E S Sproles

YIELD POINT PHENOMENA IN METALS AND ALLOYS

Yield Point Phenomena in Metals and Alloys

E. O. Hall

Plenum Press
NEW YORK

US Edition published by PLENUM PRESS
a division of Plenum Publishing Corporation,
227 West 17th Street, New York, NY 10011

Library of Congress Catalog Card Number 75–120336

SBN 306–30490–2

Printed in Great Britain

Preface

Exceptions to the rule are always interesting, and the anomalies in the stress–strain curves of mild steel and in many other metals and alloys have excited the curiosity of engineers and scientists for well over a hundred years. Yet it is only during the last twenty years that significant theoretical advances have been made, and the aim of this book has been to examine these theories against the background of the considerable volume of experimental results published over the last few years, up to mid-1969.

Hence this review volume has a two-fold aim; the first chapter attempts to review the general theories of yield point phenomena, using sufficient examples only to illustrate the theories. This chapter is intended to be complete in itself, and could be read by undergraduates who wish to appraise rapidly the general background to the problem. The remaining chapters deal, in turn, with the various alloys exhibiting yield point phenomena. Thus, chapter 2 on mild steel, is a more extensive study of quench and strain ageing, while Chapter 3 is on the refractory metals and discusses theories of the low-temperature strength. The next concerns hydrogen in metals. Chapters 5 and 6 discuss the face-centred cubic alloys, particularly the cases of the unloading yield point and intermetallic compounds. Chapter 7 covers hexagonal and ionic structures. A brief final chapter considers the areas where further research may be fruitful.

Some knowledge of dislocation theory and stereographic projection must be assumed, but the aim is to make the book as self-contained as possible, and of interest to a wide range of solid-state physicists, metallurgists and engineers.

Acknowledgments

I started this book while on sabbatical leave at the University of Cambridge in 1966 and therefore wish to thank both the University of Newcastle, for granting the leave, and Professor R. W. K. Honeycombe, of Cambridge University, for the use of facilities in his department. In addition, the help given by colleagues in supplying either unpublished data, or data prior to publication, is acknowledged in the text. Finally, I wish to thank Mrs. J. Saunders and Mrs. E. Burns for their careful typing of the draft and final copy respectively.

Contents

List of Plates

1

Yield Point Phenomena and their Theoretical Background

1.1 *Introduction*

When certain materials such as mild steel are deformed in tension, it is found that the stress–strain curve is not smooth, but shows marked irregularities, with negative slopes occurring at or near the initial yield on the curve. The actual shape of the stress–strain curve is dependent, to some extent, on the type and characteristics of the tensile testing machine used; nevertheless one may include all cases where $\delta\sigma/\delta\epsilon$ is negative as examples of yield point effects deserving attention.

Again using mild steel as an example, the progress of deformation may be divided into three stages, as shown in Fig. 1.1. The normal elastic extension AB is terminated at a stress level known as the upper yield stress σ_U. Deformation then proceeds at a decreased stress level known as the lower yield stress σ_L, but the deformation at this stage is not homogeneous: the specimen is divided into regions, known as Luders bands (after Luders (1860)†), where the strain has the value ϵ_L shown in Fig. 1.1, and other regions which are not yet deformed with zero strain. These bands are also known as Hartmann lines, after Hartmann (1896) or as 'stretcher strains'. Since this Luders strain can in steel be as high as 5%, dependent on grain size, the deformed regions on a test specimen may be clearly shown under conditions of critical illumination. Plate 1.1(a) shows a Luders band in a steel strip specimen, and Plate 1.1(b) in a stiffer, heavier specimen. Although the morphology differs in the two cases, in both specimens the upper yield stress may be regarded as a nucleation stress, and the lower yield stress as the growth stress, of the Luders bands themselves.

† In fact, these bands were first noted by Piobert (1842) – French publications often refer to these bands as the Piobert-Luders phenomena.

Thus, at the lower yield stress, deformation proceeds by the growth of Luders bands, which spread along the specimen, until at the point D (Fig. 1.1) the entire surface of the test specimen is covered, and all areas of the test length have been strained by an amount ϵ_L. Beyond this point, from D to the ultimate tensile stress at E, deformation is essentially homogeneous and thereafter necking develops, leading to normal ductile fracture at F.

As will be seen, there are numerous variants of this stress–strain curve, dependent on material, temperature, grain size and other metallurgical variables; nevertheless these general principles may apply. The technological importance of yield points is great; in pressed mild steel components for example, the Luders bands may

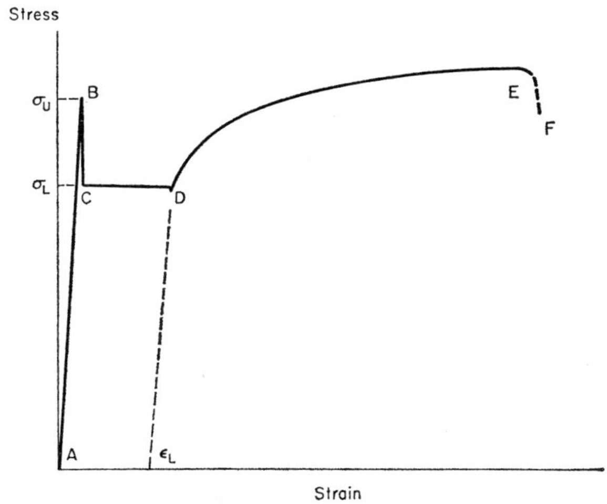

FIGURE 1.1. Diagrammatic stress–strain curve of mild steel

lead to markings resulting from the inhomogeneous deformation – these are commonly known as stretcher strains. Plate 1.2 shows a typical example in a pressed steel pie dish, where in most cases customers are not concerned with irregularities on the underside of pies, but in other examples of large pressed components, such as motor car bodies, stretcher strains make it difficult to achieve the high degree of surface finish required prior to painting. Elaborate procedures are adopted, such as deforming the sheets by temper rolling, or roller levelling by a total amount somewhat less than ϵ_L, so that on subsequent pressing, deformation will occur virtually homogeneously from the numerous Luders band nuclei so produced (see, for example, Butler and Wilson (1963) and Verduzco and Polakowski (1966)).

Although in certain alloy systems the Luders strains may exceed several hundred per cent, in others the value of ϵ_L may be exceedingly small. The variations involved and their dependence on the metallurgical variables is the core of this monograph.

1.2 *The effects of tensile machine and specimen stiffness*

Before studying yield point phenomena any further, it is necessary to dispose of two elementary, yet important, aspects of the measurement of yield points; the effects of the tensile machine and specimen stiffness.

Tensile machines are divided into two types, the so-called 'soft' and 'hard' machines. In the former, the load is considered connected to the specimen by a soft spring, so that if the specimen yields suddenly the load is virtually unaffected. Deadweight loading, hydraulic machines and pivoted beam type tensile machines come in this category, although in the latter, some allowance can be made by noting the drop in the beam as the yield point is reached. But for following rapid changes in load, such as is found with mild steel at elevated temperatures, these are of limited value. For accurate measurements, and to follow rapid changes in load, hard machines are necessary. Here the load is measured and transmitted to the specimen by a load cell and stiff members, so that very small sudden elongations in the specimen result in a large drop in load, and accurate and rapid recording of load is likewise essential. Tensile machines of the Instron type or, for lighter loads, the inverted Polanyi type described by Adams (1959), are convenient for this study. Load cells, with outputs recorded on fast (1 s F.S.D.) recorders, are also essential.

The effects of machine rigidity may be simply illustrated by reference to Fig. 1.2(a). Here, the tensile specimen shown is imagined to have a Young's modulus E, while the machine and supporting members have an effective spring constant K. Thus, under a load L, the extension of the system is $L/K + Ll/(AE)$ where l is the specimen length and A its cross section.

If the specimen extends by an amount δl the overall extension is constant; the load measured changes by δL so that

$$\delta L(1/K + l/(AE)) + L\,\delta l/(AE) = 0$$

$$\delta L/L = -\delta l/(AE/K + l)$$

For a given value of δl, it can easily be seen that as $K \to 0$ for very soft machines, $\delta L/L \to 0$.

The spring constant K of the machine may be determined quite simply by a dial gauge using a heavy specimen or by determining of the slope of the elastic region of the load–time curve, which, as the above analysis shows, will always be less than the normally accepted modulus of elasticity.

The elastic parameters of the machine will also affect the magnitude of the yield point drop. This has been studied in a series of papers by Welter and Gochowski (1938, 1939) and by Welter (1945). As the effective stiffness of the machine decreases the load relaxation decreases and will become less abrupt, until, as shown in Fig. 1.2(b), only a rounded yield is seen. Here the stress barely falls below the

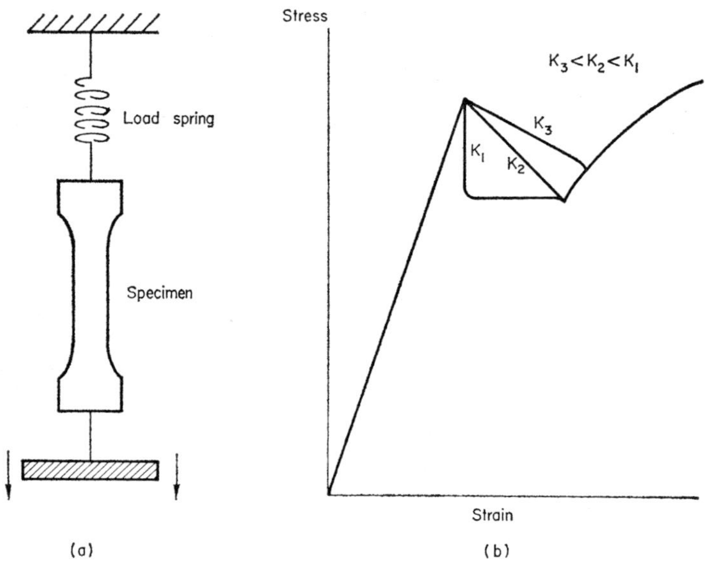

FIGURE 1.2. (a) Elastic elements of a tensile machine. (b) Effect of the spring constant K on the stress–strain curve

nucleation stress, and, as will be seen later, the Luders band will be forced through the specimen at much higher velocities. Welter and Gochowski also show the effect of axial loading on the magnitude of the yield drop.

The other factor which may affect the measurements on yield point effects is the stiffness of the specimen, which may alter the pattern of the growth of the Luders band. The reason for this is seen in Plate 1.3, which shows two Luders bands in a strip specimen which prior to deformation had a series of straight lines scribed on the surface. Examination of the lines shows that a marked shear has occurred at the Luders band front, and this shear must be accommodated by the specimen and machine. For short, stiff specimens, with a length to breadth ratio of 12† or less and with breadth 5 mm or more, it appears these shears are best accommodated by nucleating a series of Luders bands as shown in Plate 1.1(b). This has a marked effect on the lower yield stress, for as each sub-band is nucleated, a drop is recorded in the stress–strain curve, and the lower yield stress is markedly irregular. Accurate measurements of lower yield stress are made possible with wire specimens (Sylwestrowicz and Hall, 1951) or with strip specimens containing single Luders bands, as illustrated above.

1.3 *Types of yield point effects*

The yield points observed with these hard machines are found to take many different forms of instability, dependent on material and

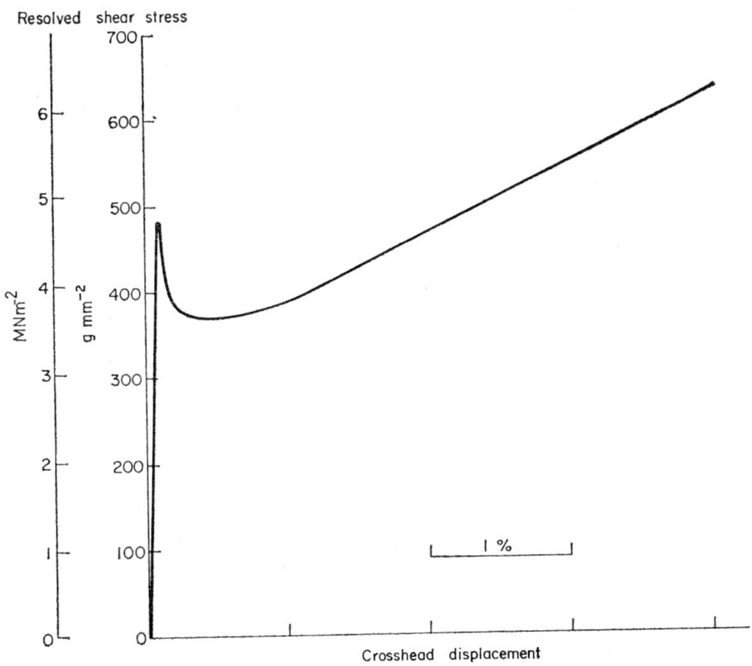

FIGURE 1.3(a). Stress–strain curves for lithium fluoride crystal at room temperature (Johnston and Gilman, 1959)

† Lomer (1952) suggests this figure, but points out that the type of test used will influence the result – for example, a tube of $\frac{1}{2}$ in. o.d. and $\frac{1}{16}$ in. wall will give a single Luders band if deformed in torsion, and a complex series of bands if pulled in tension.

testing temperature, and some of these are illustrated in Fig. 1.3.(a)–(d). They should of course be compared with the comments on Fig. 1.1 which is typical of mild steel at room temperature.

Figure 1.3(a) represents the stress–strain curve of single crystals of certain ionic crystals. In this case, the specimen is lithium fluoride containing only a small number of free dislocations. It will be seen later that this is a necessary (but not a sufficient) condition that yield points should exist. The yield point here is a short region of negative slope on the stress–strain curve, before the yield stress rises again. This type of curve is also noted in materials subjected to radiation damage by high energy particles where the dislocations become pinned by the point defects generated in bombardment (Section 1.4.5). Similar results are found in the hexagonal metals, such as zinc single crystals containing nitrogen, or in zirconium under certain conditions of grain size. In these cases a true Luders extension may not exist.

Figure 1.3(b) is another single crystal stress–strain curve; this shows a curve from copper at very low temperatures. Here there is no upper yield point, but after a certain amount of normal slip, yield points appear. In certain single crystals, this is accompanied by the

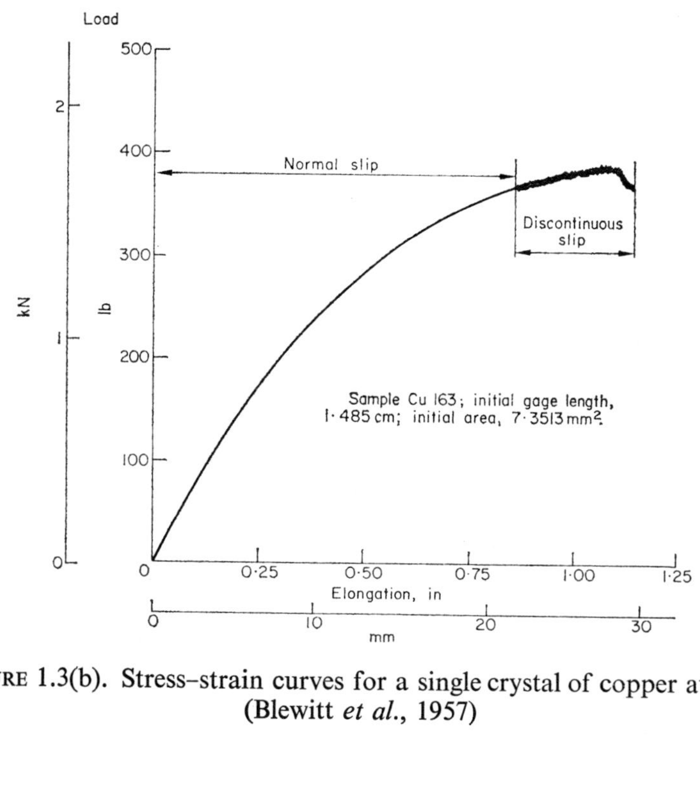

FIGURE 1.3(b). Stress–strain curves for a single crystal of copper at 4·2 K (Blewitt *et al.*, 1957)

growth in density of slip band traces which usually commence at one end of the gauge length, and spread in a parallel easy-glide mode along the gauge length (see Chapter 6).

Figure 1.3(c) shows a series of successive yield points obtained in mild steel at elevated temperatures, at about 200°C, and should be compared with Fig. 1.1 for a room temperature test. The multiple yield points seen here as deformation begins are the result of interrupted motion of the Luders band along the specimen. The movement

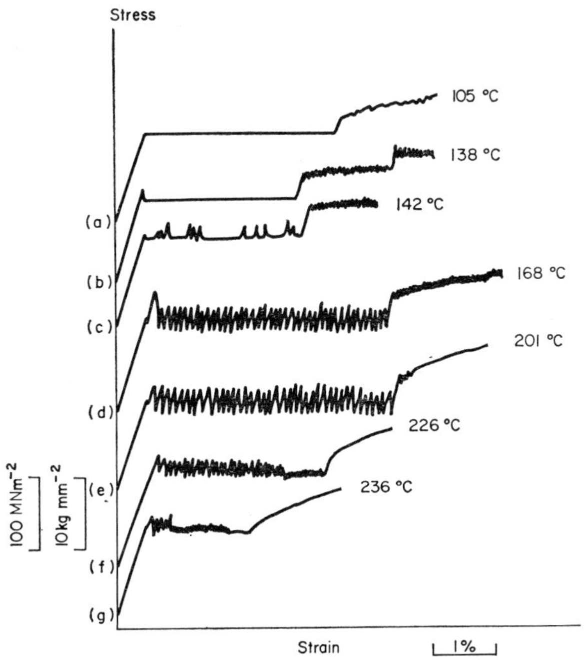

FIGURE 1.3(c). Stress–strain curves for polycrystalline mild steel at elevated temperature (Blakemore and Hall, 1966)

of dislocations near the band front becomes locked – a phenomenon known as strain ageing (Section 1.7) – and as a result the stress has to rise to release the band front again. The ductility is thereby reduced – a phenomenon known as blue brittleness – a result of simultaneous straining and ageing.

Figure 1.3(d) shows a case of austenitic stainless steel at high temperatures. Here the general stress level continues to rise as deformation proceeds – and in certain cases the curves are smooth at the commencement of yield. As the strain increases, serrations build up slowly and reach their maximum at ultimate tensile strength. This mode, characteristic of duralumin and bronzes, nickel–hydrogen and even some magnesium-base alloys is properly called the Portevin–le Chatelier effect after its discoverers (1923) who first noted it in duralumin. The name of the effect is often applied to curves such as

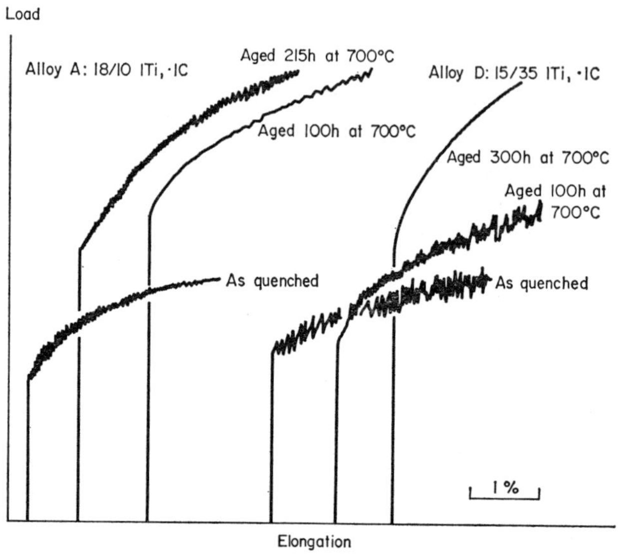

FIGURE 1.3(d). Stress–strain curves for austenitic stainless steels at 500°C (Harding and Honeycombe, 1966)

Fig. 1.3(c), although, as we shall see, the mode of locking may differ in these two cases; but both these forms of serration are due to simultaneous straining and ageing.

1.4 *The upper yield point – experimental*

1.4.1 *Determination of the upper yield stress*

The upper yield point is the stress required to initiate the deformation, yet the accurate measurement of this stress has proved a difficult problem in physical metallurgy. Despite the frequency with which figures are quoted in engineering manuals, it may fairly be said that true upper yield stresses have never been measured on any conventional shaped heat-treated test piece. Luders bands form first at the shoulders, at a stress only a few per cent above the lower yield level, and from there spread out as an array of sub-bands. To get true values the band should nucleate along the gauge length, and specimen misalignment must be reduced to a minimum.

The effects of misalignment may be greatly reduced by the use of thin wire specimens, but again results are only of value if the wires are not crushed in the grips. This may be done by holding the wire in cups and bonding with a low melting point solder† (Hall, 1951a). The results, however, still depend critically on the straightness of the wires, and it is necessary to take the maximum value recorded as representing the values closest to ideal alignment. A marked grain-size dependence is noted. The finest grained samples can show an upper yield almost twice the lower, and it can in fact exceed the U.T.S. of the sample.

The true value of the upper yield stress may be approached by a method due to Paxton (1953). If a sample is stressed to a level above the lower yield stress (but below the upper yield stress) and subjected to a low-temperature heat treatment in situ, the phenomenon of strain ageing will occur (see Section 1.7) and it becomes still more difficult to nucleate the Luders band. This method will not however get rid of the stress concentrations around the grips or the shoulders of the test piece completely.

The most accurate values of the upper yield point are obtained by centre annealing, a method devised by Hutchison (1957, 1963). Samples of hard-drawn wire are inserted in a furnace so that the ends of the sample protrude from the furnace tube and remain at a low temperature. Thus, they may safely be gripped, and presuming the wire is straight, deformation will be induced in a region free from stress concentration or gradients.

Figure 1.4 shows Hutchison's (1963) results for mild steel. This shows plots of both upper and lower yield stress against the inverse square root of grain diameter ($d^{-1/2}$) and in both cases may be written as

$$\sigma_U \text{ or } \sigma_L = \sigma_0 + kd^{-1/2}$$

where σ_0 and k are constants, the former temperature sensitive, the latter (the slope of the lines) virtually temperature insensitive. This equation, and the interpretation of these constants will be taken up in Section 1.6.4.

The disadvantage with this method is that it can only be used for upper yield point determinations; the other characteristics of the curve will be found to be so heavily dependent on grain size that deformation will soon spread into regions of small grains and the specimen can no longer be considered homogeneous.

1.4.2 *Delayed yield*

If, for the moment, the upper yield stress is considered to be a nucleation stress for the formation of the first Luders band, there are two

† Matrix alloy (29% Pb, 14% Sn, 9% Sb and 48% Bi) has a suitably high shear strength, expands on solidifying and will 'tin' steel, while melting at 105°C.

consequences. First, it should be possible to detect the formation of the yielding nuclei by careful measurements of strain prior to the upper yield, and second, it may take some little time for an effective nucleus to form. Thus, if impulse loading is adopted there will be a small but finite delay between the application of the load and the yielding of the specimen.

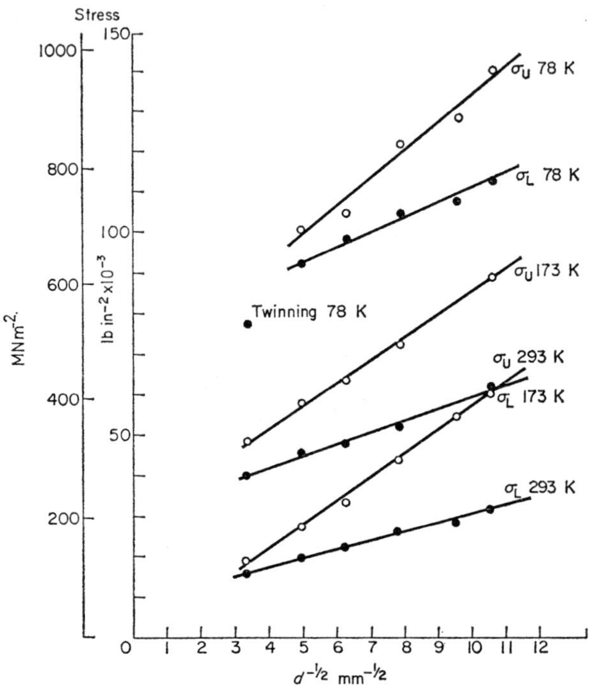

FIGURE 1.4. Variation of upper and lower yield stresses with grain size (Hutchison, 1963)

One common method of examining delayed yield is by the Hopkinson bar technique (see, for example Krafft and Sullivan, 1959, 1962). An alternative method of applying the load is to use a balanced hydraulic loading machine (Wood and Clark, 1951; Campbell and Marsh, 1962) where the balancing pressure on a hydraulic ram is suddenly removed. In this way, with critical damping of the fluid, full load can be applied to the specimen in less than 6 ms. Shock loading may also be used (Holland, 1967).

Delayed yielding has been examined in β-brass single crystals (Kramer and Maddin, 1952), in molybdenum polycrystals (Hendrickson *et al.*, 1956b) and in aluminium–magnesium alloys (Shephard and

Dorn, 1956) but most work has been on mild steel of various grades (Wood and Clark, 1951, 1952; Vreeland *et al.*, 1953a, b). Delayed yielding is not apparently notable in materials which show no yield point, as shown by Kramer and Maddin (ibid.) in aluminium and α-brass at room temperature.

Generally, the results are similar in these different materials. A plot of applied stress against delay time shows, as in Fig. 1.5 for a 0·09% C mild steel, a region where the delay time increases with falling stress

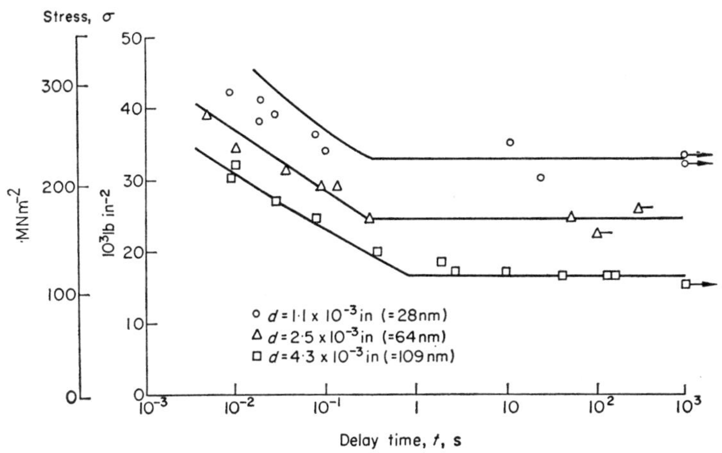

FIGURE 1.5. Delay times for yielding in mild steel for varying grain sizes (Russell *et al.*, 1961)

until it reaches a stress level where, even in a static test, no yield will occur. The stress-dependent part of the curve could be represented by the equation

$$t_d = t_0 \exp(-\sigma/\sigma_c)$$

where t_d is the delay time for an applied stress σ, and t_0 and σ_c are constants. In some early studies σ_c was tentatively identified with the upper yield stress, but since this was measured on engineering type test pieces, does not mean a fundamental correlation exists. Neither does this formula include the effect of temperature but in an aluminium–2% magnesium alloy, Shephard and Dorn (1956) found that both temperature and stress could be combined as a single function,

$$t_d = t_0 \exp(E/RT) \exp(-\sigma/\sigma_c)$$

where, in this alloy $E = 6{\cdot}8$ kcal mole^{-1} (28 kJ mole^{-1}). It is interesting to note that at least in this alloy, the same activation energy applies to the propagation rate of the Luders band.

Interest in this field has been steadily growing. Most of the papers have been concerned with theoretical interpretation of the results, and these will be discussed in the next section and in Chapter 2. However, in the case of steels, important variables such as the effect of grain size, and the more detailed effects of composition, particularly carbon and nitrogen levels, have not passed unnoticed. For example, Krafft and Sullivan (1959, 1962) have made a detailed study of the effect for steels over the range of 0·1–0·3% C, using a Hopkinson bar method, and Campbell and Marsh (1962), as well as Russell, Wood and Clark (1961) have investigated the effects of grain size on low carbon steels using hydraulic methods. Figure 1.5 comes from the paper by Russell, Wood and Clark. Notice that for a given stress, the delay times are shorter in the coarse-grained material. These points will be taken up again in Section 1.5.

1.4.3 *Microstrain*

It will be seen in a later section (1.5), that if the lower yield stress is grain-size controlled, and if the upper yield stress obeys a similar law (Fig. 1.4) then the nucleation of the first Luders band must depend on deformation occurring **before** the upper yield point σ_U. In other words, there should be evidence of some slight plastic flow in the so-called elastic region.

Evidence of flow can sometimes be seen in departure from linearity of the stress–strain curve before the yield; this is particularly so when the difference between σ_U and σ_L is large, as when using solder cups as grips (Section 1.4.1, Sylwestrowicz and Hall, 1951). However the difficulties of measurement are great, and sensitive extensometers are required. Roberts *et al.* (1952) reported observing true elastic limits at about $\frac{2}{3}$ to $\frac{4}{5}\sigma_L$, while Vreeland *et al.* (1953a) actually measured this extension, which varied up to $3{\cdot}7 \times 10^{-6}$ (0·0037%) dependent on circumstances.

In general, two methods are available, either variable stress or constant stress methods. The variable stress method is perhaps more informative and is best described in the papers by Roberts and Brown (1960) and used by Brown and Ekvall (1962) in a study on α-iron. Figure 1.6 illustrates the position diagrammatically. Brown and his co-workers distinguish the stresses marked on the curves. σ_E is the stress at which a loop observed on loading and unloading, and is the true elastic limit corresponding to dislocation movement. As the stress is increased, a level is reached (σ_A) where the loop remains open. This is the anelastic limit, and is a little below the true proportional limit σ_{PL}, which in turn is close to the upper yield σ_U. With a lower detectable strain limit of about 4×10^{-6}, the values of these parameters have been determined for iron, and in another series of papers, for zinc, aluminium, silver, copper and α-brass. Again the results will be discussed in a later section, but it may be stated here that σ_E appears temperature insensitive while σ_A is temperature sensitive in pure iron. Both depend on the impurity level.

The second method – that of deformation under constant load – has been discussed by Owen *et al.* (1958), and in effect, shows some of the difficulties in defining the so-called delay time. Figure 1.7 illustrates their results. Stage 1 is the microstrain region, which is succeeded by the microcreep region (Stage 2). Finally, gross yielding occurs, and the Luders deformation is observed (Stage 3). Owen *et al.*

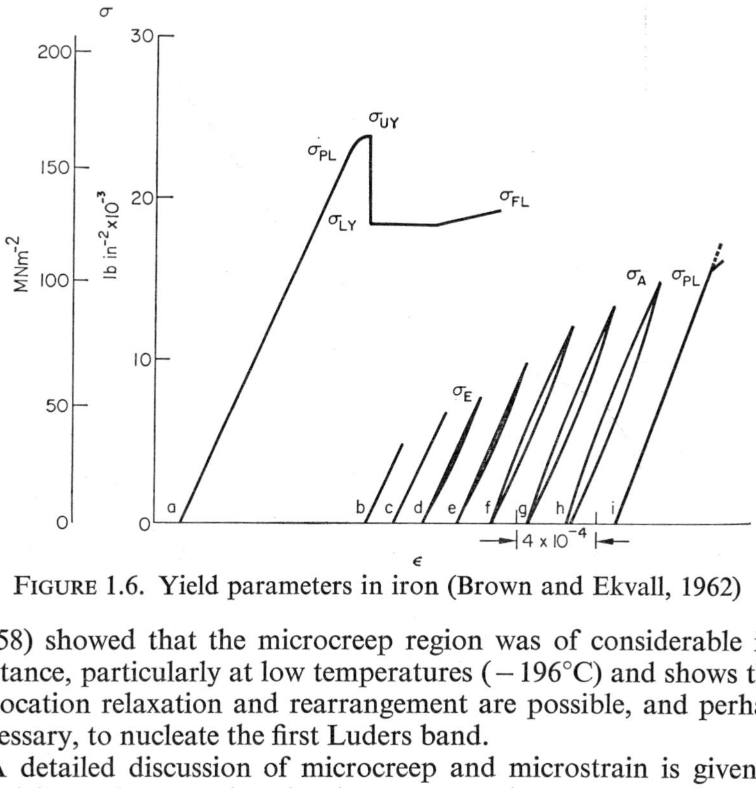

FIGURE 1.6. Yield parameters in iron (Brown and Ekvall, 1962)

(1958) showed that the microcreep region was of considerable importance, particularly at low temperatures (−196°C) and shows that dislocation relaxation and rearrangement are possible, and perhaps necessary, to nucleate the first Luders band.

A detailed discussion of microcreep and microstrain is given in Friedel's *Dislocations* (1964), Chapters 15 and 16.

1.4.4 *The effects of pressurization*

Another interesting effect, which complements those described in the previous section, was the discovery by Bullen *et al.* (1964a) that the result of applying hydrostatic pressure (removed before testing) was to diminish or even completely remove the yield point in iron. This result was extended in two later papers (Besag and Bullen, 1965, 1966 – the latter paper concerned with twinning) and was also notable in chromium (Bullen *et al.*, 1964b) – there is in fact no reason why it should not be of general applicability in all metals showing similar yield points.

The essential results are shown in Fig. 1.8. Hydrostatic pressure, up to 10^3 MN m^{-2}, is applied to the tensile specimens before testing. This produces no macroscopic deformation, but on tensile testing the yield is shown to be markedly reduced and finally disappears. The reason for this interesting effect, to anticipate Section 1.5, is that the

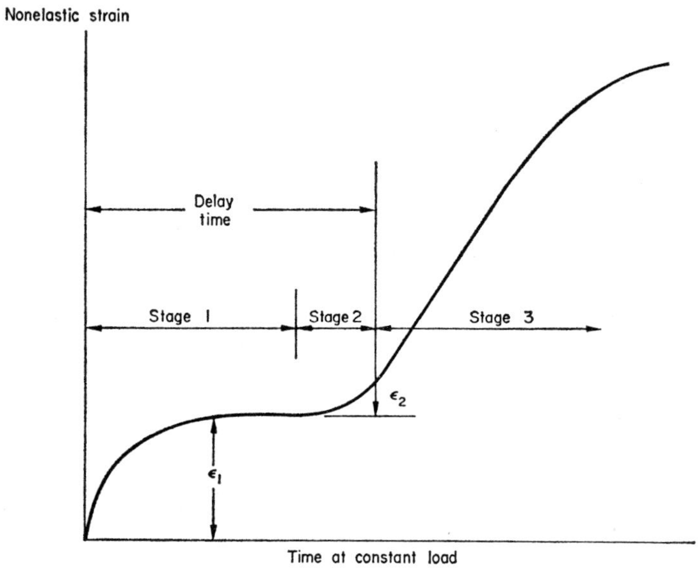

FIGURE 1.7. Stages in yielding under constant load (Owen *et al.*, 1958)

pressurization produces new dislocations around inclusions and other defects. These dislocations, unlike those in the specimen after heat treatment, are free and not locked in position; they therefore multiply readily and under lower stresses, and the yield point disappears.

It is not theoretically possible to create fresh dislocations in a homogeneous solid by hydrostatic pressure; the dislocations must have therefore arisen in centres of inhomogeneity, i.e. around inclusions. The effects were indeed much less in a high purity iron which had fewer inclusions.

Yajima and Ishii (1967) and Radcliffe (1969) have claimed a reduction in the lower yield stress of coarse-grained iron following pressurization. However, Bullen *et al.* (1967) suggest these results should be treated with reserve, and stress that it is the inclusion count, rather than the analysis of the material, which will determine the magnitude of the effect (see also Section 1.6.4).

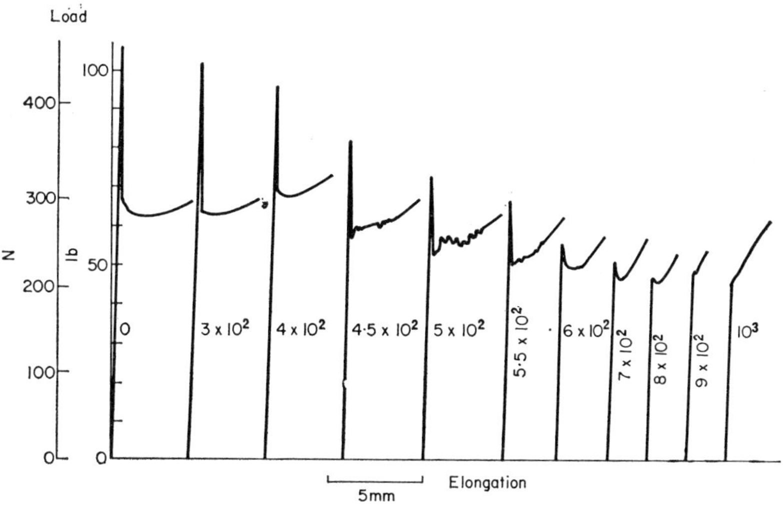

FIGURE 1.8. Suppression of the yield point in Armco iron after pressurization (applied pressures are in MN m^{-2}) (Bullen *et al.*, 1964a)

Another result of pressurization, this time resulting from shock loading, has been noted by Appleton and Waddington (1965). Samples of copper, nickel and aluminium subjected to explosive loading, show a yield point when subsequently deformed in tension. The shock-loaded material naturally contains a high density of dislocations, but Appleton and Waddington show that only a few of these are free to move. The reasons for the pinning of these dislocations are obscure.

1.4.5 *Yield points following irradiation*

A different type of yielding is thought to arise in material subject to radiation damage. Here, incident fast particles will collide with the atoms of the crystal lattice, and where the energy imparted to the atom in the crystal is high enough (generally greater than about 28 eV) the atom will be ejected from its site. Thus, a vacancy-interstitial pair will be created. Both these point defects are mobile; a certain fraction will initially recombine, but at low temperatures and for sufficiently high radiation doses, the concentration of point defects may easily reach 0·1 at.% or more.

If it is now assumed these defects are mobile, some will undoubtedly migrate to and condense on dislocations, where they cause a step or 'jog' in the dislocation line. If these dislocations are forced to move, jogs in many cases will be forced to move non-conservatively and a trail of point defects, further vacancies and interstitials, will be left behind. This is again difficult, and the dislocations remain pinned at the jogs, resulting in a remarkable lowering of the anelastic damping of the material (see, for example, Barnes and Hancock, 1958). If, after high irradiation doses, the jogs are sufficiently close, the stress needed to bow dislocations between the jogs is equal to $\alpha\mu b/l$, where μ is the shear modulus, $\alpha \sim 1$ and l is the distance between the jogs. If at first l is small then σ is high, but the dislocations bow out, combine, and the effective l suddenly increases. σ can thus fall, exhibiting a

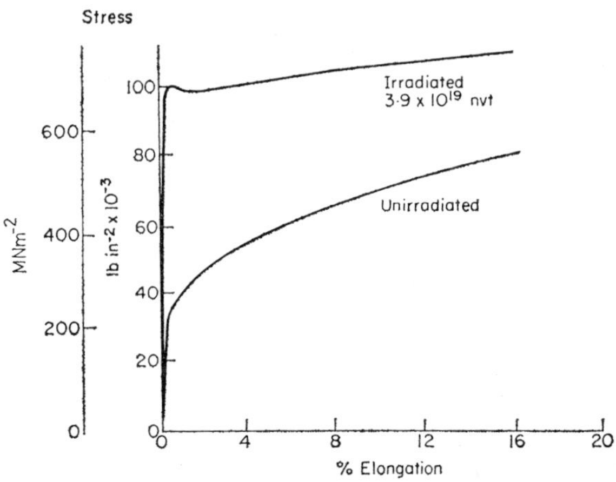

FIGURE 1.9. Yield point following irradiation of type 347 stainless steel (after Wilson and Berggren, 1955)

yield drop (Fig. 1.9). This type of yield point effect may readily be removed by a high-temperature anneal, which causes rapid recombination of the defects (Broomfield *et al.*, 1965). A similar effect may arise if a high concentration of vacancies is retained by quenching metals from temperatures close to their melting points. These vacancies can again interact with dislocations, giving a yield point, and occasionally serrated yielding (Mori and Meshii, 1969).

1.5 *The upper yield point – theoretical*

1.5.1 *The Cottrell–Bilby theory*

Since the yield point effects in mild steel have been known for over a century, and are, moreover, of considerable technological importance, it is not surprising that theories until recently arose from work on the ferrous field.

The first theory which gained credence was the so-called 'grain boundary' theory of Dalby (1913) and Kuroda (1938). As will be seen in the next chapter, the effects in mild steel are critically dependent on the amounts of carbon and nitrogen present, while in particular decarburized mild steel shows no yield point at all. Dalby and Kuroda independently suggested that each grain of the ferrite matrix was surrounded by a thin film of carbide (presumably cementite, Fe_3C) which restricted deformation of the ferrite, until, at the upper yield stress, the carbide gave way, and the nucleus of a Luders band spread across the specimen.

This theory was criticized in detail by Cottrell (1948) who pointed out that since the effects were pronounced with as little as 0·02 wt% carbon present, the carbide films must be extremely thin, and yet, after the yield stresses of the individual grains had been passed, this film would be called on to bear the whole load on the specimen. This is clearly impossible, and Cottrell proposed (and later amplified in a paper by Cottrell and Bilby (1949)) that instead the carbon atoms in solution in the ferrite segregated to dislocations, so locking them in position. The source of this segregation was the relief of strain energy caused by the migration of carbon from their interstitial sites where they cause a substantial dilation, to the tensile region around an edge dislocation where the lattice is itself dilated. Cottrell and Bilby estimate this binding energy to be about 0·5 eV per atom plane, and that a concentration of about 10^{-6} wt% C is sufficient to place one carbon atom on each dislocation per atom plane, assuming the normally accepted dislocation density of 10^8 lines cm^{-2} in well-annealed material. For yielding to occur, the applied strain energy, assisted by thermal activation, must exceed this binding energy, when the dislocations break away from their atmospheres, multiply, and so form the first Luders band.

This theory has been developed in some detail, as befits its importance. The binding energy (U) arises from the change in volume Δv caused by the solute atoms, so that

$$U = p\,\Delta v$$

where p is the hydrostatic stress field. In the case of interstitial atoms, segregating to a positive edge dislocation, the concentration will occur beneath the slip plane; in substitutional solid solutions, the concentration of solute atoms will depend on the size differences, and, in the case of a smaller solute atom, will take place above the slip plane of a positive edge. The equation above becomes

$$U = 4\mu b \epsilon' r^3 \,(\sin\theta/R)$$

where $\epsilon = (r^1 - r)/r$, the fractional size difference, and R and θ are the co-ordinates of the solute atom. The sign of the term is important, and in the case of certain substitutional alloy systems (e.g. aluminium–silver) it will in fact be very small.

Concerning ourselves for the moment with interstitial systems, it is also possible to show that the equilibrium concentration of solute atoms around a dislocation will be

$$C = C_0 \exp (U/kT)$$

where C_0 is the average concentration in the lattice.† Various means are available to measure the binding energy U; one of the simplest being to determine the temperature (known as the **upper critical temperature**) at which yielding effects disappear. Substituting with $C = 1$ (i.e. all dislocations having just one solute atom per plane), we obtain

$$U = T_{max}k \ln (1/C_0)$$

More recently, however, Keh *et al.* (1968) have suggested this relation yields $U + Q$, rather than U, where Q is the activation energy for diffusion.

In his original (1948) paper, Cottrell only considered the case of hydrostatic stresses, and this has led some research workers to state that solute atoms will not be attracted to screw dislocations; indeed Cottrell (1953a) states that aspherical distortions around a solute atom are necessary before interaction with screw dislocations will take place. However, a full analysis by Cochardt *et al.* (1955) did take into account the tetragonal distortions resulting from a carbon atom, and they were able to show that the interaction energy was at least of the same order of magnitude as an edge dislocation. This is confirmed in more recent examinations by Hirth and Cohen (1969a, b), Schoeck (1969) and Bacon (1969). There is equal reason to believe a similar interaction can occur in face-centred cubic crystals (van Bueren, 1960; Fiore and Bauer, 1967).

Any successful theory of the yield point will have to explain several features; the variation of yield stress with temperature, the effects of strain rate and the existence of delayed yielding, and the phenomena of serrated yielding (Fig. 1.3(c) and (d)) which arises with strain ageing (Section 1.7).

The Cottrell–Bilby theory of yielding considers material with all dislocations initially locked. At low stresses and low strain rates the solute atmospheres can diffuse with the dislocations, but at high stresses and/or high strain rates the dislocations escape from their

† This equation assumes a Maxwellian distribution of solute atoms. Louat (1956), Beshers (1958) and Fiore and Bauer (1967) consider that a Fermi–Dirac distribution is more appropriate, and this will affect the high-temperature distribution and hence the upper critical temperature. For example, Rudee and Huggins (1964) show that in certain circumstances C may increase as the temperature is raised.

atmospheres, and multiply. This 'breakaway' theory will now be considered.

Cottrell (1953a) gives the following formula for the critical strain rate for breakaway to occur:

$$\dot{\epsilon} = 4D\rho b/l_0$$

where D is the solute diffusion coefficient, ρ the dislocation density, b the Burgers vector of the dislocation and l_0 a characteristic length, which can be taken as the radius of the atmosphere. Since $l_0 \sim 30b$ for iron, or more characteristically, $l_0 \sim 10b$ for substitutional cases, we may write

$$\dot{\epsilon} \simeq D\rho$$

This concept was used by Bonizewski and Smith (1963) in their study of serrated yielding in the nickel–hydrogen system (Section 4.5).

The variation of yield stress with temperature was also estimated by Cottrell and Bilby (ibid.). First, the zero point yield stress (i.e. at $T = 0$ K) was determined. This gives

$$\sigma_{T_0} = \frac{3\sqrt{3}\,A'}{4\lambda^2 r_0{}^2}$$

where A' is a parameter involving the elastic constants and is $\sim 3 \times 10^{-30}$N m^2, λ the slip distance, and r_0 the equilibrium distance

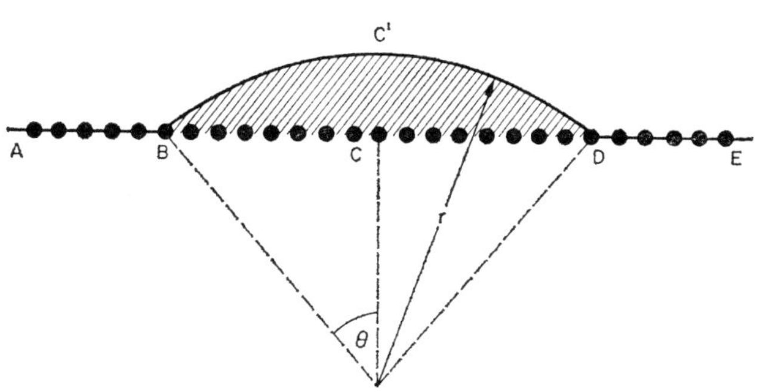

FIGURE 1.10. Release of a free dislocation loop BC'D from its original locked position (Fisher, 1955)

of the carbon atom from the dislocation core. Extrapolating known values of the yield stress to 0 K gave $r_0 \sim 7$ Å, which is reasonable. Next, the effects of thermal fluctuation were included. These fluctuations are seen as pulling a loop of dislocation line clear of its carbon atmosphere (Fig. 1.10); if the stress in the neighbourhood of the dislocation lies in the range $\sigma/\sigma_{T_0} \rightarrow \sigma + d\sigma/\sigma_{T_0}$, the time before a successful fluctuation occurs is proportional to exp $(U(\sigma/\sigma_{T_0})/kT)$ where U is an activation energy. On the other hand, during a tensile stress, the time spent in this stress range is inversely proportional to the rate of stressing $\mathrm{d}\sigma/\mathrm{d}t$. Thus yielding should occur when the quantity

$$S = \frac{\mathrm{d}\sigma}{\mathrm{d}t} \exp\left[-\frac{U(\sigma/\sigma_{T_0})}{kT}\right]$$

reaches a fixed value, or alternatively the yield stress should vary with temperature so that U/kT remains constant. Reasonable agreement with experiment was obtained using their expression for $U(\sigma/\sigma_{T_0})$ which, as Cottrell (1957b) showed later is better given by

$$U = 9\sigma_{T_0}b^3(1 - \sigma/\sigma_{T_0})^3$$

A further treatment, in rather simplified form, is given by Fisher (1955), and this will be discussed again below in terms of delayed yielding. Lothe (1962) has also discussed the effect of unoccupied impurity sites. Assuming for the moment the delay time of yielding is constant, as in a normal tensile test, Fisher (ibid.) shows that

$$\sigma T/\mu^2 = \text{constant} \quad (\mu = \text{shear modulus})$$

or, to a good approximation $\sigma \propto T^{-1}$.

It may be mentioned here that with more recent experimental determinations of the yield stresses of iron and other b.c.c. refractory metals at low temperatures, neither of these theories is entirely satisfactory. Conrad (1963) has developed a theory in terms of effective stresses and activation volumes, since very pure samples of iron (which show no yield point) show an equally strong dependence of yield stress on temperature. This is put down to changes in the Peierls–Nabarro forces which are part of the lattice energy, and control the width of the dislocations. This will however, be taken up again in Chapters 2 and 3.

Returning now to the question of delayed yielding, it is easy to see from the yield conditions of Cottrell and Bilby, given above, that the concept of delayed yield may be covered by their theory, since there is a finite time interval t before a sufficient number of loops can be formed.

Yokobori (1954), for example, using Cottrell's formulae explicitly, showed

$$t_{\mathrm{d}} = t_0(\sigma/\sigma_{T_0})^{-1/nkT}$$

where n is a parameter derived from the constants in Cottrell's activation energy equation, and the other constants have already been defined. The general expression is similar to that of Clark and

Wood (Section 1.4). Fisher (ibid.) also adopted the Cottrell approach in a rather simplified form,

$$\ln t_d = A + B\mu^2/\sigma T$$

where μ is the shear modulus, and A and B are constants. The variation with temperature appears closer; however, Cottrell (1957b) did critically comment on these simplifications.

The other missing parameter, that of grain size, has already been noted (Fig. 1.5). Russell, Wood and Clark (1961) invoked the concept of piled-up groups of dislocations at the grain boundaries initiating a Luders band nucleus. The delay time experimentally varies as $1/d^3$, where d is the main grain diameter, and theoretically could be determined as

$$t_d = \frac{A''}{d^3} \exp\left(\frac{U(\sigma/\sigma_{T_0})}{kT}\right)$$

where A'' is a constant involving grain-boundary energy. This topic will be discussed further in Section 2.4.3 where an alternative expression will be derived.

Before leaving this section, we may summarize by saying the Cottrell–Bilby theory was a remarkable step forward in our understanding of the yield point phenomena in general. As will be seen in later chapters, the concept of dislocation locking is one which has wide applicability. It is in some of the detail that difficulties arise, particularly in the variation of yield stress with temperature, and in the details of the 'breakaway' theory. Many millions of free dislocations can be introduced into a steel sample by scratching or even lightly filing the surface, and yet an upper yield point is still present. Notice also the observed microstrains before the Luders band is formed. The formulation of the proper conditions for Luders band nucleation is still not possible without considerably refining the theory.

A useful review of the interaction of solute atoms and dislocations has been published by Fiore and Bauer (1967).

1.5.2 *Suzuki locking*

The original Cottrell theory could be readily extended to the case of substitutional alloys, by assuming again that the substitutional atoms will migrate to particular parts of the dislocation so as to relieve the maximum stress, i.e. if the solute atom radius is greater than that of the solvent, these will go to the upper bounds of a positive edge dislocation. If, however, the difference in radius is small, then the interaction may be smeared out by local thermal fluctuations, and atmospheres will not form.

An alternative mode of interaction, chemical in place of elastic, has been proposed by Suzuki (1952, 1957, 1962, 1963). This theory was in fact put forward as a reason for the increased strength of dilute solid solutions; in its simple form, it can lead to yield points, and, as will be seen later, there is some evidence of the theory as a mechanism for dislocation locking, particularly in some face-centred cubic metals.

In these metals, it is well known that the normal slip dislocations can dissociate into partial dislocations of various types; one of the most common reactions being the formation of Shockley partials with a release of energy;

$$a/2[110] \rightarrow a/6[121] + a/6[21\bar{1}]$$

In certain circumstances, these partials may separate, forming a ribbon of stacking fault; the width (l_0) of the stacking fault will be determined by the stacking fault energy f and given by

$$l_0 = \mu b^2 B/f$$

where B is a constant dependent on Poisson's ratio and the angle between the Burgers vectors, b, of the partial dislocations.

In a face-centred cubic material, the fault will have an hexagonal structure, and Suzuki suggests that the solute atom concentration in this 'phase' will be different from that in the matrix C_0. The change in concentration affects the fault energy, usually lowering it, and hence the fault increases in length from l_0 to l_1.

In the presence of an applied stress the fault will shrink again, and Friedel (1964) has summarized the result in Fig. 1.11. For an applied stress $\sigma \leqslant \sigma_c$ a critical stress is given by

$$\sigma_c = [f(C_0) - f(C_1)]/b$$

where $f(C_0)$ and $f(C_1)$ are the free energies corresponding to the concentrations C_0 and C_1 in matrix and fault respectively. If $\sigma < \sigma_c$, the fault will shrink (Fig. 1.2(a)) from length l_1, to

$$l_c = 2l_0l_1/(l_0 + l_1)$$

Where $\sigma = \sigma_c$, one half of the dislocation penetrates the matrix and the fault moves with a constant length l_c, until it leaves the area of solute atom concentration, when it can move under a smaller stress.

Most of the work on this theory (Suzuki, ibid.; Flinn, 1958) has been concerned with the increased yield strengths arising from ordering, although Hirschhorn (1963a) has also considered the kinetics of stacking fault growth. It is, however, important to realize that the theory could potentially lead to a drop in load – i.e. a true yield point effect – although this may in most cases be masked by the continuing increase in yield stress from work hardening. Of even more importance to the present work is its value as a locking mechanism; many examples are known of precipitation within stacking faults, particularly in austenitic steels and possibly in palladium–hydrogen. Once precipitation has occurred, the extended dislocation will be permanently locked. The generation of new dislocations (Section 1.5.6) gives rise to the yield anomalies, and the process can be repeated to give serrated yielding (Fig. 1.3(d)).

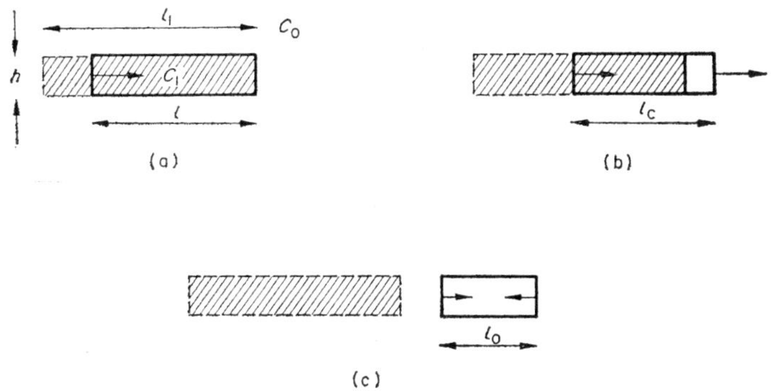

FIGURE 1.11. Release of a free dislocation blocked by the Suzuki mechanism (Friedel, 1964)

In body-centred cubic metals, it has not been expected that the stacking-fault energy could be sufficiently low for this theory to apply. In a few cases, faults have been observed in some of these metals which could be a type of stacking fault; for example, Fourdeux and Berghezan (1960) and Segall (1961) with niobium, and Nakayama *et al.* (1962) with tungsten, the latter claiming a fault energy of only 14·5 mJ m^{-2}. No faulting has been observed in other metals of this group, and indeed with wavy slip lines indicating cross slip, they would not be expected (see also Silcock (1959), Hirschhorn (1963b)). Those faults observed are probably due to relatively high impurity levels, and Wasilewski (1967) has recently estimated the energies of faults on {112} planes in these metals.

1.5.3 *Short-range order locking*

Another possible locking mechanism, capable of producing yield points in ordered alloys is known as the short-range order theory. It was originally discussed by Fisher (1954), who showed that the existence of short-range order in a crystal will increase the flow stress by an amount

$$\Delta\sigma = \gamma/b$$

where γ is the increase in surface energy of the slip plane, since a dislocation of vector b in its passage will lower the order of the lattices. If in an avalanche of dislocations this order is destroyed rapidly, a drop in flow stress (i.e. a yield point) will result. This effect could arise in some of the more concentrated alloys, and will be considered in more detail in Chapter 6, in connection with face-centred cubic intermetallic compounds.

This theory was later extended to other alloy systems by Schoeck (1956) and by Schoeck and Seeger (1959) who pointed out that from the earlier work of Snoek (1941) the carbon atoms in iron would occupy positions midway along the cube edges. This so-called Snoek atmosphere (in contrast to the Cottrell atmospheres) is in effect a localized short-range order. The flow stress of this configuration was solved and shown to be

$$\tau = 20{\cdot}5\ (A/ba^3)\ c$$

where b is the Burgers vector, a the lattice parameter and c the atom fraction of solute interstitials. A is an elastic constant. It should be noted that this equation is independent of temperature, and dependent on concentration. Certainly σ_0 in the grain-size equations (Sections 1.4 and 1.6) is linearly dependent on c, and the lower yield stress in iron is only weakly dependent on temperature over the range 0–150°C. However the latter result may be fortuitous in view of the occurrence of strain ageing (Section 1.7: see also Wilson, 1961). Nevertheless, this theory has adherents since the formation of short-range order can occur very rapidly, and some of the yield point effects with low observed activation energies may be related to this. For example, Rose and Glover (1966) have used this theory to explain strain ageing in austenite at room temperature where the diffusion rates are very low.

1.5.4 *Electronic locking*

A further form of dislocation locking has been suggested by Cottrell *et al.* (1953). Edge dislocations in a lattice should carry an electric charge, and this is particularly noteworthy in discussing ionic solids (Chapter 7). However, even in metallic lattices there should be some effect, and Cottrell *et al.* suggest that the electrical contribution to the binding energy of a solute atom and dislocation should be $\frac{1}{30}(\hbar^2 k_0^2/2m)$, where $\hbar$, k_0 and m have their usual meaning. The theory has been extended by Sugiyama (1966).

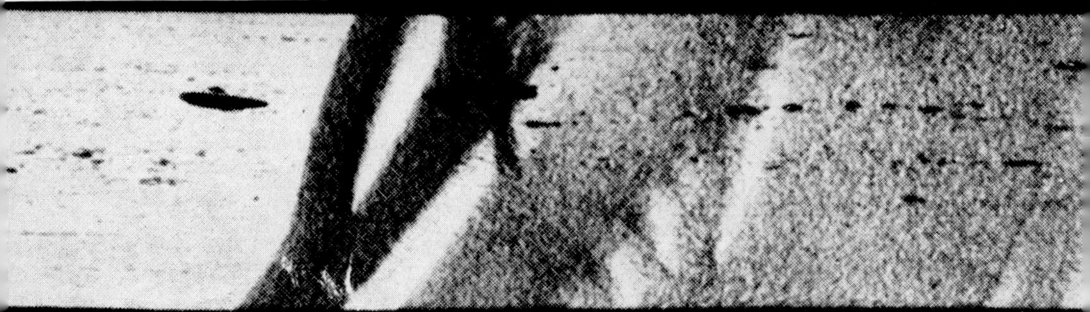

PLATE 1.1 (a) (*top*) Single Luders band in mild steel strip and (b) (*above*) complex Luders band in similar material. (Lomer, 1952)

PLATE 1.2 (*right*) Luders bands in the base of a pressed steel pie dish

PLATE 1.3

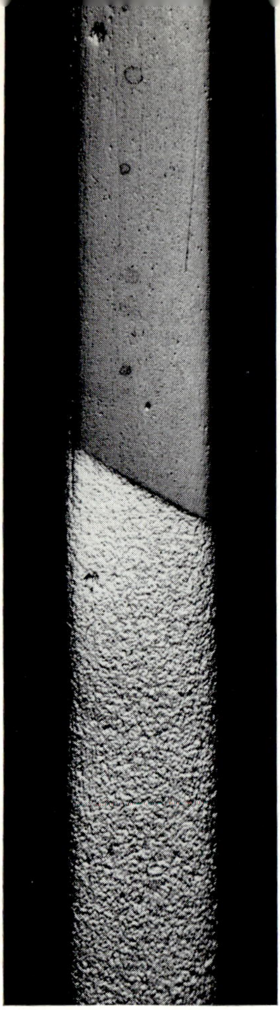

PLATE 1.4 (a)

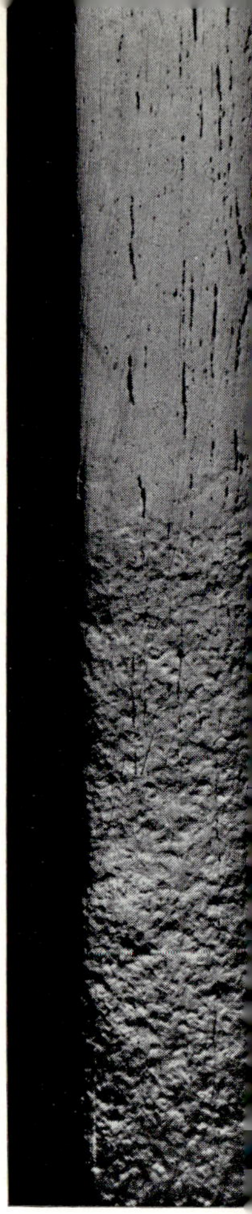

PLATE 1.4 (b

PLATE 1.5

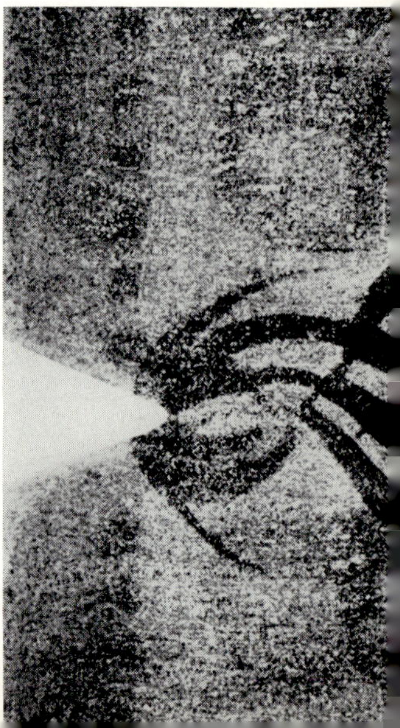

PLATE 1.3 Shear at the front of a Luders band. (Hall, 1951a)

PLATE 1.4 (a) Sharp and (b) (*above right*) diffuse Luders bands in mild steel

PLATE 1.5 Luders bands in a notch-bend specimen. (*By courtesy of* Dr. J. Knott)

In general, the contribution so calculated is small; Cottrell *et al.* calculate it as 0·02 eV for zinc in copper, compared with 0·125 eV for binding using the Cottrell–Bilby theory. To the date of writing, no yield point effects have been noted experimentally in metals which could be solely related to this mode of locking, although in nickel-base alloys with hydrogen as an interstitial impurity, this may well be the force driving the solute to dislocations (Section 4.5.3).

1.5.5 *Precipitation on dislocations*

Yet another mode of dislocation locking would appear possible, as a natural extension of the Cottrell–Bilby theory. If segregation occurs at dislocations, might it not be possible for segregation to produce nucleation, and hence precipitation? This feature should be most readily observed in cases where heterogeneous nucleation is occurring, as in age-hardening alloys being treated above the G.P. solvus temperature, or where diffusion rates are low and supersaturation high.

Many practical examples are now known, using thin film electron microscopy, such as the work of Hornbogen (1962) on iron–gold alloys, Hale and McLean (1963) on iron–carbon and iron–nitrogen alloys, Tekin and Kelly (1965) on vanadium steels, Dollins and Wert (1969) on niobium–nitrogen alloys, Kent and Kelly (1965) and Hall (1968) on magnesium base alloys, and in the 'decoration' of dislocations in alkali halide crystals, and in aluminium–magnesium alloys. Not in every case are yield points noted, but the effect is sufficiently general when coupled with precipitation on stacking faults to make it worthy of comment.

Surprisingly, there is little theoretical work on this aspect of metal physics, and what little there is simply underlines the difficulty of the problem. A thermodynamic treatment was given by Cahn (1957) who added a strain energy term to the usual thermodynamic free energy change on nucleation:

$$\Delta F = \Delta F_{\text{v}} + \Delta F_{\text{s}} + \Delta F_{\varepsilon}$$

where the subscripts v, s and ϵ refer to the volume, surface and strain respectively. The free energy is developed in terms of a parameter

$$\alpha = 2Af/\pi\gamma^2$$

where γ is the surface energy, $f = -\Delta F_{\text{v}}$ and A is an elastic constant associated with the dislocation. Typically, $\alpha \sim 0{\cdot}6$, and in the case of a continuous precipitate along the dislocations, Cahn estimates that the nucleation rates on dislocations might be $10^{14}\ \text{m}^{-3}\ \text{s}^{-1}$, when the homogeneous nucleation rate is only $10^{-64}\ \text{m}^{-3}\ \text{s}^{-1}$.

It should be mentioned here that continuous precipitation along dislocations, although found in quench-aged α-iron, is not the only possible mode of precipitation. Kent and Kelly (see Chapter 7) found discrete particles along dislocations in magnesium–thorium alloys, and Hall (1968) found rods perpendicular to dislocations in magnesium–zinc. Indeed, Kelly and Nicholson (1961) have pointed out another factor which will influence precipitate morphology. According to them, the misfit vector is another controlling variable, and they show precipitation will not occur when the Burgers vector of the dislocation is perpendicular to the misfit vector of the precipitate.

The only other theoretical paper on this topic is an extensive one by Bullough and Newman (1962b). They treat the cases of both discrete and continuous (rod-like) precipitation on dislocations. In their view, transfer velocity across the core–matrix interface is the controlling step, and if the elastic binding energy of the impurity atoms exceeds the chemical binding energy of the precipitate in the free matrix then precipitates will appear down the dislocation line. Their main interest was in the kinetics of the precipitation, and this will be discussed later in Section 1.7. The question of precipitation on stacking faults, mentioned in a previous section, is also discussed by Kelly and Nicholson (ibid.); this will again be dealt with in detail in Section 2.8.

1.5.6 *Multiplication mechanisms*

It will be obvious from the preceding discussion that if a dislocation is locked in position by a precipitate, it will be permanently locked. On straining such a specimen, yielding will have to occur by the generation of new dislocations, which will rapidly multiply under the applied stress, and possibly produce a situation where these new dislocations can continue to move under a lower stress.

This theory is based on the work of Johnston and Gilman (1959) and Johnston (1962a, b) on lithium fluoride, and applied to metals by Hahn (1962) and Cottrell (1963). The unlikely material lithium fluoride supplied two important clues; first, the material exhibited a yield point in the asgrown condition (Fig. 1.3(a)), and second, the crystals could be grown with a notably low dislocation content of only a few thousand lines per square centimetre. Furthermore, selective etching techniques and pulse loading enabled dislocation velocities to be determined, and Johnston and Gilman showed that the velocity/stress relationship was given by

$$v = (\sigma/\sigma_r)^{m^*}$$

where σ is the applied stress, σ_r is a reference stress, and the exponent

m^* is characteristic of the material. Values of m^* determined to date are given in Table 1.1.

TABLE 1.1. Velocity factor for dislocation movement
(Values at room temperature unless otherwise stated)

Crystal class	Metal	m^*	Reference
B.c.c.	Fe (high purity	5·0–6·0	Michalak (1965, 1966), Stein (1966)
	Fe (0·05% C)	15·0	Stein (1966)
	Fe–Si	20·0	Grånäs and Aronsson (1968)
	Fe–Si	35·0–40·0	Stein and Low (1960) Johnston and Stein (1963)
	Fe–Si	44·0	Floreen and Scott (1964a, b)
	Fe–Si	~60·0	Guard (1961), Noble and Hull (1964)
	Cr	5·0–9·0	Reid *et al.* (1967)
	Nb	7·0–18·0	Gubermann (1968)
	Mo	6·4	Prekel *et al.* (1968)
	W	4·8	Schadler (1964)
	NiAl	~40·0	Pascoe and Newey (1968)
F.c.c.	Cu	0·7	Greenman *et al.* (1967)
	Pb	1·0	Parameswaran and Weertman (1969)
	Al	1·0	Gorman *et al.* (1969)
	Ni	16·0	Rhode and Pitt (1967)
Cubic	Si	1·85 (600°C)	Suzuki and Kojima (1966)
	Ge	1·5 (600°C)	Chaudhuri *et al.* (1962)
	GaAs	2·0 (450°C)	Chaudhuri *et al.* (1962)
	InSb	1·9 (218°C)	Chaudhuri *et al.* (1962)
Cubic	LiF	16·25	Johnston and Gilman (1959), Johnston (1962b)
	NaCl	0·8	Gutmanas *et al.* (1963)
H.c.p.	Zn	1·0	Pope *et al.* (1967)

The following simple treatment will show how the upper and lower yield points arise. The number of mobile dislocations n in the specimen is related to the imposed strain rate and the dislocation velocity by the relation

$$\dot{\epsilon} = \rho b v$$

where b is the Burgers vector of the dislocations. If $\rho = \rho_1$ prior to yielding when the stress is σ_U, and $\rho = \rho_2$ immediately after yielding when the stress is σ_L, then

$$\rho_1 v_1 b = \rho_2 v_2 b$$

or

$$\rho_2/\rho_1 = v_1/v_2 = (\sigma_U/\sigma_L)^{m^*}$$

i.e.

$$\sigma_U/\sigma_L = (\rho_2/\rho_1)^{1/m^*}$$

Taking mild steel with $m^* \sim 15$, and if we take the number of **free** dislocations prior to yielding as $\rho_1 = 10^6$ lines cm^{-2}, and after yielding $\rho_2 = 10^9$ lines cm^{-2}, then

$$\sigma_U/\sigma_L = 1{\cdot}6$$

which is the order of magnitude observed (Fig. 1.4).

By making reasonable assumptions about the variation of dislocation density with strain, or using a coefficient of work hardening, it is possible to produce an expression for the initial portion of the stress–strain curve, as has been done by Hahn (1962) and by Cottrell (1963). The theory is also able to give a reasonable correlation between the observed delay time and the stress (see Section 2.4.3); Johnston (1962b) has used the small number of initial free dislocations to explain certain microstrain effects.

The velocity factors listed in Table 1.1 may be determined not only by pulsed loading, but also by studying the variation in yield stress with strain rate. Since

$$\dot{\epsilon} = \rho b(\sigma/\sigma_r)^{m^*}$$

$$\ln \dot{\epsilon} = m^* \ln \sigma + K$$

where K is a constant, and hence

$$m^* = (\Delta \ln \dot{\epsilon})/(\Delta \ln \sigma)$$

provided that the dislocation density ρ remains constant. This will not be so, but as has been discussed by Michalak (1964, 1966) it is still possible to derive reasonable estimates of m^* by this method by extrapolating to zero strain (see also Floreen and Scott (1964a, b), Li and Michalak (1964) and Ono (1966)). An alternative approach, with slightly different results, has been proposed by Yada (1967).

The effect of temperature on m^* has also been reviewed and examined by Michalak (1965), Bernstein (1969), Balasubraminian (1969) and Evans (1969). m^* generally increases the lower the temperature, but in high purity iron, the rate of increase is more rapid than in silicon iron, and below about 100 K, is higher than Fe–Si. Certain investigators, notably Johnston and Gilman (1959), Christian (1967) and Krausz (1968), have suggested the velocity equation be written

$$v = C_1\sigma^{m^*} \exp(-C_2/T)$$

which may make it possible to apply classical reaction rate theory to obtain this empirical result. Christian (1967) suggests that on this basis m^*T should be constant; however experiments do not bear out this point except possibly at low temperatures. In particular, it would be desirable if this wholly empirical relation could be related to the activation analysis which has been so valuable in discussing the low-temperature properties of b.c.c. metals, and this may yet be possible (Li, 1967).

The major advance given by this theory is that it avoids having to deal with thermal unpinning as a necessary part of the Cottrell–Bilby theory. The initial number of free dislocations must be low, so that a discussion of the elements responsible for pinning, whether by precipitates, atmospheres, point defects or other means, is relevant. The second necessary condition for yield points is that the velocity exponent (m^*) is not too high, otherwise the yield point may be masked. The main weakness in the theory lies in the expression for dislocation velocity, which has a purely empirical basis. Christian (1964, 1967) in particular has criticized this view, and discussed the significance of observed variations in the value of m^*, but is not yet possible to substitute another velocity equation which has a sound theoretical, rather than an empirical, basis.

1.5.7 *Effects of composition on yielding*

It should, in principle, be possible to distinguish between the modes of locking previously discussed by studying the variation of yield stress with varied amounts of solute element.

Suzuki (1957) analysed the expected variation of Cottrell locking with composition, and showed that it should follow $C(1 - C)$, where C is the atomic concentration of solute. Koppenaal and Fine (1961), following Flinn (1956, 1958) suggested that short-range order locking should vary as $[C(1-C)]^2$. Suzuki (ibid.) also showed that chemical locking should vary as $C(1-C)$, but later papers by Hendrickson (1962) and Guard and Fine (1965) revealed that this was not correct. Suzuki locking reached a maximum, dependent on the alloy system considered, but provided the thermodynamic parameters are known, correlation should still be possible.

In practice, the distinction may not be so easy. In interstitial systems, the low concentrations make these terms linear in C (Cracknell and Petch, 1955a). The effects of concentration on electronic locking are difficult to estimate, while in the case of precipitate locking, the magnitude of the yield drop will depend more on the density of unlocked dislocations, and this again is an estimate which is difficult to predict. Direct studies of dislocation densities are of more value here than an indirect study of yield stresses.

1.6 *The lower yield point*

The lower yield point can be described as either the growth stress of the Luders band (as in the case of mild steel) or as the stress below which the band ceases to grow. The distinction is not trivial, and is in fact necessary to discuss cases of serrated yielding (Fig. 1.3(c)), where as will be seen in the following chapters, the band is moving in a jerky manner. A further distinction is probably also needed for the case of single crystals, where the Luders band is simply the progressive spread of parallel slip from one end of the gauge length to the other. In this case, there is often only a minimum in the curve (see Fig. 1.3(a)), although in some cases extensive Luders strains are obtained (Chapters 6 and 7). Hahn (1962) refers to the former as 'continuous' yielding, and in cases where an extensive Luders strain is present as 'discontinuous' yielding. It is felt here it is better to draw no clear distinction between the two cases. Hereafter in this section it is assumed that specimen conditions are such that only a single band with a unique front is propagating along the specimen.

1.6.1 *The Luders strain*

The Luders strain (ϵ_L) is simply the strain experienced by the deformed region of the test piece, and can readily be determined from the stress–strain curve of the material, for the point D (Fig. 1.1) is always clearly marked, often by a slight drop in load, and the rate of strain hardening immediately afterwards is rapid.

The Luders strain is a function of grain size, and this feature is particularly marked in the case of mild steel (Fig. 1.12). A similar relation appears to hold in niobium, and in aluminium–magnesium alloys. While the grain size is fine, however, the bands appear sharp as shown in Plate 1.4(a) and deformation virtually proceeds grain by grain along the specimen (Morrison and Glenn, 1968); with coarse grain sizes the sharp front to the bands may be lost, leading to the so-called 'diffuse' Luders bands illustrated in Plate 1.4(b) (Hall, 1950, 1951a). These diffuse bands are common in aluminium alloys, presumably because of the larger grain sizes; because of the smaller Luders strains, they are more difficult to observe optically. The strain rate at the band front is markedly reduced, and the full Luders strain is only developed at a distance of several tens of grain diameters behind the Luders front (Carrington and McLean, 1965). In most commonly produced grades of mild steel, however, grain refining gives an added yield strength, and sharp Luders bands are commonly observed.

Observation and recording of the Luders band pattern is relatively easy using critical macro-illumination, as described by Boxall and Hundy (1955). Liss (1957) has suggested light copper plating as a means of improving their visibility. Strain-sensitive etchants, such as Fry's reagent (45 g cupric chloride in 180 cm^3 HCl and 100 cm^3

water) are also helpful in the case of complex Luders band patterns around notches, or in bend tests (Plate 1.5). (It only works well on high-nitrogen steels according to Fell (1937).) However, the transition from sharp to diffuse band propagation can only be followed with difficulty by normal metallographic practice, using, for example, slip line counts, and diffuse bands themselves are rendered more readily visible by the Moiré fringe technique of Theocaris and Koroneos (1963), or by other optical techniques, e.g. Hall (1950). The use

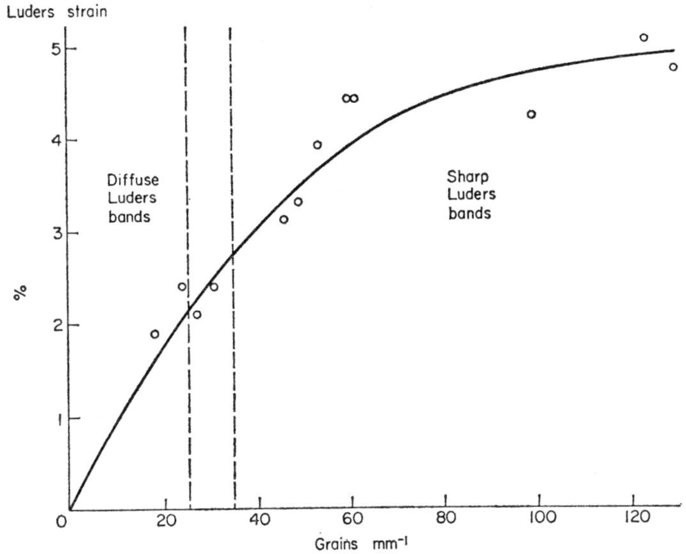

FIGURE 1.12. Variation of Luders strain with grain size in low carbon mild steel

of special lacquers such as Stresscoat, which chip or peel away from the deformed regions, is only of limited use with the higher values of ϵ_L, i.e. with sharp Luders bands.

1.6.2 *Shear at the band front*

It has already been mentioned (Section 1.2) that a macroscopic shear can be readily observed at the front of the Luders band, and very often the band front appears to make an angle of 45° to the tension axis (Plate 1.1(a)). In other words, the band front follows the plane of maximum shear. This is only a first approximation, for closer study by Lomer (1952) using five-sided specimens, showed that the true angle between tensile axis and band front was 48° ± 1·5°. Jaoul (1961) also pointed out that Luders bands take up angles which minimize stress at their edges, and one slip system is mainly developed in each grain, the band front following this plane of maximum resolved shear stress, averaged over the grains of the cross section. Textures in the sheet will thus also affect the observed angle (Grimes, 1967).

Friedel (1964) has pointed out the role of the nucleus of the Luders band in determining the magnitude of the yield drop. In Fig. 1.13 a

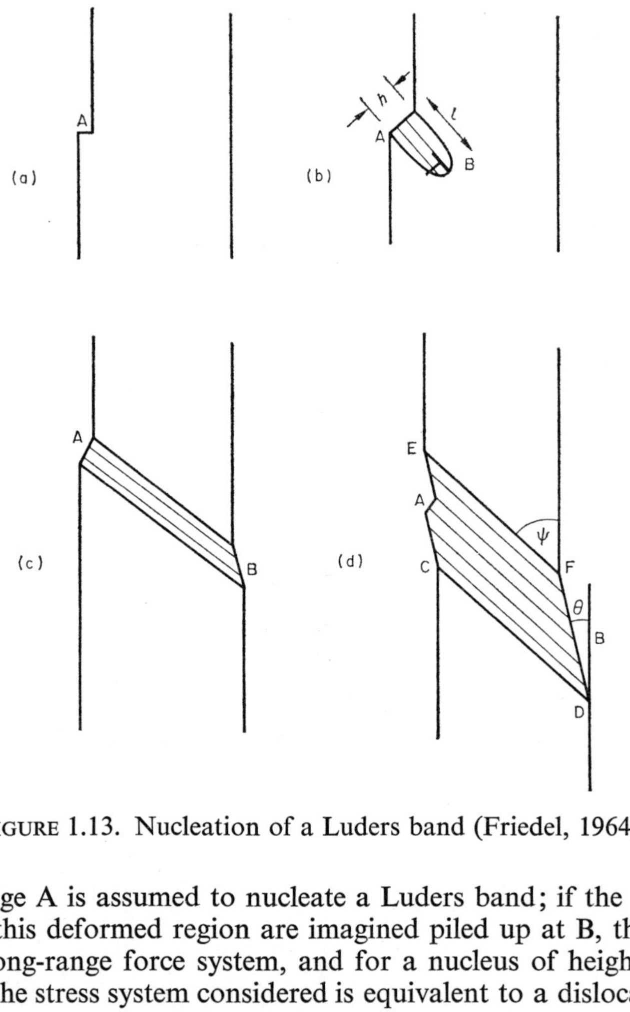

FIGURE 1.13. Nucleation of a Luders band (Friedel, 1964)

small ledge A is assumed to nucleate a Luders band; if the dislocations in this deformed region are imagined piled up at B, they produce a long-range force system, and for a nucleus of height h and angle θ, the stress system considered is equivalent to a dislocation of Burgers vector $h\theta$. If the nucleus is semicircular, of radius l, the supplementary stress to continue its growth is

$$\Delta\sigma \simeq \mu(h\theta)^2/h\theta l = \mu h\theta/l$$

This is a notable stress difference, since $h \sim l$ and θ is a few per cent, and undoubtedly contributes to the yield drop.

Crussard (1963a, b; 1964) suggests that the initial Luders bands form across the front of a strip specimen, i.e. in the plane of minimum stiffness of the material, giving a frontal angle of close on 90°; thereafter, tongues of deformation spread out (see Plate 1.1(b)) and the front finally grows to form an angle of 45–50° (Plate 1.1(a)).

The macroscopic effects on specimen shape are nevertheless quite remarkable – circular wires become elliptical in cross section following the passage of the shear front, while thin strip, deformed at a suitable temperature can in fact be bent in the thick plane of the specimen (Hall, 1951c).

1.6.3 *Velocity of the Luders band front*

The Luders strain ϵ_L is closely related to the velocity of the Luders band front. If V is the velocity of the crosshead of the testing machine then in time δt a length of material δl at the band front becomes $(1 + \epsilon_L)\,\delta l$ as the band front passes through it. In this time, the crosshead has moved $V\delta t$ so that

$$(1 + \epsilon_L)\,\delta l - \delta l = V\delta t$$

$$\epsilon_L = V/v_L$$

where $v_L = \delta l/\delta t$ is the velocity of the Luders front (Sylwestrowicz and Hall, 1951; Moon and Vreeland, 1968). Where more than one front is operating, this equation becomes

$$\epsilon_L = V/Nv_L$$

where N is the number of active fronts, and their velocity is each reduced by a factor $1/N$ (Butler, 1962a).

Since the deformation increases from zero to ϵ_L over a very few grain diameters behind the front, the strain rate there is correspondingly very high. Cottrell (1964) has shown that the localized rate is

$$\dot{\epsilon} = v_L \epsilon_L / S$$

where S is the thickness of the yielding region. If $v = 10$ mm s^{-1}, $\epsilon_L = 0{\cdot}03$ and $S = 0{\cdot}1$ mm, then $\dot{\epsilon} = 3$ s^{-1}. Thus, a bar deforming homogeneously at this rate would double its length in about $\frac{1}{3}$ s.

Part of the difficulty of dealing theoretically with Luders band propagation lies in our lack of knowledge of the work-hardening rate at the band front. Hart (1955) has attempted a model for a Luders band front, using material with assumed properties given by the equation

$$\sigma = \sigma_0 \exp\,(K_1\epsilon + K_2\dot{\epsilon})$$

where σ_0, K_1 and K_2 are parameters. This equation, the exact form of which is not considered critical, gives

$$\epsilon = \dot{\epsilon}_0/K\,[1 - \exp(-Kl/v_L)]$$

where $K = -(1 - K_1/K_2)$ and $\dot{\epsilon}_0$ is the value of $\dot{\epsilon}$ at $l = 0$, the front of the band.

The simple relation between Luders strain and band velocity and crosshead velocity is true at all temperatures and stresses, but it cannot be used to predict variations in ϵ_L with temperature. This factor will alter both the yield stress, and the work-hardening coefficient and these in turn affect ϵ_L.

Fisher and Rogers (1956) measured the velocity of the bands at 24°C and −78°C using deadweight loading and Stresscoat on the specimens, and applied theory of delayed yielding to the material at the band front. The formula $\ln t_d = A + B\mu^2/\sigma T$ (Section 1.5.1) was applied, and gave

$$\ln v_L = C - (D/\sigma_L T)$$

In other words, the band velocity depended only on stress and temperature. Other measurements at room temperature were made by Butler (1962a, b). There were a number of unexplained effects occurring with the propagation, but at the single temperature of testing, he showed

$$\sigma L = A' + B' \log \epsilon_L$$

which is equivalent to Fisher and Rogers. The parameters A' and B' depend on grain size and composition. The generalized equation $\epsilon_L = V/Nv_L$ was shown to hold.

The theoretical basis for the Fisher–Rogers treatment was the thermal unpinning of dislocations. With the development of theories of dislocation multiplication and the feeling that thermal unpinning rarely arises, these theories have become untenable; nevertheless, they have their value and de Kazinczy (1959b) used the Fisher–Rogers treatment to prove that hydrogen does not affect the yielding behaviour of α-iron (see Chapter 4).

Hahn (1962) and Conrad (1963c) have endeavoured to apply more general theories to the front velocity. Hahn for example, proposed that the velocity of the band front is determined by the velocity at which dislocations move near the band front, i.e.

$$v_L \propto (\sigma_L/\sigma_r)^{m*}$$

As will be seen in the next section, there are difficulties associated with the varying dislocation density; nevertheless it will also give the Butler relationship above. Conrad suggested from his earlier work (1961) that

$$\dot{\epsilon} = \nu \exp [-H(\tau^*)/kT]$$

where ν is a frequency factor, and $H(\tau^*)$ the activation energy, itself a function of the effective shear stress τ^*. Assuming the same dislocation mechanisms control propagation of the band front (see Conrad (1963c)) then

$$v_L = v_{L_0} \exp [-H(\tau^*)/kT]$$

Figure 1.14 shows Conrad's plot of results, plotted on a stress axis with arbitrary zero for $v_{L_0} = 10 \mu\text{m s}^{-1}$. From his theory can be derived an activation volume v^*; the value of v^* from the Luders band

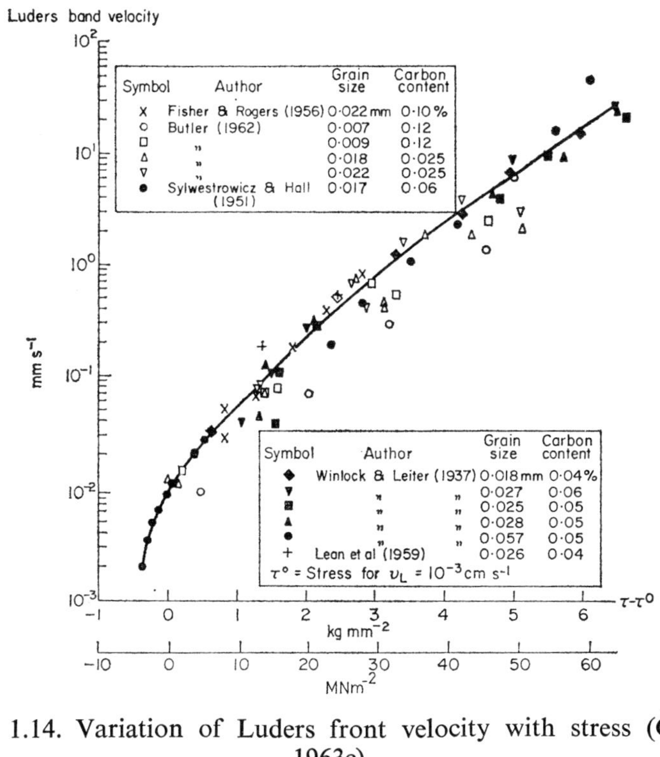

FIGURE 1.14. Variation of Luders front velocity with stress (Conrad, 1963c)

velocity measurements agrees well with Conrad's more detailed theories on the temperature dependence of the yield stress. A similar approach has been used by Violan *et al.* (1967) (see Chapter 3).

A similar study was carried out on niobium by Conrad and Stone (1964). Using 'Stresscoated' specimens, the band front velocities were measured on $\frac{1}{16}$ in. diameter rods. The results were very similar to mild steel, with the addition that rotation of the band front was noted. This is not seen on thin strips; it may arise from the relatively short length: diameter ratios used. The results, again, fitted Conrad's formula quoted above.

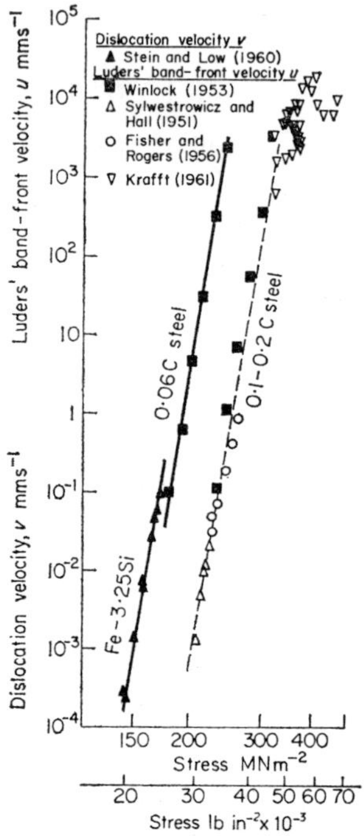

FIGURE 1.15. Stress dependence of dislocation velocity and Luders front velocity (Hahn, 1962)

1.6.4 *The lower yield stress*

As explained in Section 1.2, under conditions where a single Luders band is propagating through the specimen, the lower yield stress may be measured very accurately. The variables which will concern us here are the effects of strain rate, grain size and temperature.

The strain rate variation comes from the equations given earlier in Section 1.5.6. Since

$$\dot{\epsilon} = \rho b v = \rho b(\sigma_L/\sigma_r)^{m*}$$

$$\ln\dot{\epsilon} = m^* \ln \sigma_L + C$$

and a plot of $\ln \dot{\epsilon}$ against $\ln \sigma_L$ should give a slope of m^*. This method of determining the velocity parameter was used by Hahn (1962), his results being reproduced in Fig. 1.15. It is seen the slope of this line is parallel with the direct determination of dislocation velocities in silicon iron by Stein and Low (1960). It now appears that this analysis is in error, for Hahn's values imply an equality with m^* for Fe–Si alloys. This is now known not to be so (cf. Table 1.1). The difficulty arises because the dislocation density has not been held constant, and because of the manner of the Luders extension, results cannot be extrapolated to zero strain. Thus, values of m^* derived from this analysis are too large: Moon and Vreeland (1968) show $m^* \sim 60$ if derived from Luders band values, while direct measurements give $m^* \sim 40$. Christian (1964, 1967) has discussed these difficulties, and points out that a properly developed, thermally activated stress theory such as that of Li (1967) is a better approach. This point will be discussed again in Chapter 3.

The effects of grain size on the upper yield point have already been noted in Section 1.4.1, and are equally marked on the lower yield stress. Although the advantages of grain refinement have been known for many years (Arrowsmith, 1924), the relation between mean grain diameter (d) and σ_L was first given by Hall (1951b), and later studied in detail by Petch and his co-workers (e.g. Cracknell and Petch (1955a) and Heslop and Petch (1956)), giving the result as

$$\sigma_L = \sigma_0 + k_y d^{-1/2}$$

Although Baldwin (1958; see also Kochs, 1959), Conrad (1963a) and Anderson *et al.* (1968) have suggested alternative relationships, this result has wide applicability and holds not only for steels of a variety of grades, but for many other metals. It has also been found to hold, at least at small strains, in metals which show no yield (Armstrong *et al.*, 1962), in hardness measurements (Hall, 1954; Jindal and Armstrong, 1967, 1969) and in intermetallic compounds. It may also apply for strains well beyond the yield, or beyond the Luders extension if one is present (Evans and Rawlings, 1968). Results are summarized in Table 1.2.

This grain-size relationship (often known as the Hall–Petch equation) was originally developed on the basis of the unpinning of dislocations, and since in fine-grained mild steel the edge of the Luders band appears sharp (Lomer, 1952), it seemed reasonable to suppose that deformation was spreading, grain by grain, through the specimen. In Fig. 1.16, AB represents a grain boundary dividing the yielded region from the unyielded region. Dislocations in the yielded grain A run up to the boundary, and form piled-up arrays of dislocations, whose presence increases the stress in grain B, until a dislocation

TABLE 1.2. Typical values of σ_0 and k for various metals and alloys

(Measurements at room temperature unless otherwise stated)

Crystal structure	Metal	Condition	σ_0		k		Reference
			kg mm^{-2}	MN m^{-2}	kg mm$^{-3/2}$	MN m$^{-3/2}$	
B.c.c. (A2)	Mild steel 0·03% C	Upper yield point	0	0	4·01	1·24	Hutchison (1963)
	Mild steel 0·12% C	Upper yield point	0	0	5·25	1·63	Petch (1964)
	Mild steel 0·06% C	Lower yield point	4·2	41	2·1	0·66	Hall (1951)
	Mild steel 0·12% C	Lower yield point	7·2	71	2·39	0·74	Armstrong *et al.* (1962)
	Mild steel 0·15% C	Lower yield point	9·2	90	2·5	0·79	Hull and Mogford (1958)
		Irradiated (1·5 × 10^{19} nvt)	12·4	122	2·5	0·79	
	Swedish iron 0·02% C	Lower yield point	4·8	47	2·28	0·708	Cracknell and Petch (1955a)
	Swedish iron decarburized	Propnl. limit	3·7	36	0·66	0·21	Armstrong *et al.* (1962)
	Mild steel 0·04% C	Lower yield point	4·6	45	2·29	0·712	Wilson and Russell (1960a)
	Mild steel 0·04% C	Strained 4%	19·7	193	0·37	0·12	
	Mild steel 0·04% C	Strained 4% and aged 10^4 min at 60°C	22·1	217	0·94	0·29	
	Fe–0·23% C–0·65% Mn	Yield stress	18·7	183	2·12	0·658	Morrison (1963)
	Fe–0·23% C–0·65% Mn + 0·16% Nb	Yield stress	23·9	234	2·13	0·662	
	Fe–3·25% Si	Yield stress	32·5	319	2·1	0·66	Suits and Chalmers (1961)
	Fe–3·25% Si, 73 K	Twinning stress	28·0	275	11·1	3·45	Hull (1961)
	Vanadium	Yield point	30·9	303	0·72	0·22	Lindley and Smallman (1963a)
	Vanadium, 20 K	Twinning stress	87·9	862	4·84	1·50	

	Chromium, 0°C	Yield point	18·9	185	2·93	0·908	Marcinkowski and Lipsitt (1962)
	Chromium, −195°C	Twinning stress	54·1	531	14·2	4·40	
	Niobium	Yield point	7·0	69	0·13	0·041	Adams *et al.* (1960)
	Molybdenum	Yield point	11·0	108	5·7	1·8	Johnson (1959)
	Molybdenum	10% strain	40·0	392	1·7	0·54	Johnson (1959)
	Tantalum	Yield point	12·6	124	1·0	0·31	Koo (1962)
	Ta–10% W	Yield point	77·3	758	−0·28	−0·089	Tedman and Ferris (1962) (low accuracy)
	W (250°C)	Yield point, bending	14·7	144	7·6	2·4	Farrell *et al.* (1967)
F.c.c.	Aluminium	0·5% strain	1·6	16	0·22	0·070	Carrecker and Hibbard (1957)
(A1)	Al/3·5% Mg	Yield point	5·0	49	0·85	0·26	Phillips *et al.* (1952)
	Copper	0·5% strain	2·6	26	0·36	0·12	Feltham and Meakin (1957)
	Cu–4 at.% A	Yield point	4·1	40	0·75	0·23	Köster and Speidel (1965)
	Cu–8 at.% A	Yield point	6·0	59	1·38	0·427	Köster and Speidel (1965)
	Cu–5.9 at.% As	Yield point	3·8	37	0·87	0·27	Köster and Speidel (1965)
	Cu–15 at.% Ga	Yield point	5·0	49	1·13	0·351	Köster and Speidel (1965)
	Cu–3·2 at.% Sn	Yield point	11·6	114	0·42	0·13	Russell (1965a)
	Cu–8 at.% Sn	Yield point	12·0	118	0·75	0·23	Köster and Speidel (1965)
	Cu–20 at.% Zn	Yield point	4·3	42	0·78	0·24	Köster and Speidel (1965)
	Cu–30 at.% Zn	Yield point	4·6	45	1·00	0·310	Quoted in Armstrong *et al.* (1962)
	Cu–30 at.% Zn	Strained 20%	34·4	337	1·10	0·342	Quoted in Armstrong *et al.* (1962)

TABLE 1.2. (Continued)

Crystal structure	Metal	Condition	σ_0		k		Reference
			kg mm^{-2}	MN m^{-2}	kg mm$^{-3/2}$	MN m$^{-3/2}$	
	Cu–36 at.% Zn	Yield point	3·7	36	1·24	0·386	Köster and Speidel (1965)
	Cu–12·5 Ni–12·5 Zn	Yield point	5·3	52	0·83	0·26	Köster and Speidel (1965)
	Cu–29 Ni–29 Zn	Yield point	8·0	79	1·87	0·579	Köster and Speidel (1965)
	Nickel	Yield point	3·5	34	0·71	0·22	Sonon and Smith (1968)
	Silver	0·5% strain	3·8	37	0·22	0·070	Carrecker (1957)
H.C.P. (A3)	Zinc	0·5% strain, 0°C	3·3	32	0·71	0·22	Armstrong *et al.* (1962)
	Zinc	17·5% strain, 0°C	7·3	72	1·17	0·364	Armstrong *et al.* (1962)
	Magnesium	0·2% strain	0·7	7	0·9	0·3	Hauser *et al.* (1956)
	Magnesium	0·2% strain, −196°C	1·5	15	1·5	0·47	Hauser *et al.* (1956)
	Titanium	Yield point	8·0	79	1·3	0·41	Guard (1955)
	Zirconium	0·2% strain	3·0	29	0·8	0·3	Keeler (1955), Coleman and Hardie (1966)
Cubic ($L1_2$)	Ni_3 Mn	Yield point, ordered	10·7	105	3·49	1·08	Johnston *et al.* (1965)
		Yield point, disordered	10·7	105	1·96	0·607	Johnston *et al.* (1965)
Cubic ($L2_0$)	FeCo	Yield point, ordered	5·1	50	0·92	2·8	Marcinkowski and Fisher (1965)
		Yield point, disordered	38·5	378	0·36	0·11	
Cubic (A_{12})	$Mg_{17}A_{12}$	0·5 strain	20·7	203	−1·02	0·31	Grierson and Parkins (1965)
Hex. (B8)	$TlBi_2$	Yield point	2·5	25	0·38	0·12	Robinson and Bever (1966)

source there becomes unpinned and grain B yields. From dislocation theory, the effective stress acting on the dislocations is $\sigma_L - \sigma_0$, where σ_0 is a friction stress impeding the movement of dislocations and equal in fact to the single crystal yield stress. At a distance r_c ahead of the piled-up group, the stress is given by $(\sigma_L - \sigma_0)\sqrt{(d/r_c)}$ (Eshelby *et al.*, 1951), where d is the length of the slip path, assumed to be the grain diameter. If then grain B yields, i.e. the dislocations there become unpinned when this stress concentration reaches a critical value, σ_c, then writing $k_y = \sigma_c r_c^{1/2}$, the Hall–Petch equation is obtained.

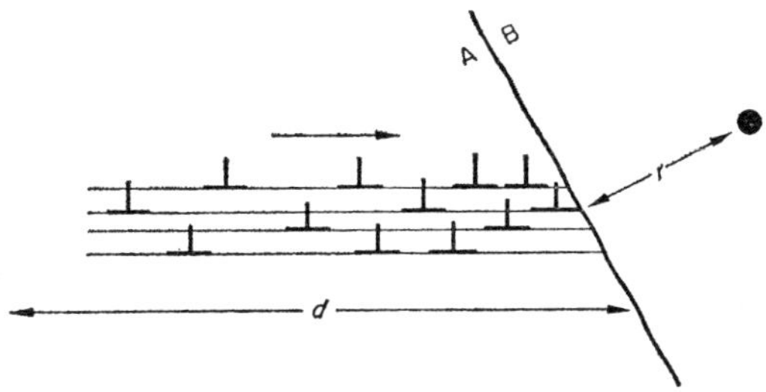

FIGURE 1.16. Dislocation pile-ups (Cottrell, 1964)

The determination of the constants in the grain-size equation is a tedious and lengthy affair, and considering the apparent widespread applicability, it is natural that ways and means have been considered to reduce this. One method, proposed by Rosenfield (1962) and Johnson (1962) and used by Gilbert *et al.* (1962), Mogford and Hull (1963) and others, is to equate the homogeneous deformation section of stress–strain curve to the usual equation

$$\sigma = K\epsilon^n$$

Extrapolation of this curve plotted logarithmically back to intersect the elastic part of the curve (AB, Fig. 1.1) gives a stress intercept equal to σ_0. From a knowledge of the lower yield and grain size of the specimen, both σ_0 and k_y may be obtained.

Recently, however, this method has come under criticism. Lindley and Smallman (1963b), Phillips and Chapman (1965), Holzmann and Man (1966), Dingley and McLean (1967) and Christ and Smith (1967) have all criticized this approach, due to the variation of the parameters in $\sigma = K\epsilon^n$ with grain size, and it now seems there is nothing for it but to carry out stress–strain curves and grain size counts on a number of specimens to obtain the parameters accurately.

The effects of variables on the two parameters in the Hall–Petch equation for mild steel have been admirably investigated by Petch and his co-workers. Petch himself (1953) noted that the slope of the lines in the $d^{-1/2}$ plot is virtually independent of temperature; Cracknell and Petch (1955a) examined the effects of carbon and nitrogen, Heslop and Petch (1956) examined the effects of manganese and testing temperature and Codd and Petch (1960) the effects of boron.

The main results from these, and the other references listed in Table 1.2, may be summarized as follows:

(a) k_y is not very sensitive to temperature. This is certainly true in mild steel (Figs. 1.4 and 1.17), and increases only slightly at low temperatures in copper-base alloys (Köster and Speidel, 1965).

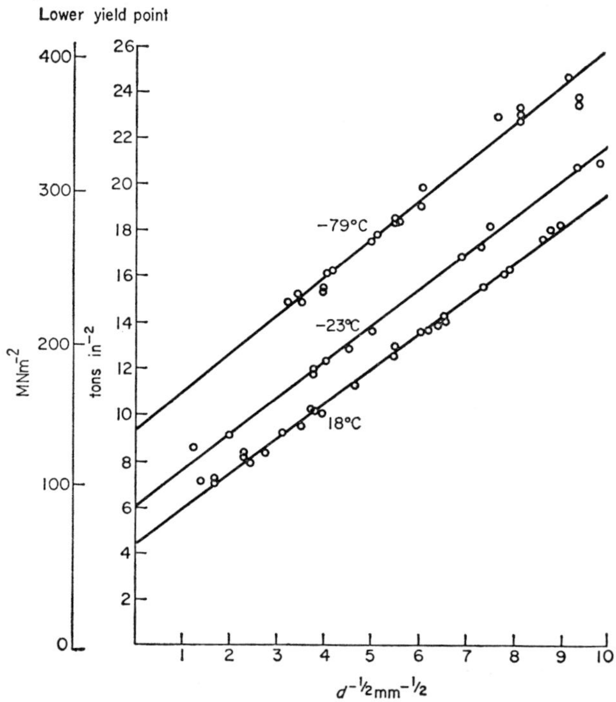

FIGURE 1.17. Dependence of lower yield stress on grain size for mild steel (Petch, 1958)

(b) In mild steel, k_y is not very sensitive to composition (Fig. 1.18), but falls appreciably on decarburization (Table 1.2). In copper-base alloys, k_y again seems to increase with added alloying element, but in certain cases a maximum is found (Köster and Speidel, ibid.; Hutchison and Pascoe, 1969). A similar effect occurs in nickel–sulphur alloys (Floreen and Westbrook, 1969).

(c) k_y is not very sensitive to irradiation (but see also Section 2.6).

(d) In iron, k_y is reduced by pressurization (Yajima and Ishii, 1967, 1969, Radcliffe, 1969).

(e) k_y is insensitive to strain rate (Harding, 1969).

(f) Beyond the Luders extension, the grain-size law still holds, but k_y is generally smaller than at the yield stress. If strain ageing occurs (Section 1.7) k_y increases again, but does not regain its original value.

(g) σ_0 is markedly temperature sensitive.

(h) σ_0 is markedly sensitive to composition, strain, strain rate, strain ageing and irradiation.

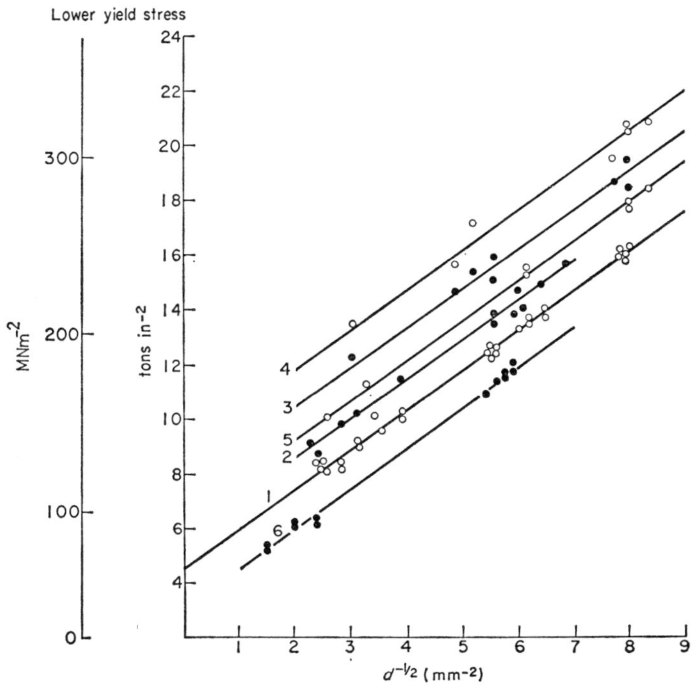

FIGURE 1.18. Dependence of lower yield stress on grain size, for (1) annealed mild steel En 2; (2) En 2, nitride; (3) En 2 quenched from 650°C; (4) En 2, quenched, aged 1 hour at 150°C; (5) En 2, quenched, aged 100 hours at 200°C; (6) annealed Swedish iron (Cracknell and Petch, 1955a)

We will discuss first the variations in σ_0 – the so-called friction stress – because it represents the force restricting the motion of free dislocations on the slip plane. Heslop and Petch (1956) suggested that because of the sensitivity of σ_0 to both temperature and composition, it could be regarded as made up of two components:

$$\sigma_0 = \sigma_0' + \sigma_0''$$

where σ_0' is dependent on temperature only (Fig. 1.19) and σ_0'' dependent on composition only (Fig. 1.20).

It is interesting to note that σ_0 (Fig. 1.19) shows the same steep rise at low temperatures which is exhibited by all the b.c.c. refractory metals, and this has led Heslop and Petch (ibid.), Conrad and Schoeck (1960) and in particular Conrad (1963b, 1966) to suggest that the Peierls–Nabarro force – the lattice force determining the width of a dislocation – is responsible for this rapid rise in yield stress at low temperatures. This point is discussed again in Chapters 2 and 3, and is not relevant to our discussion of yield point effects at this stage.

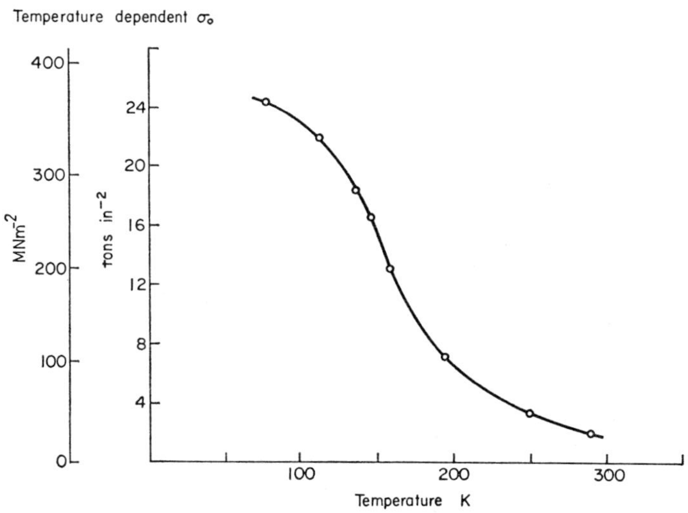

FIGURE 1.19. Variation of σ_0' for mild steel on temperature (Heslop and Petch, 1956)

What now remains for discussion is the variation of the parameter k_y, which, from the derivation of the grain-size equation, is given by

$$k_y = \sigma_c r_c^{1/2}$$

Since σ_c is a stress needed (on Cottrell–Bilby theory) to free the dislocations in the neighbouring grain, k_y was frequently called the 'unpinning parameter', and as such should be temperature-dependent and follow the Cottrell–Bilby equation (Section 1.5.1). However, as seen above, this is not the case, and this surprising result had led Conrad and Schoeck (1960) and others to take this as evidence that unpinning (in the Cottrell–Bilby sense) rarely, if ever, occurs.

The dislocation multiplication theory of yielding discussed above also required permanently pinned dislocations, and the question therefore arises where the new dislocations are nucleated. Cottrell and Fisher (quoted in Cottrell, 1963) suggest that these new dislocations are created by loops being thrown out from the grain boundaries themselves, and Fisher (1961) has reported observing these in furnace-cooled samples. (Plate 1.6.) (Hornbogen has a similar effect in Fe–P alloys (Hornbogen, 1963).) Fisher also points out that it is possible to vary the degree of pinning by heat treatment. In quenched, and lightly

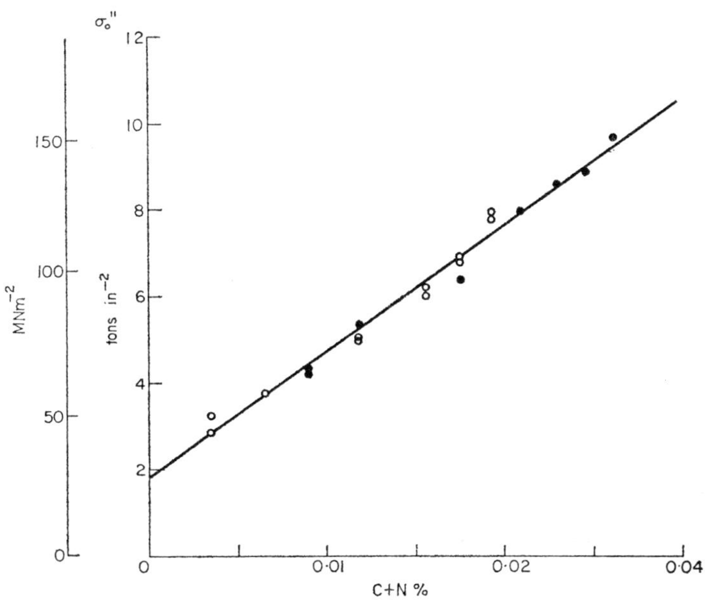

FIGURE 1.20. Variation of σ_0'' for mild steel on composition (Cracknell and Petch, 1955a)

quenched-aged samples, it may be possible to unpin dislocations, in which case

$$k_y = \sigma_p r_p^{1/2}$$

where σ_p is an unpinning stress $<\sigma_c$, and which will give smaller values of k_y than the usual (0·70 $MNm^{-3/2}$) observed in mild steel. Figure 1.21 shows this to be the case. k_y is very temperature-dependent in these lightly aged alloys, but twelve hours heat treatment is enough fully to lock the dislocations, and thereafter the value of k_y is virtually independent of temperature. In quenched material, dislocations are never seen being emitted from grain boundaries. Wilson (1967) confirms this view, and considers that grain-boundary segregation largely determines the value of k_y.

These observations have led to discussion on the magnitude and variation of k_y with temperature and strain. Armstrong *et al.* (1962), in reviewing determinations of k_y, show that the value of k_y should be given by

$$k_y = m^2 \tau_c r^{1/2}$$

where m is the Taylor orientation parameter, being related to the number of operative slip systems in the polycrystalline aggregate, necessary to keep grain boundaries contiguous. This will vary from a minimum of 2 for an infinite number of slip systems through 3.1 for a face-centred cubic aggregate, to ~6·5 for hexagonal materials. τ_c in

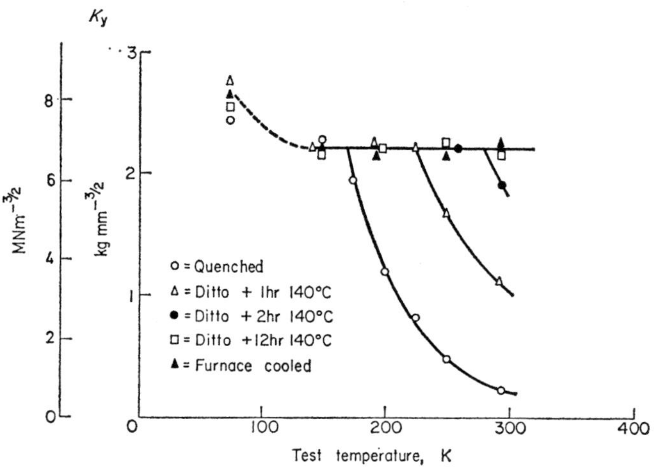

FIGURE 1.21. Dependence of k_y on heat treatment and testing temperature (Fisher, 1961)

the expression above is now the critical **shear** stress for the operation of new sources; r has its original meaning.

Thus, with changes in temperature, it is possible that the value of m decreases as the temperature falls, while τ_c increases, and the results give a roughly constant value of k_y. The values of k_y also roughly correlate; the h.c.p. metals should be about nine times greater than f.c.c. metals – the actual value (Table 1.2) is about 4. The same argument has been invoked in ordered alloys: Marcinkowski and Fisher (1965) show that k_y for ordered FeCo is theoretically about twice that of disordered, since in the latter case cross slip is easy (i.e. m = 2), whereas it is difficult in the ordered case and m ~ 3·1. The actual ratio is about 3.

For strains beyond the initial yield, as mentioned previously, the Hall–Petch equation is still found to apply. σ_0 naturally rises, due to strain hardening and the increased dislocation density, but k_y (or rather k_ε at strains beyond the yield) will fall.

The value of k_ε has been estimated by Meakin and Petch (1963) and by Marcinkowski and Fisher (1965) as

$$k_\varepsilon = m^2 \tau_c r^{1/2} + \frac{m^{3/2} \alpha \mu}{\sqrt{2}} (\epsilon b)^{1/2}$$

where μ is the shear modulus, α a constant $\sim 0{\cdot}4$ and the other symbols have already been defined. This analysis implies that the stress–strain curve is roughly parabolic in nature.

This picture is, however, only a first approximation, and Armstrong *et al.* (ibid.) cite several complicating factors, due to strain hardening and the formation of obstacles within the grains, as well as possible changes in the parameters. In α-iron, dislocation tangles soon form within the grains (Plate 1.7). In the ordered alloys, such as FeCo, quite marked changes in k_ε may occur (Marcinkowski and Fisher, ibid.), the values going through a maximum for both the ordered and disordered states, but at different strains. On the other hand, this cannot be characteristic of ordered alloys, since Ni_3Mn gives values virtually independent of strain, while α-iron and other metals and alloys also seem to show few anomalies, either decreasing or increasing in a smooth manner.

The presence of the three parameters which now combine to form k_y, and the difficulties of discussing their variation with temperature, has led Li (1963) to suggest other possible ways of nucleating dislocations within the undeformed grains. The role of sub-boundaries, and in particular, the possibility of grain-boundary ledges creating free dislocations is considered in some detail. This is consistent with Fisher's observations mentioned earlier. The value of k_y is then:

$$k_y = \alpha \mu b (8M/\pi)^{1/2}$$

where α is again $\sim 0{\cdot}4$, and M is the number of ledges per unit length of grain boundary, and is independent of grain size. It will be seen this is virtually independent of temperature, and the analysis can also be developed to allow for effect of impurities condensing on the grain boundary and also for plastic strain. There are some difficulties remaining, however. The values of ledge densities for hexagonal material do seem rather high, the values of k_y for these metals having been explained in the theory of Armstrong *et al.* (ibid.) as an orientation effect. Secondly, Marcinkowski and Fisher (1965) have shown in ordered FeCo, the additional stress needed to nucleate a dislocation by this model is

$$\tau = E_0/b$$

where E_0 is the antiphase boundary energy and b the Burgers vector. This is very much higher than the yield stress of the ordered alloy, which experimentally, is softer than the disordered case (Chapter 6). Nevertheless, Li's model is an interesting development and merits further study, particularly in conjunction with further investigations on k_y in other ordered systems.

Other derivations of k_y are possible. Gouzou (1964) has derived the form of the Hall–Petch equation by considering the energies needed to move a dislocation from an atmosphere of carbon atoms in the grain boundary, as well as in the grains. Köster and Speidel (1965), who noted numerous pile-ups in electron micrographs of their copper-base alloys, also derived k_y as

$$k_y = \mu/50\sqrt{l_i}$$

where $2l_i$ is the slip-line spacing.

Recently, closer attention has been given to the pre-yield micro-strain, as a means of considering the Luders band nucleus. This will be covered in some detail in Section 2.4.2, and it is sufficient to note here that, in the special case of silicon-iron, etch pitting is a very effective extensometer. This technique has enabled Worthington and Smith (1964, 1966) to distinguish three stresses affecting their etch-pit patterns, (i) where the first slip line is formed, (ii) where the first slip breakthrough occurs at a grain boundary and (iii) where the Luders front is formed. (ii) should follow a $d^{-1/2}$ law, but with $m = 2$, so its value of k_y should be less than for the lower yield stress. In fact, in fine-grained material ($d^{-1/2} \sim 8$) there is little difference between these two latter stresses. This again shows the difficulty of dealing with material which in its relatively coarse-grained state, is prone to show diffuse Luders bands. Worthington (1969) has also used this technique to show that pressurized material has a higher proportion of slipped grains prior to macroyielding.

Finally, Dingley and McLean (1967) have determined a value of k_y simply by considering the maximum stress needed to open up a loop of fresh dislocation line, and equating this to τ_c as defined previously. Ignoring any misorientation factors, the value of k_y so obtained agrees closely with experiment.

All these theories have two other experimental facts to deal with. In the case of niobium, k_y is very small (Adams *et al.*, 1960) or even zero (Churchman, 1959; Johnson, 1960a). A similar result applies to nickel (Wilcox and Smith, 1965). Armstrong *et al.* (1962) suggests that contamination during heat treatment may be interfering with the determination of an accurate value of k_y in niobium. Another anomaly is the intermetallic compound $Mg_{17}Al_{12}$, which has an ordered cubic (A12) structure, and a **negative** value of k_y, as yet wholly unexplained (Grierson and Parkins, 1965). They also show In_2Bi has a negative value of k_y, although the number of results is not sufficient to quote a value with any real accuracy. This compound is thus not included in Table 1.2.

All theories discussed above however still require that grain boundaries act in some way as effective barriers to dislocation movement; and that dislocations pile up against them. Since dislocations in iron may readily cross slip, arrays might not be expected to form, and in fact dislocation arrays are not often seen in thin film electron microscopy (Brandon and Nutting, 1960; Morrison and Glenn, 1968) possibly because they move out of the foils during thinning, but Suits and Chalmers (1961) have clearly seen arrays in silicon iron; as have Carrington and McLean (1965) and Smith (1960, 1962). Köster and Speidel (1965) have seen numerous dislocation pile-ups in copper-base alloys showing yield point effects. Fisher (1961) suggests a slip band may be as effective as a dislocation array.

In single crystals, of course, these concepts of the spread of deformation along the specimen cannot strictly apply. Here the deformation spreads as a sequence of parallel slip bands along the specimen (see, for example, Piercy *et al.*, 1955; Blewitt *et al.*, 1957, 1960), the progress of deformation probably engendered by the local constraints and stress concentrations set up between the deformed and undeformed regions.

Finally, it should be mentioned that Hutchison (1963) has found the same grain-size relation for the upper yield point in mild steel, and others have obtained the same law for twinning stresses (see Table 1.2). In both cases, however, k_y is much larger than for deformation by slip. This result may be interpreted to mean that the movement of slip dislocations is necessary to cause a Luders band or twin nucleus, as the case may be; the higher value of k_y may result from yielding taking place under triaxial stress conditions, or for other reasons in the variation of r_c.

It will also have been obvious from the comments through this chapter that, at least in the case of mild steel interstitial impurities (carbon and nitrogen) are in some way responsible for these yield points. This fact will be discussed in detail in the next chapter, but the identification of elements responsible is a vital part of any alloy system showing yield points.

1.7 *Strain ageing*

1.7.1 *Introduction*

One of the further characteristics of yield points in metals and alloys is their ability often to reappear after the material has been strained and subjected to a low-temperature (ageing) heat treatment. This is the phenomenon of strain ageing, apparently first systematically examined by Martens (1879). (Quoted by Kenyon and Burns, 1939.)

Suppose the material has been strained to an amount in excess of the Luders strain ϵ_L, the stress relaxed and the material aged. After short ageing times, the stress–strain curve on restraining may take the form shown in Fig. 1.22(a). In this state the curve is often known as the 'unloading' yield point, as it can occur in a variety of pure materials, as well as in mild steel.

The unloading yield point deserves a separate discussion, which will be postponed until Chapter 5, where it is seen to be commonly observed in f.c.c. metals; it is however distinct from the types of dislocation locking discussed in this chapter. For those alloys showing Luders deformation, and undergoing longer periods of ageing, the yield point becomes more obvious (Fig. 1.22(b)), until, as can be seen in this diagram, there is a true Luders extension for the second time. This band, which will be termed the **secondary** Luders band, has a diffuse front in the early stages (Hall, 1950, 1951a), but progressively sharpens as ageing proceeds, and at the same time hardness and yield stress also increase. In mild steel, the problem of strain ageing is of considerable importance in the production of sheet, as ageing can proceed at a slow but observable rate during storage at room temperature, and sheets need temper rolling before being used in deep-drawing operations to avoid stretcher strain formation.

The process of ageing can, in the case of mild steel, be repeated to produce tertiary Luders bands, but the ductility of the material rapidly deteriorates, and ductile failure occurs.

1.7.2 *Kinetics*

The study of the kinetics of strain ageing is an important and difficult part of the theory of yield point effects, and is rendered even more difficult by the variety of parameters employed. In many cases, the kinetics of ageing is one of the few methods available of identifying the solute atoms responsible for the yield point effects.

At short ageing times, when no Luders extension has developed the arbitrary choice of parameters makes comparison difficult. Figure 1.22(a), taken from Titchener and Davies (1965) illustrates this point. In mild steel, the existence of the yield plateau makes the

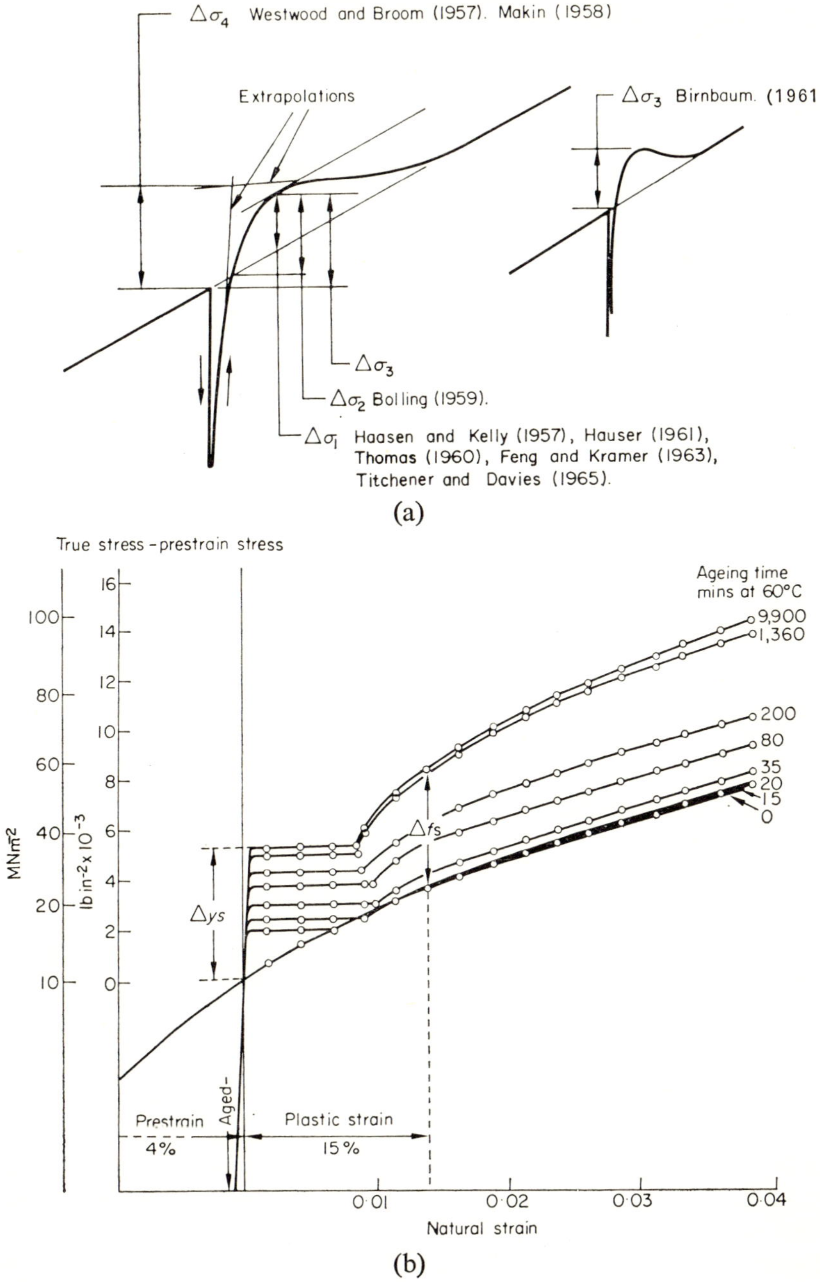

FIGURE 1.22. Development of yield points in strain-aged material after (a) short and (b) longer ageing times ((a) is diagrammatic only and shows the various parameters employed) ((a) Titchener and Davies, 1965; (b) Wilson and Russell, 1959)

choice less difficult, but even here there is some cause for confusion. Most investigators (e.g. Mura *et al.*, 1961) use a factor

$$f = (\sigma_U - \sigma_L)/\sigma_L$$

where σ_U is the upper yield point after ageing, σ_L the yield stress before ageing. σ_L will be a function of pre-strain, as well as composition, and in turn will depend on dislocation density. Other investigators have used hardness measurements, giving an ageing factor

$$f = \frac{H_t - H_0}{H_m - H_0}$$

where H_m is the maximum, H_0 the initial, and H_t the hardness after time t. This value of f cannot strictly be compared with that given

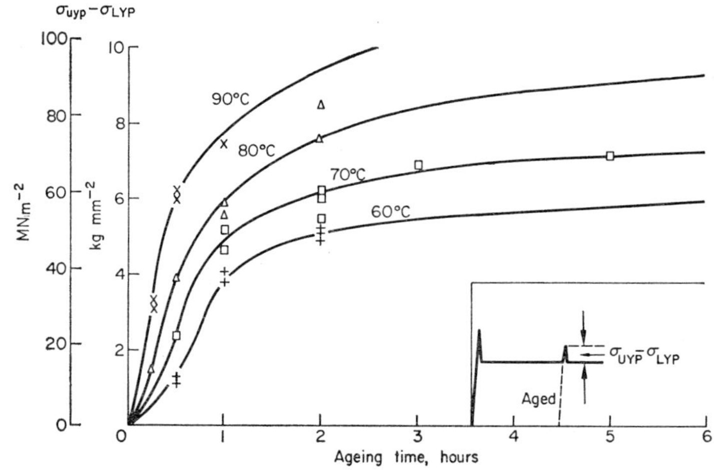

FIGURE 1.23. Ageing curves for mild steel (Sylwestrowicz and Hall, 1951)

above, because the hardness value around the indent will depend on the Luders band pattern, which will depend on the upper and lower yield stresses. In yet other cases, particularly in hydrogen embrittlement studies, some have used ductility, or even the Luders extension ϵ_L as the parameter.

One ageing parameter which is free of most of these objections was developed by Sylwestrowicz and Hall (1951). If a primary Luders band is run partway along a specimen, and the test stopped, then after ageing, an additional stress must be applied to start the band in motion again (Fig. 1.23, insert). The value of $\sigma_U - \sigma_L$ is dependent only on conditions at the band front, and will not change with the position of the front, so that several tests may be carried out on the one specimen. Figure 1.23 shows a series of curves for mild steel aged in the temperature range 60–90°C, and this same method has recently been used by Blakemore and Hall (1966) in studies at higher temperatures.

If the ageing process is thermally activated, and, moreover, is a singly activated process, then the activation energy (E) may be determined as follows. If the times t_1 and t_2 are found for the ageing parameter (f) to reach a given value at temperature T_1 and T_2, then it may be easily shown that the activation energy for the process is given by

$$\ln\left(\frac{t_2}{t_1}\right) = \frac{E}{R}\left(\frac{1}{T_1} - \frac{1}{T_2}\right)$$

where R is the gas constant.

The value of E so obtained may then be compared with the activation energies for diffusion of the various elements in the matrix, and the solute atoms responsible for ageing may be identified.

At this stage it should also be noted that all ageing effects do not increase with time; some, particularly in the f.c.c. metals, decrease with ageing. For example, some of the unloading yield points in copper are in the category (Chapter 5), and the effects of serrated yielding in aluminium base alloys are often less in fully aged than in freshly quenched or air-cooled material (Chapter 5). Whatever the physical situation, the appearance or reappearance of a yield point after straining must mean that dislocations are now more difficult to move than before; in other words, they are now locked in position.

1.7.3 *Mechanisms of locking and their kinetics*

It is not intended in this section to cover again the various modes of locking, as these have been introduced above, but instead to consider the kinetics of locking so that, in later chapters, these may be compared with experiment.

(a) *Cottrell–Bilby locking.* Cottrell and Bilby (1949) extended their theory of locking of dislocations in iron by carbon atmospheres (Section 1.5.1) by also determining the rate at which these atmospheres form. Figure 1.24 shows a positive edge dislocation, and plotted around this are the lines of constant interaction energy $A \sin \theta/r$. Under this interaction field, carbon atoms will diffuse along paths which form the orthogonal set (dotted); those circles of smallest radii will drain their carbon (and nitrogen) atoms first, and the carbon in solution will be depleted from an ever-increasing radius as time goes on. The limiting capture radius R is related to the dislocation density ρ by

$$R = (\pi\rho)^{-1/2}$$

For times that are not too long, Cottrell and Bilby showed that the number of carbon atoms removed from solution in α-iron after time t (N_t) would be given by

$$N_t = 3(\pi/2)^{1/3} N_0[ADt/kT]^{2/3}$$

where N_0 is the initial concentration, A an elastic constant, D the diffusion coefficient, t the time and T the temperature.

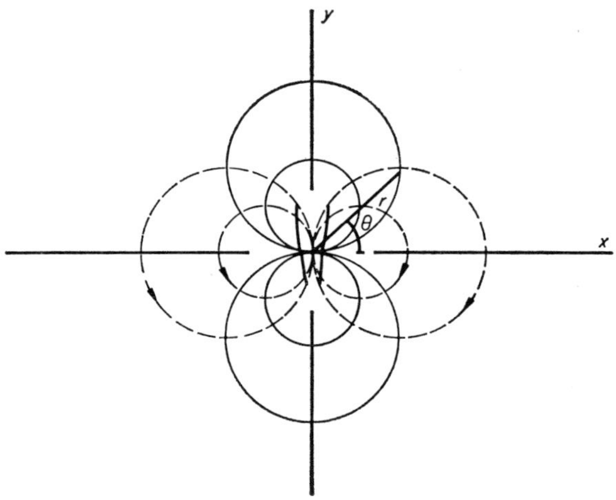

FIGURE 1.24. Equipotential (full lines) and lines of flow (dotted) for solute atoms migrating to a positive edge dislocation (Cottrell, 1953a)

This formula seemed to be strikingly confirmed by Harper (1951) who showed, using internal friction methods, that the rate of removal of carbon from solution was, in fact, proportional to $(Dt)^{2/3}$. In his experiments, however, he used specimens quenched from high temperature, and the conditions are not necessarily the same as in strain-aged material. For example, a number of later studies on furnace cooled material using internal friction show time exponents of $n \sim 1{\cdot}5$ at early stages of ageing, and $n \sim 0{\cdot}5$ at later stages (see, for example, McLennan (1965)).

All theories involving dislocations are difficult to handle at the dislocation core, among them Ham (1959), Mura *et al.* (1961) and Bullough and Newman (1959, 1962a, b). Bullough and Newman (1962c) have in fact criticized both the work of Mura *et al.* and the original paper of Cottrell and Bilby. In the former case, they contend the theory is identical with Cottrell and Bilby, and in the later case they contend the diffusion equation is incorrect, because it ignores back diffusion due to the concentration gradients built up near the core. In their view the differential equation to be solved should read

$$\frac{1}{D}\frac{\partial c}{\partial t} = \frac{1}{r}\frac{\partial}{\partial r}\left[r\frac{\partial c}{\partial r} + \frac{rc}{kT}\frac{\partial \phi}{\partial r}\right]$$

the first term in square brackets being omitted from the original treatment. (In this expression c is the concentration, and ϕ the interaction energy.) Thus, the Cottrell–Bilby treatment is too rapid in absolute terms, but the solution of the differential equation above leads to complex expressions which are difficult to compare with experiment.

(b) *Suzuki locking.* At high temperatures, the effect of thermal energy is to cause the Cottrell atmospheres to evaporate, and the disappearance of the yield point in iron tested above about 300°C is considered to be due to this effect.

In certain face-centred cubic alloys, however, such as the Nimonic series and certain austenitic steels, yield point effects are often observed at high temperatures. The Suzuki theory of solute migration to stacking faults could be invoked here, because in many cases precipitation has occurred on the faults and are clearly visible in thin film electron microscopy (Section 2.8.3). Unfortunately, while the actual locking stresses have been calculated and applied with some success to solid solution hardening in alloys, the kinetics of migration to the faults have not been studied. However, it is reasonable to assume that if the numbers of free dislocations can be rendered small enough by this means, then yield points will appear, and these at high temperatures.

Honeycombe (private communication) contends that true Suzuki locking is not yet proven, and that the yielding of austenitic steels where the dislocations are pinned by precipitates is due to climb of the dislocations over the barriers presented by the particles (see Chapter 2) leading to breakaway and multiplication.

(c) *Short-range order locking.* The mechanism of Snoek ordering, as formulated by Schoeck and Seeger (1959) has already been discussed in Section 1.5.3. It can readily be seen in this analysis that the formation of ordered regions will occur in very brief time intervals, as the impurity atmospheres have, on average, only a few atomic jumps to make before they settle down in their octahedral sites.

Although Schoeck and Seeger did not estimate any kinetics for their locking model, this theory has therefore been invoked to explain certain ageing effects occurring at very short times, as has been done by Wilson and Russell (1959) in α-iron, and Rose and Glover (1966) in austenite.

(d) *Precipitation on dislocations.* In this, extending the concept of dislocation pinning covered in Section 1.5.4, there are two papers of assistance which deal with the kinetics of precipitation or dislocations.

Bullough and Newman (1962a) have pointed out, in their revised treatment of the formation of Cottrell atmospheres, that a $t^{2/3}$ law holds only up to a short time at the beginning of ageing, up to about $N_t/N_0 \sim 0{\cdot}3$, and that the fraction of total solute atoms removed from solution is very much greater than predicted by their revised theory. This in turn implies that the dislocations act as sinks for solute atoms in other ways, i.e. precipitates may form there.

Haasen (1959) suggested a simple means of predicting the solute atom flow in this case; the mass flow rate to a line in one direction is proportional to $(Dt)^{1/2}$, thus, in three dimensional flow to a dislocation, the rate of depletion of solute should be given by

$$N_t/N_0 \propto (Dt)^{3/2}$$

and this flow rate is observed in certain cases.

The treatment of Bullough and Newman (1962b) is more sophisticated. Two conditions are possible (and are observed in practice)

(i) the chemical binding energy of the solute atoms in the precipitate is greater than the elastic binding energy to the dislocation cores – discrete precipitate particles form.

(ii) the elastic binding energy of the solute atoms in the core exceeds the chemical binding energy – continuous (rod-like) precipitates form down the dislocations.

Bullough and Newman (ibid.) tended to reject the former of these two models, because the kinetics of the precipitation could not be made to fit Harpers $t^{2/3}$ law (see Section 1.7.3(a)). As we have seen, however, the law may only have limited validity, and discrete particles are known to form in some cases. There are, naturally, other unknown parameters in the solution, including the velocity transfer coefficient across the core/matrix interface, but if the interaction potential between the impurity atoms and the dislocation is zero, an exact solution is possible, given by

$$\frac{N(t)}{N_\infty} = \frac{4}{X^2 - 1} \sum_{n=1}^{\infty} \frac{[1 - \exp(-MX^2\alpha_n^2)]J_1^2(X\alpha_n)}{\alpha_n^2(J_0^2(X\alpha_n) - J_1^2(X\alpha_n))}$$

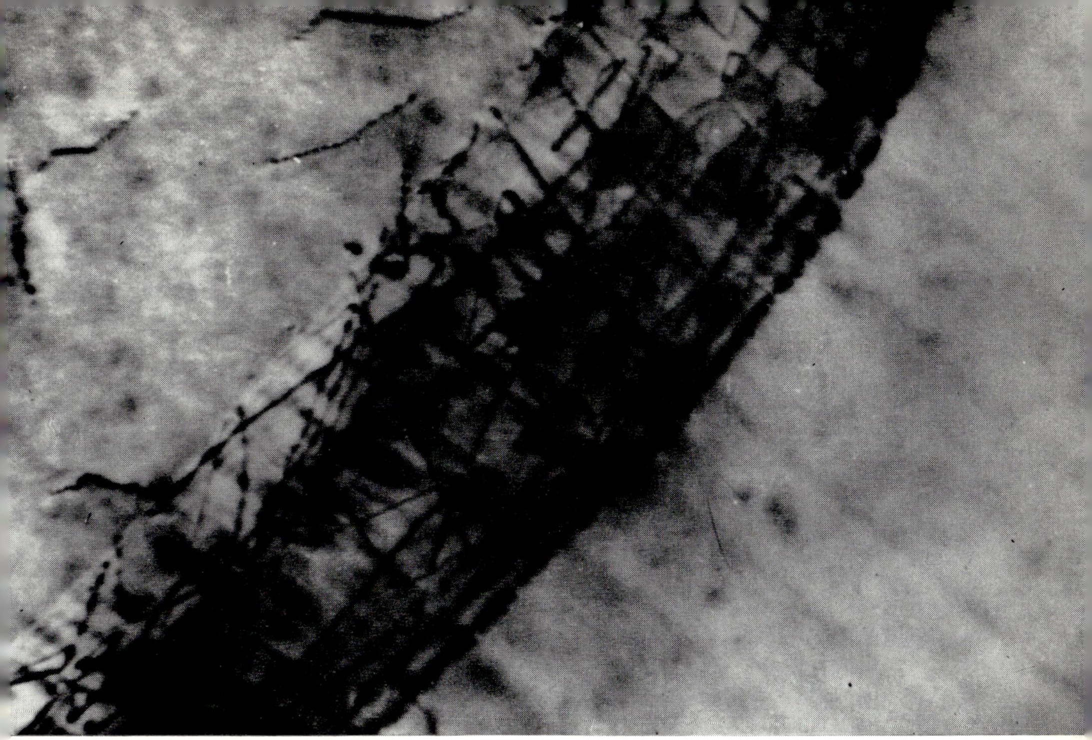

PLATE 1.6 Formation of dislocation loops at a grain boundary ahead of a Luders front in mild steel (× 50,000). (Fisher, 1961)

PLATE 1.7 Dislocation cell structure in deformed iron (× 50,000). (Fisher, 1961)

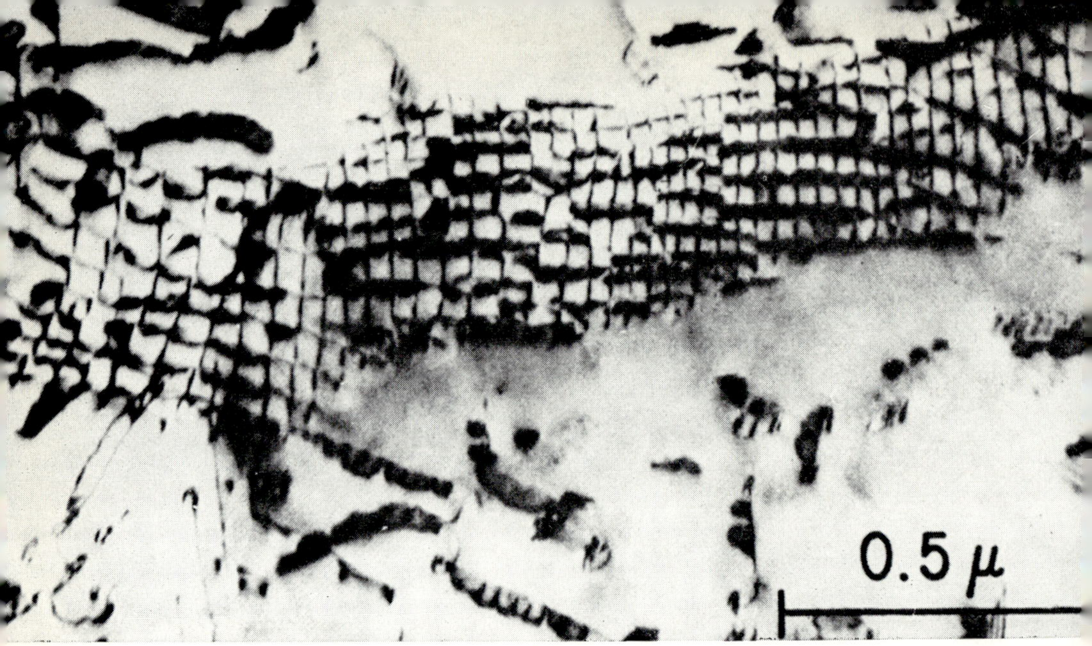

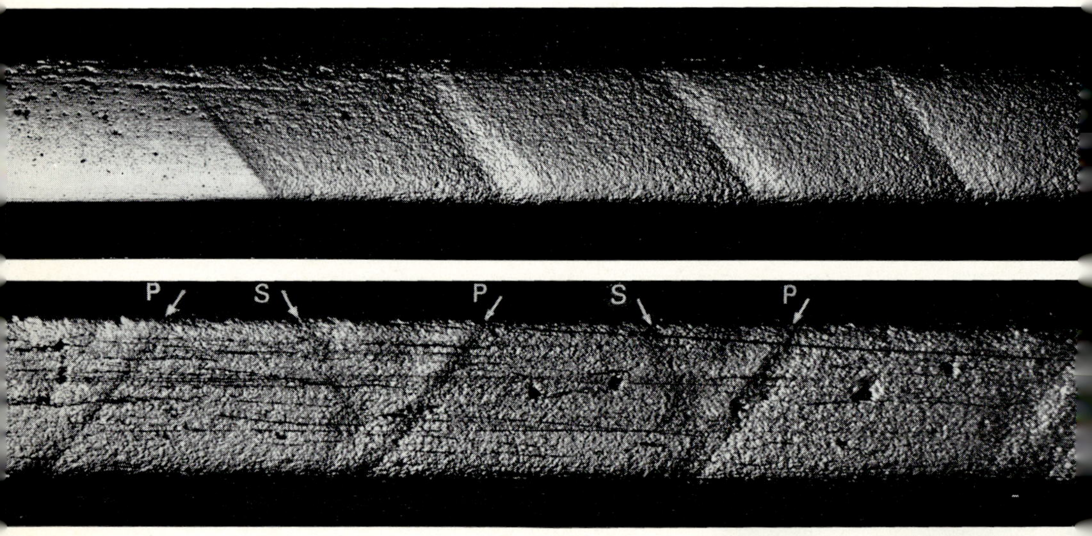

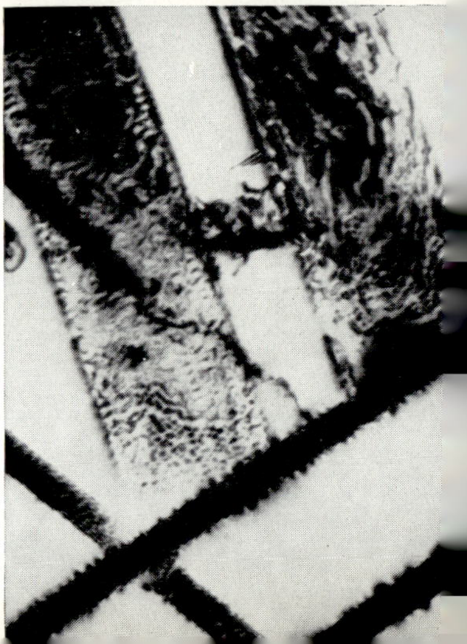

PLATE 2.1 (*top*) Precipitation of α''–$Fe_{16}N_2$ on dislocations in a 0·02% N–Fe alloy quenched and aged two hours at 100°C. (Keh and Leslie, 1963)

PLATE 2.2 (*above*) (a) Primary band markings and (b) primary (P) and secondary (S) band markings in mild steel deformed at 200°C.

PLATE 2.3 (*right*) Precipitation on stacking faults in 18 Cr–10 Ni–1 Nb austenitic steel (×80,000). (*By courtesy of* Prof. R. W. K. Honeycombe) where α_n are the roots of the Bessel function equation

$$J_0(\alpha)Y_1(X\alpha) - J_1(X\alpha)Y_0(\alpha) = 0$$

and

$$M = Dt/R^2, \quad X = R/r_0 \quad \text{and} \quad N_\infty = \pi(R^2 - r_0^{\,2})C_0$$

Blakemore and Hall (1966) have used this approach to discuss yielding in mild steel, for if discrete particles form along the new dislocations arising from the yield, then on ageing and straining the new yield stress will be given by

$$\sigma = \alpha\mu b/l$$

when $\alpha \sim 1$, and l is the mean distance between the precipitate particles.

Bullough and Newman also analysed the kinetics of continuous precipitation, appropriate to the case of quench ageing (Chapter 2). Here a $t^{2/3}$ law does apply, and Bullough and Newman (ibid.) have obtained a satisfactory agreement with the Harper formula, but at the expense of some rigour because it was necessary to suppose a transient variation in transfer velocity across the core/matrix interface. Later theoretical developments will be of immense value in studies in this field.

1.7.4 *Blue-brittleness – the Portevin-le Chatelier effect*

Having covered the basic concepts of ageing it is necessary now to consider the possible consequences of simultaneous straining and ageing. Depending on the system under investigation, and the imposed strain rate, a situation may arise where the diffusion rate of the impurities responsible for locking the dislocations matches the velocity of the dislocations themselves. When this happens, the dislocations may suddenly become locked in situ, deformation stops, the stress rises, and an upper yield point is observed when deformation recommences.

This process thus gives rise to a serrated stress–strain curve, similar to Figs. 1.3(c) and (d), and as a result of continuous ageing in the deformed regions, there is a marked decrease in ductility. This phenomenon in mild steel is known as ‘blue-brittleness’ because it occurs in the blue-heat range of 450–550 K. It was first noted by Adamson in 1878 (quoted by Kenyon and Burns (1939)), but is now known to occur in many aluminium- and copper-base alloys, in nickel charged with hydrogen, and indeed in most metals showing yield point and ageing effects. Since the appearance of serrated yielding has a characteristic temperature range (aluminium–magnesium around room temperature, nickel–hydrogen around 210 K), it is probably better to call the phenomenon of serrated yielding the Portevin–le Chatelier effect, after the Frenchmen who first noted that effect in duralumin (1923).

It has further been noted that the temperature at which these serrations first appear is a function of strain rate: the higher the strain rate, the higher the temperature at which serrations first occur. Cottrell (1953b) has used this fact to deduce an equation for the conditions for serrated yielding. If this condition is a thermally activated one,

$$\dot{\epsilon} = C \exp(-E/kT)$$

where C is a frequency factor, and E the activation energy for diffusion. Applying the equation to α-iron, Cottrell shows that the minimum diffusion rate D to produce serrated yielding is given by

$$\dot{\epsilon} = 10^9 D$$

This equation has been widely used, to interpret the elements responsible for strain ageing (see Chapter 3) or to determine the activation energies in nickel–hydrogen (Chapter 4). There are, however, some dangers in this. In the first place, the constant C in the equation above contains a term proportional to the amount of active solute. This is often not accurately known. Secondly, as Cottrell himself points out, the effective diffusion rate increases during a tensile test, because of the creation of numerous vacancies. This is presumably the reason why aluminium–magnesium alloys do not show serrated yielding at the start of a tensile test, but rather serrations develop as the test proceeds (Chapter 5). Freshly quenched or irradiated material, which does have a higher vacancy concentration, develops serrations earlier.

1.8 *Pseudo yield points*

1.8.1 *Introduction*

It is now possible to define more exactly the type of yield points which will be studied in more detail in the following chapters. In particular, we will be dealing with a system of dislocations which can be or are pinned or locked in position from a variety of causes, and which, as a result, produce an instability in the stress–strain curve on straining. There are, however, other cases where similar instabilities may arise, but which, as will be explained, cannot be classified as true yield points.

1.8.2 *Geometric softening*

The first case to be considered is the well-known one of geometric softening. Single crystals deform on certain slip planes and directions, the initial system being the one where the Schmid factor $\sin \chi \cos \lambda$ is the largest (here χ and λ are the angles between the tension axis and the slip plane and direction respectively). As deformation proceeds, the crystal planes will rotate so that the specimen axis tends to coincide with the slip direction. If the initial orientation is within a certain range, and the rate of work hardening low, this slip-plane rotation may cause the Schmid factor to increase, and a lower stress is needed to support the continuing deformation. An apparent yield point results.

The classical example is in the hexagonal metals (Boas and Schmid, 1929; see also Byrne, 1961). Here the single slip plane gives rise to a marked pseudo yield if the initial orientation is such that χ_0 and λ_0 are both greater than 45°.

Price and Kelly (1964a) have applied this analysis to face-centred cubic crystals, where it is equally valid provided a single slip system is operating. The applied load L is related to the yield load L_0 and tensile strain ϵ by the relation

$$L = L_0 \cos \lambda_0 \left[1 - \frac{\sin^2\lambda_0}{(1 + \epsilon)^2}\right]^{1/2}$$

for zero work hardening, and they point out under these conditions a yield point should result **regardless** of initial orientation. Thus, the often quoted results on hexagonal metals given above must imply a marked change in work-hardening rate around χ_0 and $\lambda_0 = 45°$.

Price and Kelly (ibid.) also show that the formula given above can distinguish between a pseudo yield and a genuine yield. Figure 1.25 shows their results for some Al–20% Ag and Al–15% Zn alloy single crystals. The dotted curve shows the theoretical plot from the formula above; in the former case the yield is genuine, in the latter it is not. This simple analysis should always be carried out in single crystal studies; unfortunately it is not often performed.

Geometric softening can also lead to single crystal Luders bands developing (see, for example, Byrne (1961)), where the slipped region spreads into the undeformed part of the crystal in a controlled manner. This can be a further source of confusion unless the yield is shown to be a genuine one.

1.8.3 *Work softening*

A similar but more subtle geometric effect arises in the face-centred cubic system. If a metal single crystal is deformed an amount ϵ at some temperature T_1, and the temperature then lowered to T_2, then on restraining the crystal the yield stress is found to be less than if the specimen had been deformed to a strain ϵ at T_2. The stress decrease observed is known as the reversible flow stress in this type of test, because it is thermally recoverable.

Stokes and Cottrell (1954) and Cottrell and Stokes (1955) in the course of a study on the reversible flow stress in aluminium single crystals, noticed that a yield point could be engendered by first

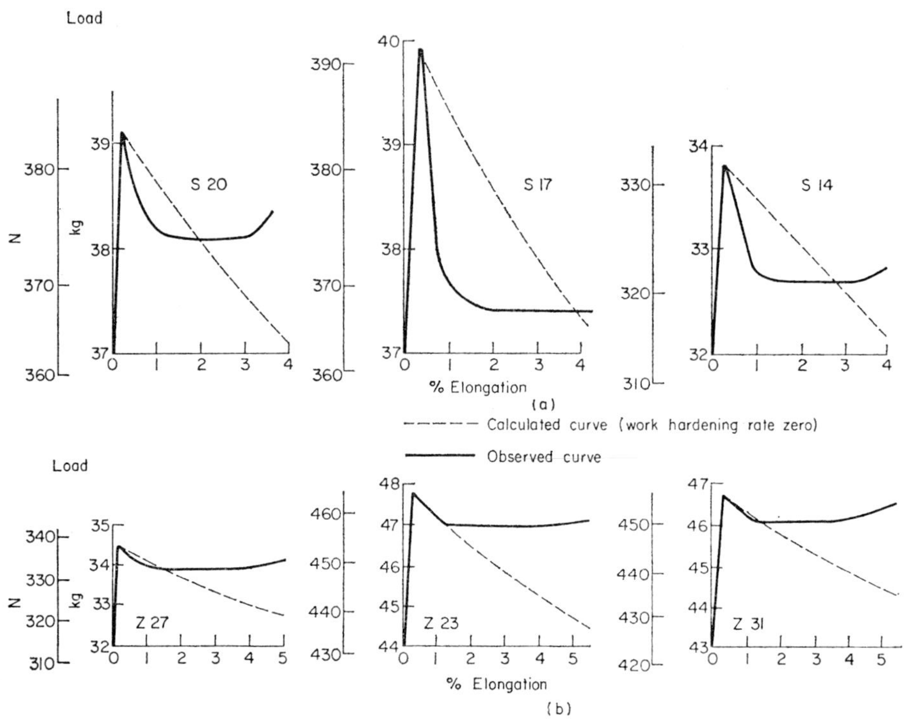

FIGURE 1.25. (a) Genuine and (b) pseudo yield points in Al–20% Ag and in Al–15% Zn alloy single crystals respectively (Price and Kelly, 1964a)

straining the crystals at 90 K, and then restraining at 300 K (Fig. 1.26). This effect only appeared after substantial extension, long after easy glide had ceased. The effect is not due to strain ageing, and Cottrell and Stokes suggest a simple explanation. There is no equation of state for deformed metals; both the density and distribution of dislocations are important in determining the flow stress. At low temperatures, dislocations can associate to form sessile dislocation barriers (the Lomer–Cottrell barriers), but on straining at higher temperatures, these barriers can dissociate again with the help of thermal fluctuations, and a massive flow begins. A yield point, as well as a Luders extension is thus seen.

In general, therefore, we will be concerned in this monograph with cases where yield points arise from the locking of dislocations by impurities, and this is manifested either by an initial yield, and the reappearance of that yield after strain ageing, or the appearance of

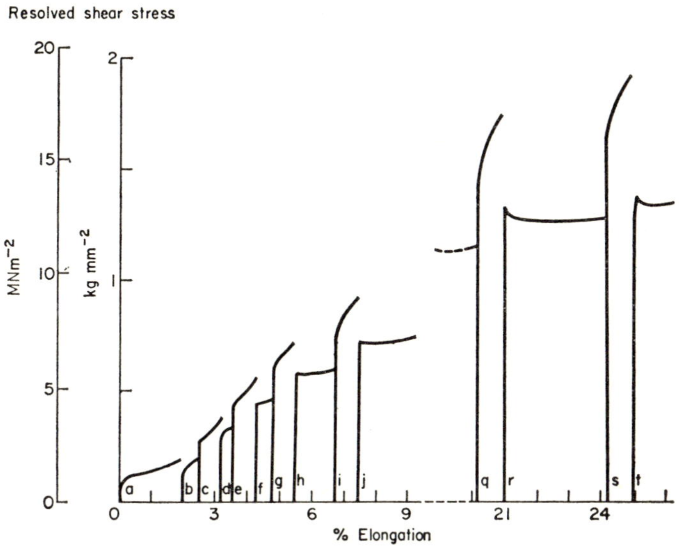

FIGURE 1.26. Work softening yield points in aluminium single crystals (each pair of curves consists of a prestrain at 90 K, ageing at 293 K for 15 min, and restraining at the higher temperature. A yield point is first seen in curve h) (Cottrell and Stokes, 1955)

serrated yielding during testing. In this sense, yields due to work softening are not within the scope of this work, although they are relevant to the discussion on unloading yield points (Chapter 5).

1.8.4 *Transient yield points*

Yet another pseudo yield can arise from varying the strain rate during the course of a tensile test. With modern testing machines, this is often done merely by push-button control, and Bolling, Hayes and Wiedersich (1961) show that this can lead to a pseudo yield point. The reason is that stress and strain rate are related, by virtue of the strain rate sensitivity,

$$B = \Delta\sigma/\Delta(\ln \dot{\epsilon})$$

which is constant for a given metal at a given temperature and stress. Increasing the strain rate (naturally) increases the yield stress, but is, in the transient stages, a function of the ratio of specimen to machine sensitivity. This then gives rise to transient yield points until the conditions equalize (see also Osborne, 1962, Guin, 1969). The yield points observed by Yada and Kiritani (1965) probably fall in this category.

The existence of this transient will thus depend, to some extent, on the machine, metal used, and testing temperature. It can readily be checked, since lowering the strain rate will induce a yield point in the reverse sense, or alternatively a metal may be tested which is known to show no unloading yield point at the testing temperature (Section 5.2). It is thus unlikely to cause confusion; but it is worth noting that some of the transient yield points in 61S aluminium observed by Lubahn (1949, 1952) may come in this category.

1.8.5 *Twinning and thermal softening*

The mode of homogeneous deformation known as twinning has already been treated in several monographs. In twinning, macroscopic blocks of a crystal are sheared to retain their original crystal structure, but are altered in orientation with respect to the matrix. The formation of a twin takes place extremely rapidly at rates approaching the velocity of sound, so that the specimen extends rapidly, often with an audible click, and with a sharp drop in load.

In general, twins form in an irregular manner at randomly dispersed points along the test piece. The stress–strain curve may show the resultant serrated yielding at the beginning of the test, or, as shown in Fig. 1.27, after a certain amount of slip deformation. Occasionally, however, the serrations due to twinning are very regular, as found by Blewitt *et al.* (1957) for copper single crystals at 4 K, Thornton and Mitchell (1962) in brass or in the more recent work of Altshuler and Christian (1966) in iron. This type of yielding is, however, unlikely to be confused with yield point effects, although Blewitt *et al.* (1957, 1960) have seen Luders type bands where the spreading twin represents the deformed region, as well as the more common slip line Luders bands in single crystals. Molybdenum -50 at.% rhenium alloys show quite marked serrated yielding at 20°C but this is now known to be the result of extensive twinning (Jaffee *et al.*, 1961) – strain ageing in an interrupted tensile test is not observed. Bolling and Richman (1965) noted a similar effect in the compound Fe_3Be. In the majority of cases, twins are usually clearly distinguishable from Luders bands, and in polycrystals may be detected by polishing and etching. It should be noted, however, that the ‘cry’ of twinning, mentioned above, can also occur in serrated yielding – see, for example, Portevin and le Chatelier (1923) – although in this case it is probably the result of vibration within the testing machine complex.

There are a few cases quoted in the literature where other anomalous examples of serrated yielding are cited. Wessel (1957a), found serrated yielding in nickel, zirconium and β-brass at 4 K (Fig. 1.28) and Powell *et al.* (1958) found serrations in the stress–strain curves of

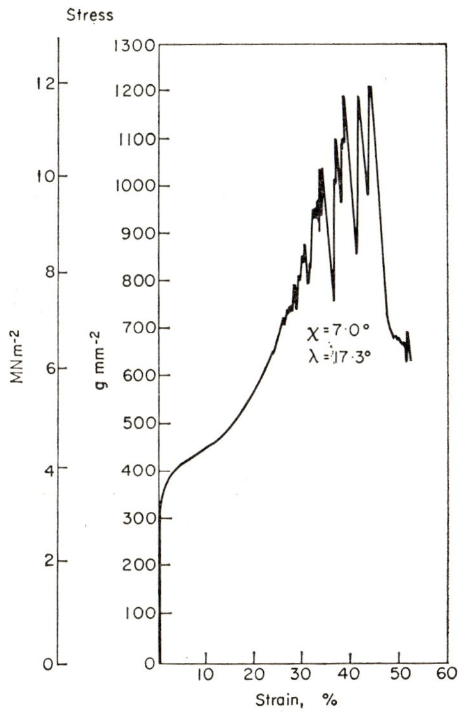

FIGURE 1.27. Twinning after slip in a cadmium single crystal (Boas and Schmid, 1929)

Type 301 stainless steel at 10 K. Here, another explanation has been put forward, namely that localized deformation heats the sample lowering the yield stress still more – the so-called thermal instability theory (Zener and Hollomon, 1944; Bazinski, 1957, 1960; Lean, 1960). In a way, this is little more than the old localized melting theory of slip-plane deformation; this general theory as applied at room temperature has long ago been discounted (see, for example, Cottrell (1953a)), but it may well be that at these temperatures, where the yield stress is a sensitive function of temperatures, and the specific heat is low, that this analysis is more generally applicable, particularly in metals with high stacking-fault energies. Strain ageing is absolutely ruled out at liquid helium temperatures.

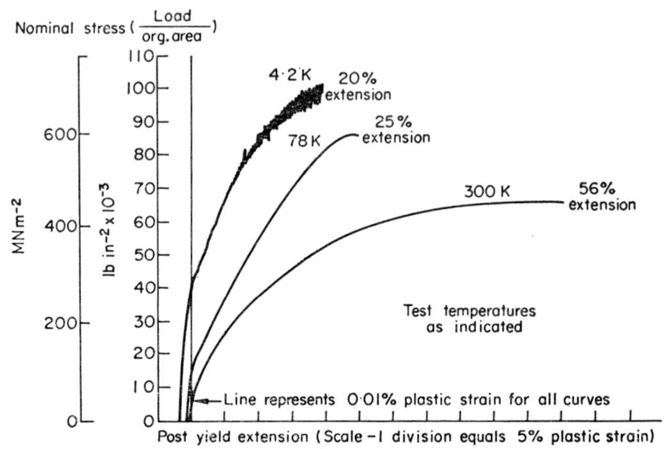

FIGURE 1.28. Serrated yielding in β-brass at low temperatures (Wessel, 1957a)

In the chapters which follow each group of metals will be examined in turn, and the validity of the theories put forward earlier in this chapter will be discussed.

2

Iron and Its Alloys

2.1 *Introduction*

From the previous chapter, it will have become clear that the presence of yield point phenomena in iron and steel is due to the presence of interstitial impurities in the metal. From the Cottrell theory (Section 1.5.1) it would be expected that substantial alloying elements could also form atmospheres to lock dislocations in position, but little has been done on, say, precipitation hardening steels which might fall in this category. In the case of the high alloy austenitic steels, it will again be found (Section 2.8) that carbides are responsible for the serrated yielding found there.

TABLE 2.1. Radii of interstitial solutes in Iron

Metal	Radius (Å)	r/r_{Fe}
α-Fe	1.28	—
C	0·77	0·60
N	0·72	0·57
B	0·94	0·73
O	0·60	0·47
H	0·46	0·36

Thus, it is the interstitial elements carbon, nitrogen, boron, oxygen and hydrogen with which we will be concerned in this chapter. These five form the only possible elements which, by virtue of their atomic size, can fit into the interstices of the α-iron, and then only with difficulty. Their approximate atomic radii are given in Table 2.1.

There are two possible sites in the α-iron lattice where these elements may reside, the so-called tetrahedral and octahedral sites

(Fig. 2.1(a) and (b)), so named because of the configurations of the surrounding atoms. The theoretical radii associated with these sites in the b.c.c. lattice are 0.291*R* for the tetrahedral hole, and 0·154*R* for the octahedral, where *R* is, in this case, the atomic radius of α-iron. These then reduce to 0·36 and 0·19 Å respectively.

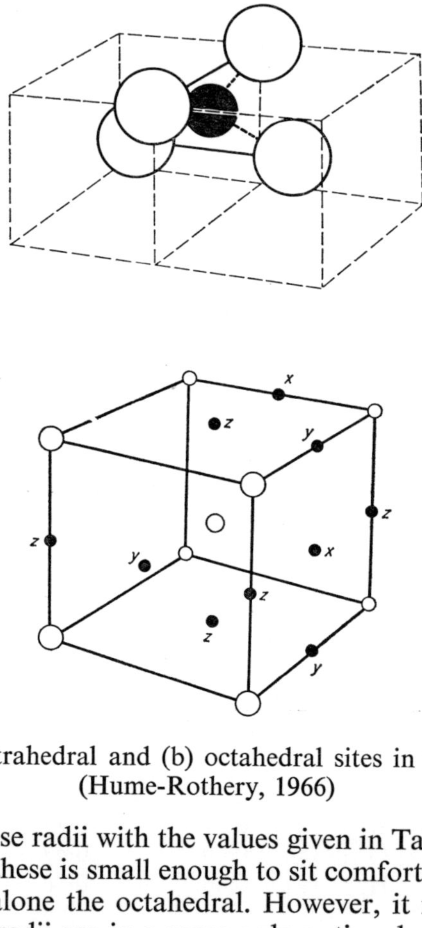

FIGURE 2.1.(a) Tetrahedral and (b) octahedral sites in the α-iron lattice. (Hume-Rothery, 1966)

Comparing these radii with the values given in Table 2.1, it is seen that only one of these is small enough to sit comfortably in the tetrahedral hole, let alone the octahedral. However, it must be remembered that these radii are in a sense only notional, and depend to a marked degree on their surroundings and their state of ionization.

Despite the larger value of the tetrahedral hole, it seems that the octahedral holes are in fact preferred. The co-ordinates of the former $(0, \frac{1}{2}, \frac{3}{4})$ show the solute atom surrounded by two pairs of equally spaced iron atoms. The co-ordinates of the octahedral sites, however, fall into three groups, $(\frac{1}{2}, 0, 0)$, $(0, \frac{1}{2}, 0)$ and $(0, 0, \frac{1}{2})$, plus body-centring translations (Fig. 2.1(b)) and if these sites contain a solute atom slightly too large, the atom may be accommodated by a displacement of only the two nearest neighbours, instead of four as in the case of the tetrahedral holes. A tetragonal distortion of the lattice results.

This fact has considerable application in the theory of internal friction in b.c.c. metals; the interstitials can be made to move from one octahedral site to another under the influence of an applied stress, and if in addition they can be made to move in resonance with an applied (alternating) stress, a notable increase in damping will occur. This damping peak is often known as the Snoek peak, and the octahedral ordering as Snoek ordering, after the discoverer (Snoek, 1941). The detailed treatment of internal friction is outside the scope of this work, but the technique has valuable applications in determining the

TABLE 2.2. Diffusion constants of interstitial elements in α-iron

Element	D_0 (mm^2 s^{-1})	E (kcal mole^{-1})	E (kJ mole^{-1})
C	0·488	18·35	76·7
N	0·394	19·16	82·0
B	10^8	62·0	260
O	3·71	23·4	97·9
H	114	8·5	35·5

amount of free solute left in solution in the α-iron lattice, and reference will be made to the results of this technique at several points in this chapter.

Different conditions will naturally apply in the face-centred cubic lattice of γ-iron. This is a close-packed lattice and, assuming the iron atoms are again spherical, the largest interstice present is an octahedral hole of radius $0{\cdot}41R = 1{\cdot}31$ Å. This is larger than either type in α-iron, and we would thus expect the interstitial element solubilities to be higher in the γ than in the α phase.

Having dealt with the size of the five interstitial elements in steel, and their position in the lattice, it is also necessary to consider their diffusion rates, since this notably affects the strain-ageing kinetics. Recent values of D_0 and E, the diffusion constant and activation energy respectively, are given in Table 2·2.

There are two points of interest in this table. First, the values imply fairly rapid diffusion even at room temperature. This is in marked contrast to most of the substitutional systems. Second, the values for carbon and nitrogen are remarkably similar, and in many cases it is difficult to distinguish between the two.

Of the five elements listed, the case of hydrogen will be taken up in Chapter 4, while there is little point in considering the case of oxygen. This is because oxygen is practically insoluble in both α- and γ-iron, and the effect of the deoxidants often added in steelmaking will greatly outweigh the small amount of oxygen (if any) left in solution. Edwards *et al.* (1939) states categorically that oxygen has no effect on strain ageing. Little work has been carried out on boron, but this will be covered as part of the next section (2.2.2).

2.2 *The effects of carbon, nitrogen and other elements*

It is first necessary to prove conclusively that yield point and strain-ageing effects in mild steel are caused by the presence of interstitial carbon and nitrogen, which can, under appropriate conditions, lock the dislocations in position. Proof of this can be given in three ways, all of which cause the yield point to disappear:

(a) by decarburizing or otherwise purifying the steel;
(b) by alloying with strong carbide or nitride forming elements; or
(c) by quenching from high temperature, which leaves the dislocations free of their Cottrell atmospheres.

The first two of these will be dealt with here; the effects of quenching will be treated separately later (Section 2.3). This section concludes with a discussion of the effects of other elements which are commonly added to mild steel.

2.2.1 *Decarburizing and denitriding*

Decarburization by annealing in wet hydrogen has been known for many years to remove the yield point extension in mild steel, Ludwik and Scheu (1925) being among the first to appreciate the yield point may be decreased by purification. Adcock and Bristow (1935) also found no yield point in their specially purified iron, but the effects of decarburization by wet hydrogen were examined independently and systematically by Edwards *et al.* (1943) and Low and Gensamer (1944). Figure 2.2 taken from the latter paper, clearly shows the effect.

Later work in this field has primarily been concerned with the effects of specific carbon and nitrogen levels on the yield stress and ageing behaviour of various grades of mild steel. For example, Tjerkstra (1961) examined the stress–strain curves of decarburized samples, showed that the Hall–Petch equation still applied, and was able to indicate that carbon (and/or nitrogen) must be absorbed into and so strengthen the grain boundary. The value of k_y in decarburized specimens is considerably lower than in carburized samples (Table 1.2).

Imai and Ishizaki (1962) examined the relative ageing rates at room temperature of a series of nitrogen-bearing steels. The rates levelled off at N_2 contents greater than approximately 0·007%, but below this, strain ageing was progressively decreased.

Enrietto (1966) has examined the separate effect of nitrogen by annealing samples in **dry** hydrogen. This will cause denitriding, but not decarburizing, so the initial yield point is present, but the susceptibility to strain ageing is markedly reduced. It is worth noting that in fact the Luders extension is increased by denitriding, since the

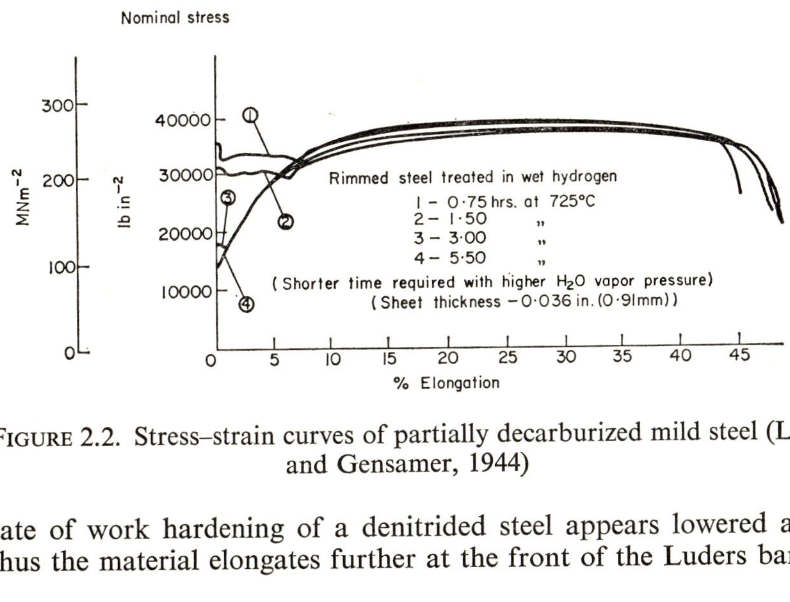

FIGURE 2.2. Stress–strain curves of partially decarburized mild steel (Low and Gensamer, 1944)

rate of work hardening of a denitrided steel appears lowered and thus the material elongates further at the front of the Luders band.

2.2.2 *Effects of carbide and nitride forms*

The possibility of producing non-ageing steels by alloying has likewise been known for many years (see, for example, Hayes and Griffis, 1934), for in deep-drawing purposes these steels will not need temper rolling or roller levelling to remove the possibility of stretcher strain formation.

Best known of additives to produce non-ageing steel is aluminium, which combines with the nitrogen present to form AlN. The nitrogen levels available for strain ageing are thus reduced to very low levels. Unfortunately, this addition acts as a deoxidant, producing a killed steel, rather than the rimmed variety which is generally preferred in deep drawing. Thus, a large amount of research has been carried out on the effects of aluminium and other additions, to correlate the ageing characteristics with composition. These additions may be conveniently classed in three groups:

(a) Nitride formers; Al, Si and B
(b) Carbide former; Mo
(c) Carbide and nitride formers; Cr, V, Nb and Ti

(a) *Nitride formers.* The early paper of Edwards *et al.* (1940) showed surprisingly, that aluminium did not eliminate strain ageing, although the use of aluminium killed steels was widespread at that time. It would appear from later work that Edwards *et al.* had used too high a normalizing temperature (950–1050°C) and too high an ageing temperature (one hour at 250°C). A more extensive study by Leslie and Rickett (1953) has resolved some of these apparent anomalies. Killed steels showed no ageing after one hour at 100°C, but ageing increased above this temperature possibly due to uncombined carbon, or to dissolution of the AlN particles.

Leslie and Rickett (ibid.) also showed that aluminium *per se* cannot eliminate strain ageing at 100°C, but for best results must be combined with silicon. Likewise, a study by Laxar *et al.* (1961) showed that up to 0·08% Al decreased ageing, but above 0·3% ageing appeared to increase again. This point deserves further consideration.

The effects of silicon, acting in conjunction with aluminium, have been mentioned above, but silicon can also form a stable nitride, thought to be isomorphous with AlN (Arrowsmith, 1963) and which can form readily in the range 600–650°C. More recent work by Baker (1967) has shown that manganese is present in this phase. In view of this recent result, SiN may co-precipitate with AlN showing the effectiveness of the two additives acting together.

The work of Leslie and Rickett (ibid.) and Imai and Ishizaki (1962) both show, again, that silicon acting alone cannot eliminate strain ageing. The latter paper in fact shows that the progressive reduction of ageing is only achieved with quite high silicon additions (up to 25%), but no attempt was made here to vary heat-treatment conditions, or to promote specially the formation of silicon nitride.

At these and higher silicon contents, one encroaches on the electrical steel area, where the steels produced show much reduced elongation and a tendency to deform by twinning rather than by slip. Hull (1961) has investigated this transition as a function of grain size and temperature, and shown that whereas single crystal and coarse-grained samples are brittle, or twin, at (say) 77 K, fine-grained samples deform by slip with the usual initial yield point. Biggs and Pratt (1958) have shown that deformation suppresses twinning in single crystals of iron, but on ageing, twinning can occur. A similar result has been found in silicon–iron by Jolley and Hull (1964) while at higher temperatures (90–150°C), Gell and Worthington (1966) showed that twinning occurs as a result of dynamic strain ageing.

The use of boron as an additive to control ageing by the formation of boron nitride (BN) has been proposed by Morgan and Shyne (1957a, b). Provided the additive level was between 0·007 and 0·020% a significant reduction in strain ageing could be achieved. Above this level strain ageing increases again, and it can be seen from Tables 2.1 and 2.2 that boron might be as effective an interstitial element as carbon and nitrogen. This has in fact been shown from the work of Codd and Petch (1960). It was found in Chapter 1 that the value of k_y in the Hall–Petch equation is to some extent a measure of the degree to which dislocations are locked. Codd and Petch showed that k_y was notably increased by the addition of approximately 0·004% B. In fact, a similar analysis applied to the other two major interstitials (C and N) showed that boron was clearly the strongest locking element, and followed by nitrogen and carbon in that order. The advantage of boron over aluminium and silicon is that the amounts required are much smaller, thus the possibility of producing a non-ageing rimmed steel does arise.

All these three elements can, as is well known, form carbides, but there is little possibility of this occurring in iron, since the chemical affinity of the iron with these additives outweighs the possibility of carbide formation.

(b) *Carbide formers* (Mo). Molybdenum is not considered a strong nitride former, but is commonly added to tool steels to assist in the promotion of a fine dispersion of carbides, including Mo_2C. The possibility thus arises that additions of molybdenum might suppress strain ageing. However, Edwards *et al.* (1940) found that additions of 0·3–2% Mo had little effect. Leslie (1959) (quoted in Baird, 1963) was equally unsuccessful. It would thus seem that best chances of success would be to use additives which are both carbide and nitride formers, as seen in (c) below.

(c) *Carbide and nitride formers* (Cr, V, Nb, Ti). Edwards *et al* (ibid.) included in their classic investigation several of the elements which form both carbides and nitrides. Chromium is the weakest of these, and a 6% Cr addition was needed before ageing at 250°C was appreciably reduced. The initial yield is also removed by this treatment (see Fig. 2.2).

Vanadium is a much more effective additive, and Edwards *et al.* (ibid.) found 0·69% V was sufficient to eliminate strain ageing. Epstein *et al.* (1950) found 0·03–0·05% V was very effective in reducing strain ageing at 100°C, and serrated yielding in the blue-brittle region (approximately 250°C) was also suppressed, but at these levels of alloying, the initial yield was still present.

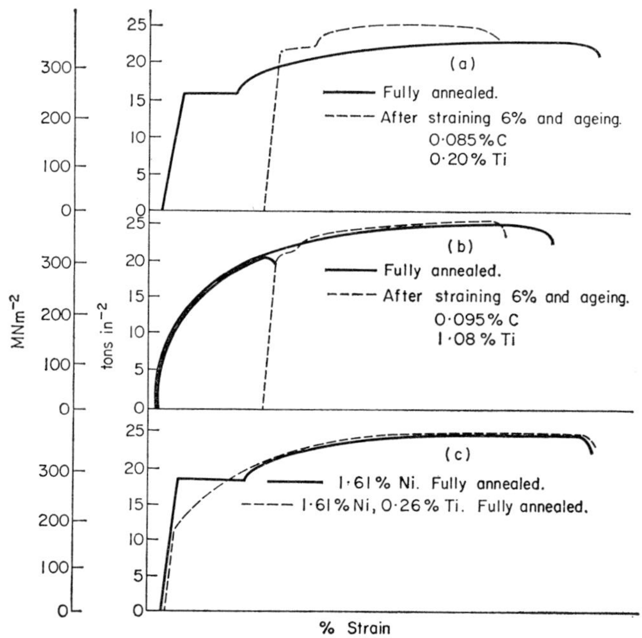

FIGURE 2.3. Stress–strain curves of some nickel and Ni/Ti-bearing steels (Edwards *et al.*, 1940)

Even more effective is niobium. Edwards *et al.* quote 0·36% Nb as sufficient to eliminate both strain ageing and the initial yield. Later work by Morrison (1963) examined a wider range of niobium contents, and also clarified some of the conditions of heat treatment. After normalizing, an intermediate anneal at 650°C was found markedly to reduce the sensitivity to strain ageing, presumably since at this temperature the nitrogen may be taken up by the growing carbide.

Finally in this section, we must consider the effects of titanium.

Edwards *et al.* (ibid.) have shown this is by far the most effective of all the group of additions; and from more recent work on alloy steels, it is expected that the removal of carbon or nitrogen from solution by titanium will not be critically dependent on heat treatment. The normal Ti : C ratio in alloy steels is about 4 : 1, but Edwards *et al.* needed rather more than this (9 : 1) to remove both ageing and the initial yield. This could have been from an (unknown) nitrogen level in their steel. The stress–strain curves from some of their Ti bearing steels are shown in Fig. 2.3. Note the absence of an initial yield in the second and third examples.

2.2.3 *Effects of other elements*

This section will be concerned with the effects, if any, of other alloying elements normally present in mild steel. These are usually Cu, Ni, Mn, S, P and As. Of these, copper and nickel usually have little effect. The work of Edwards *et al.* (1940) showed that additions of these two elements slightly increased the susceptibility to strain ageing. Imai and Ishizaki (1962), working with copper-bearing steels found small copper additions (approximately 0.3%) to be fairly effective reducers of strain-ageing susceptibility; thereafter the effect increased again.

Manganese is a weak carbide and nitride former, and it is known from internal friction investigations that the manganese and other interstitials do form some association in the lattice. Thus, a decrease in ageing rate would be expected. This is in fact seen in the work of Imai and Ishizaki (ibid.) on ageing at 100°C, and to a lesser extent in Edwards *et al.* (ibid.) on 250°C ageing.

Sulphur, like oxygen, is normally considered to have zero solubility and in general stringent efforts are made to keep the sulphur content low. Its companion element, phosphorous, does have a limited solubility and a small amount of research, using both internal friction and tensile testing, has shown some decrease in strain ageing due to this element. This is consistent with the view that phosphorous may be a weak nitride former (Baird, 1963).

Hornbogen (1963) has also shown that a dilute iron–phosphorous alloy work hardens much more rapidly than pure iron, with loops thrown out from grain boundaries, and clear evidence of extensive piled-up groups of dislocations.

Finally it should be mentioned that Irvine and Pickering (1963) have examined the general mechanical and impact properties of a wide range of experimental steels, and show quite rightly that in choosing the alloying levels of these additions, other properties can be seriously affected. The usual compromise between cost, yield stress, and I.T.T. may then have to be reassessed in terms of the advantages of a non-ageing steel.

2.3 *Quench ageing*

2.3.1 *The* Fe–C *and* Fe–N *phase diagrams*

These alloys possess one further property which helps to prove that carbon and nitrogen are responsible for the yield point and ageing behaviour in mild steel. This property, known as quench ageing, arises from the increased solubility of carbon and nitrogen in the α-iron lattice at high temperature. On quenching down to room temperature, the solute atoms are 'frozen' in their random distribution through the matrix, the dislocations are free of solute atmospheres, and the initial yield point is not present. On ageing the supersaturated lattice, much of the excess solute will be precipitated, and the dislocations will again become locked, so that the yield point will reappear.

This treatment is exactly similar to the solution treatment, quench-

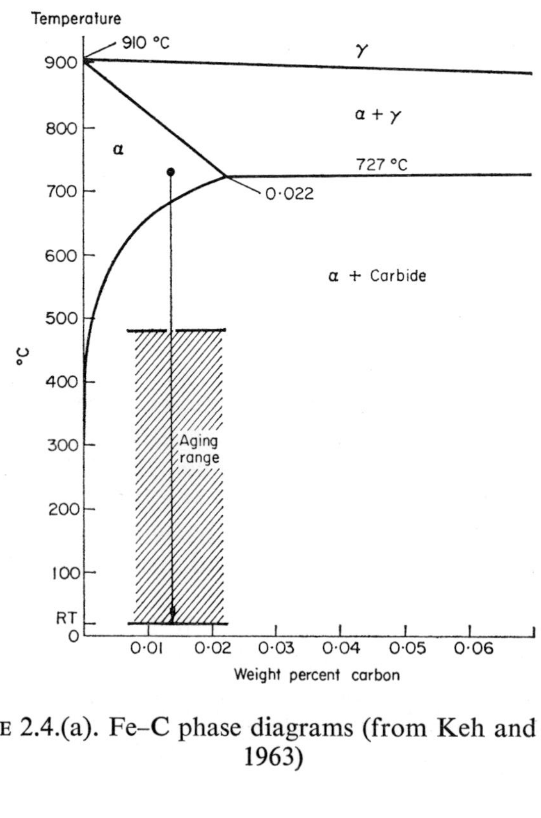

FIGURE 2.4.(a). Fe–C phase diagrams (from Keh and Leslie, 1963)

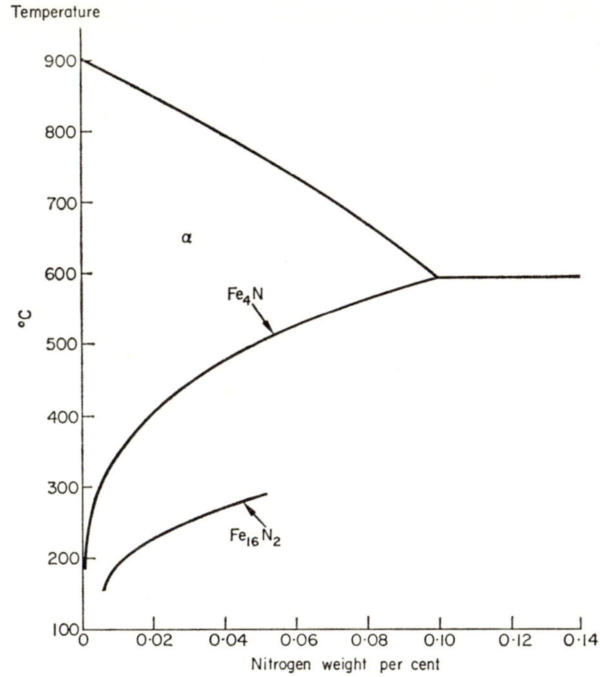

(b)

FIGURE 2.4.(b) Fe–N phase diagrams (from Keh and Leslie, 1963)

ing and ageing carried out on normal age-hardening alloys of the duralumin type and which will be discussed in Chapter 5. The difference in behaviour between the two cases is that an initial yield does not necessarily develop in the duralumin case; nevertheless there is considerable hardening of the iron by the precipitation, and over-ageing can also occur. Details of the ageing kinetics are given in the next section.

The amounts of solute available for ageing in both the Fe–C and Fe–N systems are shown in Figs. 2.4(a) and (b). It is seen that the maximum solubility is higher in the nitrogen system than in the carbon (0·10 against 0·022 wt%) and that the eutectoid temperature is lower in the nitrogen case than in the carbon (863 K against 1000 K). The room temperature solubilities are both very low, but from data given by Leslie and shown in Fig. 2.5, the equilibrium solubility is about ten times higher for nitrogen than for carbon. Thus it is always advisable not to ignore the nitrogen analysis for a steel.

The decomposition products also differ in the two cases. In the iron–nitrogen system, the equilibrium precipitate is Fe_4N (α'), a face-centred cubic structure, and having the $\{210\}_\alpha$ plane as the habit. At lower temperatures, another nitride $Fe_{16}N_2$ (α'') is formed, and it has a body centred tetragonal structure with $\{100\}_\alpha // \{100\}_{\alpha''}$. The morphology of these precipitates will follow later, but it will be the precipitation of α'' which will be concerned with quench ageing. In the iron–carbon system, there are likewise at least two carbides which may form. The high-temperature form is Fe_3C, orthorhombic, with a $\{110\}_\alpha$ habit. There is a further low-temperature form, often called ϵ-carbide, but which Keh and Leslie (1963) and Askew and Wells (1967) maintain is isomorphous with $Fe_{16}N_2$ and has the same habit as α''.

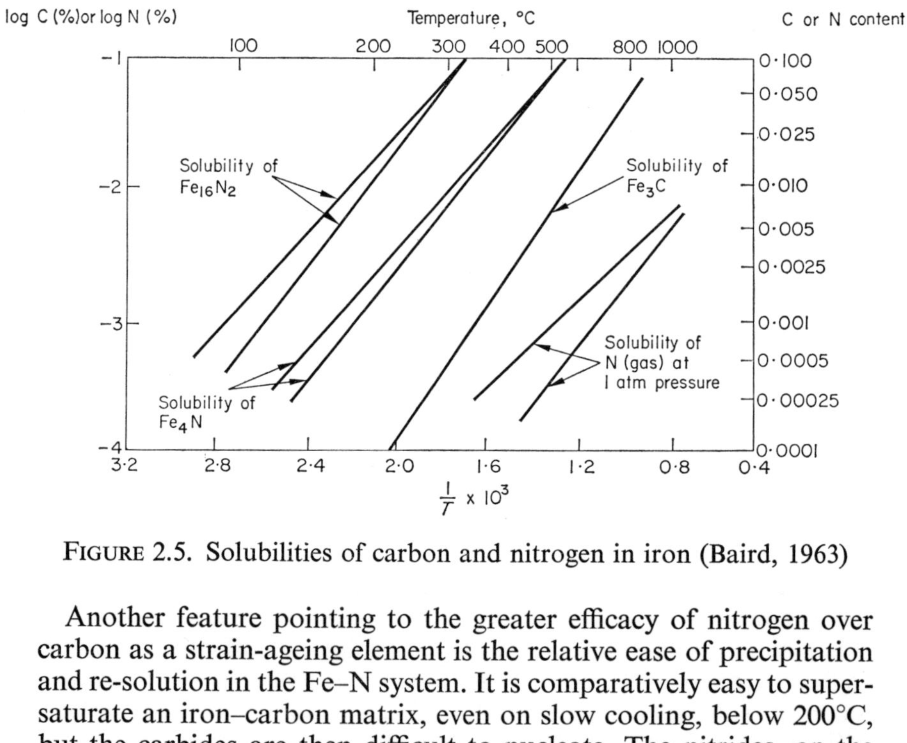

FIGURE 2.5. Solubilities of carbon and nitrogen in iron (Baird, 1963)

Another feature pointing to the greater efficacy of nitrogen over carbon as a strain-ageing element is the relative ease of precipitation and re-solution in the Fe–N system. It is comparatively easy to supersaturate an iron–carbon matrix, even on slow cooling, below 200°C, but the carbides are then difficult to nucleate. The nitrides, on the other hand, appear to have a lower stability, and re–solution takes place more readily giving a ready supply for dislocation locking, as well as the higher equilibrium solubility mentioned above. This also means that the strain-ageing characteristics of iron–nitrogen alloys will be much less sensitive to prior heat treatment than comparable iron–carbon alloys.

2.3.2 *Ageing kinetics*

The kinetics of quench ageing were first studied in a detailed manner by Davenport and Bain (1935), and their results for a 0·06% C steel are shown in Fig. 2.6. These early results are not only shown in their own right, they also illustrate several other important features. Nabarro (1948) used these curves to show that the activation energy was that of carbon in iron, by determining the time taken to reach maximum hardness over the temperature range shown in the diagram. Secondly, it will be seen later (Section 2.5) that these curves do not parallel those for strain ageing; over-ageing is readily observed under these circumstances, and as the temperature is raised above about 40°C, the maximum hardness begins to fall. This has no parallel in strain ageing. Finally, it should be noted that the hardness

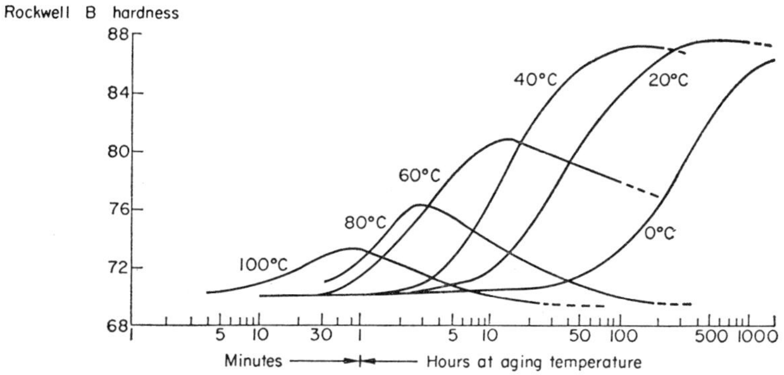

FIGURE 2.6. Quench-ageing hardness curves (Davenport and Bain, 1935)

of the as-quenched samples is higher than the furnace cooled samples; some quench hardening must therefore be present.

The curves shown above are thus typical of normal age-hardening alloys, with the exception that, in this case, hardening is achieved with about 1% of the solute used in duralumin. The explanation of quench ageing in mild steel should thus be considered in similar terms to duralumin. The fact that the initial yield point returns at the same time is largely incidental, but must be associated in some form with the segregation or precipitates that form during ageing.

To deal firstly with the effects of heat treatment on quench ageing, it would be expected that maximum hardening would be achieved by heat treatment giving a finely dispersed precipitate phase. Thus, in steels containing more than 0·022% C (see Fig. 2.4(a)), rapid cooling through the transformation temperature will produce a fine dispersion of carbide nuclei (Butler, 1962c), and, incidentally, a more complete precipitation of the carbon on ageing (Richards and Barratt, 1965). As explained earlier, nitride precipitation will be less dependent on heat treatment.

Pre-ageing is known to affect the distribution of precipitates in aluminium-base and magnesium-base alloys (Embury and Nicholson, 1965; Hall, 1968), and the same would appear to hold in the quench ageing of iron. Rickett *et al.* (cited by Keh and Leslie, 1963) showed that a marked increase in the hardness from ageing at 100°C could be achieved by pre-ageing at 25°C. As will be seen below, this produces a large number of precipitate nuclei, and the resultant precipitate dispersion on ageing at 100°C is then much finer in scale.

Pre-straining before ageing might also produce a similar effect if conditions are correct for the precipitation of carbide and nitride on dislocations. As far as is known, there are no critical hardness experiments to test this.

While much work has been done on ageing kinetics using hardness, another valuable technique for studying the kinetics of quench ageing is internal friction (I.F.) By measuring the magnitude of the anelastic damping, a relaxation peak is observed at temperatures slightly above room temperature; the magnitude of this damping peak is proportional to the amount of carbon in solid solution.

As shown in Section 1.7, the rate of removal of carbon from solution to form the dislocation atmospheres is given by

$$N(t) = 3\sqrt[3]{\pi/2}\, N_0\, [ADt/kT]^{2/3}$$

where the symbols were previously defined. By a slight modification of this law, Harper (1951) was able to rewrite the equation as:

$$\ln(1 - W) = -(t/\tau_t)^{2/3} \qquad (2.1)$$

where W is the fraction precipitated, and τ_t is a characteristic time which will depend, inter alia, on dislocation density as well as temperature.

The agreement with experiment in this analysis is probably fortuitous, since on the one hand the simple theory does not allow for saturation or back diffusion effects, while on the other the possibilities of nucleation of carbide precipitates within the matrix were not investigated. The exponent in equation (2.1) above should in fact vary with the mode of precipitation. Ham (1969) has given analytical solutions for the growth of spheroids, rods and discs; in each case the exponent at small times differs from 2/3. At later stages of ageing, depletion and impingement factors become important.

Similar kinetic studies may be carried out using electrical resistivity (E.R.) determinations, and in Table 2.3 are listed the collected results of several investigators, all of whom assumed the Cottrell–Bilby theory (or Harper's modification of it) applied. Again, there are two points of importance.

TABLE 2.3. Measurements of strain-ageing kinetics in quenched Fe–C or Fe–N alloys (*after* Baird, 1963)

Investigator	Method	Pre-strain (%)	Ageing temperature (°C)	Solute content (wt%) C	Solute content (wt%) N	Calculated dislocation density ($cm^{-1} \times 10^{-10}$)	Density of solute (atoms/atom plane)
Harper (1951)	I.F.	5	30	0·015	—	7·7	13
		10	30	0·009	—	10·2	6
		15	30	0·013	—	15·7	6
Dahl and Lücke (1954)	E.R.	4.5	120	0·015	—	0·82	75
		5·1	150	0·015	—	1·67	45
		5·3	90	0·018	—	0·95	118
Thomas and Leak (1955)	I.F.	7	24–54	—	0·015	7·7	11·5
		7	35–67	0·015	—	6·6	15·6
		7	35–67	0·0095	—	5·0	13·1
		7	35–67	0·0055	—	4·0	9·4
Pitsch and Lücke (1956)	I.F.	4·8	90	0·01075	—	2·6	40·5
		4·8	120	0·01575	—	2·2	58
		4·8	149	0·01835	—	3·0	66
		4·8	170	0·01915	—	2·4	65·5
Nacken and Heller (1960)	I.F.	12·2	20–140	0·0203	—	13·6	13
		12·2	20–140	0·0203	—	13·8	13
		12·2	20–140	0·0127	—	13·2	9
		12·2	20–140	0·017	—	13·2	11
		12·2	20–140	0·0185	—	17	10
		5·2	20–140	0·0185	—	11	15
		30·3	20–140	0·0185	—	36·4	4
Wilson and Russell (1960b)	E.R.	4	20 and 60	0·01	0·004	10	12
Köster, Bangert and Hahn (1954)	I.F.	33	40	0·018	—	30	—
Barrand and Leak (1963)	I.F.	20	43	0·016	—	5·19	20

First, all these investigators pre-strained their quenched samples before ageing, and secondly the number of carbon atoms per dislocation plane is quite large, in some cases more than 100. These figures would suggest that precipitation, rather than atmosphere formation, is extremely likely here.

The pre-strain listed in Table 2.3 would seem to have been introduced to increase the number of dislocation sites without (at that stage) any real knowledge of the dislocation substructure. In fact, there is clear evidence that the kinetics of precipitation are also changed by the introduction of plastic deformation. Quenched and unstrained samples lose their carbon at a rate corresponding to $t^{3/2}$ (Wert, 1949; Wepener, 1957) and this is interpreted by Wert and Zener (1950), Doremus (1958, 1959) and others as precipitation. Plastic deformation then alters the kinetics, often abruptly, to a $t^{2/3}$ law. The removal of carbon could thus be considered a two stage process (McLennan, 1965) but the details of the process need much further study from the theoretical end. On the experimental side, a recent paper by Butler (1966) has extended the internal friction results mentioned earlier to higher ageing temperatures. The process can then be considered in two stages, with ϵ-carbide precipitating on dislocations as the early rate-controlling step, and later, indications were present of the precipitation of Fe_3C. The time exponents varied in this case from 1·2–1·0, which was considered an appropriate value for long cylinders precipitating on dislocations according to Ham (1959).

Thus, the weight of evidence, at least in unstrained samples, is for precipitation to be occurring both within the matrix and on dislocations, while for strained samples the situation is less clear. It now remains to be seen how these views are correlated using electron microscopy.

2.3.3 *Structure of quench-aged alloys*

To resolve the structures present in quench-aged steels, as in most other steels, requires the use of electron microscopy. Even in the early models where replication had to be used, extraction replicas revealed the presence of a carbide which was not cementite (Tsou *et al.*, 1952) and which was probably ϵ-carbide. The careful replica work of Doremus and Koch (1960) gave clear indications of the precipitation on dislocation arrays, but with the advent of thin film microscopy, the situation has essentially been resolved.

The dislocation structure in cold-worked iron is now known to produce, after a few per cent deformation, an arrangement of cells within the grains, the cell walls being outlined by dislocation tangles. Simple dislocation theory had presumed that the Burgers vectors in α-iron were all of the form $b=\frac{1}{2}a$ [111], but Brandon and Nutting (1960) and Carrington *et al.* (1960) also found clear evidence that [100] dislocations exist in the iron lattice. [110] type dislocations have also been noted, and the relative proportions are roughly 60% [111] type, 20% [100] and 20% [110] (Dingley and Hale, 1966).

These results are of course of general applicability. In respect of quench-aged material, the first transmission electron microscope studies were by Leslie (1961), closely followed by Keh and Wriedt (1962), Smith (1962), Phillips (1963) and Hale and McLean (1963). General agreement has been good, and the work has been reviewed in the perspective of earlier results by Keh and Leslie (1963).

Nitride precipitation, at the lower temperatures, produces platelets of α''-$Fe_{16}N_2$, approximately 40 Å and up to 1000 Å across, lying on {100} planes (Keh and Wriedt, ibid.) This mode of precipitation should give a time exponent of 2, rather than the 3/2 found experimentally. However, the amount of isolated nitride is strongly dependent on ageing conditions. Ageing at lower temperatures produces a fine dispersion of nitrides, while at 100°C the precipitates are few and coarse. This clearly corresponds to the lower hardness maximum in material aged at the higher temperatures see Fig. 2.6). The precipitate distribution is also affected by pre-ageing at, say 25°C which must be below the 'G.P. zone solvus' temperature for this material. Most interesting of all, however, is the fact that dislocations seem to be preferred sites for nitride precipitation, as shown in Plate 2.1, and this phenomenon seems less sensitive to heat treatment than matrix precipitation. This is then good evidence that at least a high proportion of dislocations are permanently locked, or at least well pinned in quench-aged alloys.

The structure in iron–carbon alloys is very similar, as are the kinetics. The carbide formed is often classed as ϵ, whose structure may be in doubt, but it nevertheless can form both in the matrix, as well as extensively on the dislocations (Smith, ibid.; Hale and McLean, ibid). The latter authors suggest ϵ-carbide is a stronger barrier to slip than the α''-nitride, and other differences occur in that the carbide development is more strongly dependent on the presence of other elements (see below). Phillips (ibid.) has examined precipitation in quench-aged Fe–C–N samples, and finds that matrix carbide precedes matrix nitride. He also found contrast effects around dislocations, which could be due to segregation preceding precipitation.

The addition of silicon to the Fe–C system has been examined by Keh and Leslie (1963). Silicon, like aluminium and phosphorous, is an element insoluble in the carbide, and its rejection reduces the carbide growth rate. A maximum particle size, characteristic of the ageing temperature, results. Precipitation of ϵ-carbide on the dislocations still occurs, at least in a 3% Si–0.018% C alloy.

Manganese (in common with Mo, Cr and W) is more soluble in cementite than ferrite, and its presence should little affect the kinetics of precipitation. Leslie (1961) has confirmed this for 200°C ageing, and at higher temperatures Baird (1966b) has shown the presence of dissolved manganese in the nitride of a Fe–Mn–N alloy. Some precipitation on dislocations is still noted. At still higher temperatures, ϵ-Mn_2N may be formed.

While there is a certain amount left to be done on the effects of added ternary elements, the microstructure of quench-aged Fe–C and Fe–N alloys is now understood, at least in principle. The act of quench ageing, or of furnace cooling, will in general produce a precipitate dispersion which will ensure a large fraction of the dislocations are locked in position by precipitates. Thus, the initial yield point, absent in freshly quenched material, will return following quench ageing, and it would appear reasonable that the initial yield in mild steel is a result of a deficiency of free dislocations. The fraction of the solute atoms remaining in a supersaturated form in the matrix is then available for strain ageing (Section 2.5).

2.3.4 *Mechanical properties of quench-aged alloys*

According to the Cottrell–Bilby theory outlined in Section 1.5.1, it should be expected that the initial yield point of dilute Fe–C and Fe–N alloys may be suppressed by quenching from high temperature. This statement naturally supposes that the solute content is sufficiently low that all the solute is dissolved at solution temperature. Straining such a quenched alloy produces a normal rounded yield.

The reappearance of the yield point on ageing such a sample has been investigated by Wilson and Russell (1959) and Nakada and Keh (1967). In both cases, it is shown that the initial rate of return of the yield point is very rapid, comparable in fact with the single jump time of a solute atom. Thus the development of the yield can only be associated with the occurrence of Snoek ordering (Section 1.5.3; see also Eshelby, 1959).

Nakada and Keh (ibid.) show that after the development of this effect, there is a slower and more marked growth of the yield which may be ascribed to Cottrell locking followed later by precipitation on the dislocations. This is also in accord with the results of Fisher (1961) and illustrated in Fig. 1.21, which shows the very low values of k_y characteristic of these quenched materials, and its rapid increase on ageing. Finally in this section it should be remarked that the studies on ageing kinetics reported in Section 2.3.2 will all have been on material in which this initial Snoek ordering will have already occurred.

2.4 *Yielding behaviour*

2.4.1 *The upper yield point*

In Chapter 1 the experimental difficulties associated with the measurement of the upper yield stress were mentioned together with the effects of some external variables, such as pressurization. In addition, the major theories of the upper yield point have been covered, so that in this section, all that will be attempted is a more detailed review of experimental results and a consideration of how these fit in with developments of the theory.

Practically the only sets of adequate data on the variation of upper yield stress with temperature are those using centre-annealed specimens by Hutchison (1963) and Petch (1964). These show that the usual grain size equation holds, but that the value of k_y for the upper yield is $\sim 1{\cdot}4$ MN m$^{-3/2}$ or about twice that for the lower yield. Secondly, the values of σ_0 will differ, being less for the upper yield stress, so that the two curves meet at some positive grain size (Fig. 1.4).

Petch (ibid.) tried to explain this in the following terms. Since the upper yield is interpreted as the nucleation stress for the first Luders band, and since, from microstrain investigations, some plastic deformation is noted below this level, Petch assumes that some small but constant number of grains, N, are deforming plastically when nucleation of the band occurs. The fraction of total grains so deforming is then Nd^3, so that the strain rate in this volume is $1/(Nd^3)$ the applied strain rate. The friction stress σ_0 is strain rate sensitive, so that the actual friction stress on moving dislocations is

$$\sigma_0{}^{U} + \Delta\sigma_0 \log 1/(Nd^3)$$

where $\sigma_0{}^{U}$ will be the (bulk) friction stress, and $\Delta\sigma_0$ the increase in friction stress caused by the localized increase in strain rate on the yielding grains.

Thus, the upper yield grain-size relationship is

$$\sigma_U = \sigma_0{}^{U} + \Delta\sigma_0 \log 1/(Nd^3) + k_y d^{-1/2}$$

It should be noted that this result is not linear in $d^{-1/2}$, but the deviations are comparatively small, especially at room temperature, and could be masked by the scatter in Fig. 1.4. Worthington (1967) has also confirmed this relationship for a 3% silicon iron, as have Birkbeck and Douthwaite (1968), using iron containing 0·015% C.

A somewhat more specific approach to the problem of nucleation of Luders bands has been given by Crussard (1963a, b, 1964) using classical nucleation theory, the form of the embryo being that shown in Fig. 1.13(b). Estimating the total energy change, made up of elastic, surface and internal energy factors, the critical size may be determined and the stress (which will be the upper yield stress) may be estimated. The result is linear in $d^{-1/2}$, but the ratio of the slopes of the upper and lower yield stress plots $k_y(\text{uys})/k_y(\text{lys})$ is 1·5 instead of the value of 1·8 found by Hutchison (ibid.). Furthermore, the two curves should intersect on the y axis, which is not the case. However, considering the theoretical difficulties and assumptions involved, the agreement is encouraging.

The basic difficulties arise in the growth of the embryo, whether this occurs by some form of microstrain or microcreep. If this is so, the temperature dependence of the anelastic limit might be followed, but this is not the case. However, problems of the yield stress of individual crystals, and their relationship to σ_0 will be raised again later (section 2.7).

An alternative approach to the problem of nucleating the first nucleus of the Luders band is given by Russell *et al.* (1961). Although this is developed in terms of a dislocation unlocking theory, the general picture it gives is independent of this, for it considers the stress around an unlocked slip band as an inclusion, and calculates the condition for breaking through the grain boundary. This gave a value for k_y as

$$k_y = 2\sqrt{\left(\frac{\pi(2-\nu)}{(1-\nu)}\right)}\,\mu g$$

where ν is Poisson's ratio, μ the shear modulus and g the grain-boundary energy. Their experimental value of k_y (which is only slightly less than Hutchison's) gave g as $5{\cdot}8 \times 10^{-7}$ J mm^{-2} which seems rather high. However, their analysis and result might possibly be more appropriate to the lower yield stress, which would give $g \sim 1{\cdot}8 \times 10^{-7}$ J mm^{-2}; furthermore, their theory cannot explain the intersection with the lower yield stress curve, while microstrain studies (see section following) tend to indicate that slip in these nuclei is actually nucleated in or near grain boundaries.

Whatever the difficulties of producing the critical size of embryo, the upper yield stress can be developed in qualitative terms of the dislocation multiplication theory as outlined in the previous chapter. The bulk of the dislocations may be considered pinned by both carbide and nitride, and the influence of strain rate may be satisfactorily derived from Hahn's (1962) or Cottrell's (1963) treatment of the stress–strain curve. Arsenault (1963) did query the value of some of the parameters in Hahn's treatment, but this was satisfactorily answered by Reid and Hahn (1964) (see also Arsenault (1964a)).

The experimental difficulties seen here have precluded any basic studies on the variation of the upper yield stress with composition, and the effects of temperature have only been cursorily examined at room and lower temperatures. These again indicate that k_y is only slightly temperature sensitive.

2.4.2 *Microstrain*

As part of the general study of the initiation of yielding in mild steel, it might naturally be expected that the presence and detection of microstrain would be of considerable importance. This is in evidence from the numbers of papers on this aspect. The definition of terms has been given earlier (Section 1.4.3) and need not be repeated here.

The first detailed discussion on microstrain measurements relevant to mild steel appear in the paper by Roberts *et al.* (1952) in which a series of steels of up to 1% carbon was examined. Evidence was obtained of deformation below the lower yield point, and the 'elastic limit' was shown to vary as the logarithm of the mean ferrite path. Fisher (1955) examined theoretically the microstrain problem as part of the application of Cottrell's theory to delayed yielding, and associated microstrain with the breaking away of lengths of dislocation line from their restraining atmospheres. Certain of Fisher's results were checked by Hendrickson *et al.* (1956a), who found that log ($\dot{\epsilon}_{\mu_0}$) the initial microstrain rate was proportional to $1/\sigma T$, where σ is the stress and T the absolute temperature, as Fisher had predicted. However, there were discrepancies above room temperature, and the ratio of the energies of dislocations, both with and without their atmospheres, seemed to be considerably in error.

With the improvements in extensometers during the ensuing years, it is now possible to distinguish the elastic limit (σ_E), the anelastic limit (σ_A) and the proportional limit (σ_{PL}), and Brown and his co-workers in particular have made a notable contribution in this field. For example, Brown and Lukens (1961) have shown the microplastic strain, using a Fisher type model, is given by

$$\gamma = C\rho D^3 (\sigma - \sigma_0{}^0)^2/(\mu\sigma_0{}^0)$$

where ρ is the density of sources, D the grain diameter, μ the shear modulus, $\sigma_0{}^0$ the stress required to move the first dislocation in the most favoured grain and C a constant $\sim\frac{1}{2}$.

Perhaps even more important is the comparative insensitivity of the elastic limit (σ_E) to temperature. This has been clearly established by Brown and Ekvall (1962) in iron down to 90 K, and subsequently by Kossowsky and Brown (1966) down to liquid helium temperatures. In the latter case, a small rise is noticed between 50 and 5 K, but this is insignificant compared with the macroscopic yield stress which increases rapidly in this range. This has also been confirmed by Stoloff *et al.* (1965) using a series of Fe–Co and Fe–V alloys.

This result has considerable significance in theoretical discussions on the variation of macroscopic yield stress with temperature, discussed in Chapter 3, and to be mentioned again later in this chapter.

Microcreep, in contrast to microstrain studies, has been little studied. Arsenault and Weertman (1963) have reviewed this neglected field, and show that the incubation time prior to deformation is a function of ageing time and carbon content, with an activation energy of 20 kcal mole^{-1} (84 kJ mole^{-1}). The deformation following this incubation period is dependent both on carbon content and on temperature. This technique should be extended over a wider temperature range and certainly to higher temperatures, as it might be of considerable interest in blue-brittle studies.

The possibility of using optical microscopy, coupled with etch pit counting as a sensitive optical strain gauge has not escaped notice. Silicon iron is particularly suited to this technique, and there are a number of relevant papers. Holden (1960) noted slip bands starting by dislocations activated near precipitate inclusions, and this was followed by Suits and Chalmers (1961) who discovered individual deformed grains in specimens stressed beyond the friction stress σ_0. The results were consistent with the microstrain being nucleated at stress concentrations, such as precipitates. This type of result also fully confirms the work of Bullen (see Section 1.4.4) who showed that application of hydrostatic pressure could again engender dislocations and so eliminate the tensile yield point.

Later work on etch pitting has attempted to go further than this, and in particular to see if the origin of the Luders band nucleus can be determined. McLean (1963) was the first to note that practically all slip bands in the microplastic range started from grain boundaries. This observation was subsequently confirmed by Carrington and McLean (1965) and by Worthington and Smith (1964, 1966), again using a 3% silicon iron. The interesting result arises that active slip lines are found when the applied stress is less than the friction stress, showing that the stress necessary to operate a source (in the grain boundary) is less than that required to propagate the dislocations across the grain. In the later paper, this result was extended to show that slip could traverse complete grains at stresses less than σ_0, and an explanation was found in the Taylor misorientation factor m (Section 1.6.4) which acts as a stress raiser.

Finally, we revert to the question of grain-size effects. In the equation for γ given above, Brown and Lukens (ibid.) assumed, by definition, that $\sigma_0{}^0$ is independent of grain size. This has been checked by Brentnall and Rostoker (1965), who found, surprisingly, that $\sigma_0{}^0$ is independent of grain size in α-iron, but varies as $d^{-1/2}$ in both 3% silicon iron and in nickel. This trend was confirmed by etch pit counts, the differences between these metals being put down to inclusion population. It may well be, however, that the theory is inadequate, and in particular it may be placing too much reliance on results from larger grained material ($d^{-1/2} < 3{\cdot}5\ \text{mm}^{-1/2}$), where the Luders bands are diffuse and slip nucleation spreads from widely spaced areas along the specimen. This view is reinforced by Worthington (1966) who showed that the stress required to 'break' through a grain boundary is grain size dependent, and in fine-grained material is only slightly lower than the lower yield stress.

2.4.3 *Delayed yielding*

In Chapter 1, Sections 1.4.2 and 1.5.1, the general phenomenon of delayed yielding in mild steel was introduced and its general characteristics noted. In the latter section, the interpretation of delayed yielding was given in terms of the Cottrell–Bilby theory, which predicted a time dependence for the dislocations to break away from their atmospheres. The activation energy for this process was calculated (Cottrell, 1957) as

$$U = 9\ \sigma_{T_0}\ b^3\ (1 - \sigma/\sigma_{T_0})^3$$

where σ_{T_0} is the zero point yield stress, and b the Burgers vector. This gave a delay time

$$t_d = t_0 \exp\ [U\ (\sigma/\sigma_{T_0})/kT]$$

which compared reasonably well with experiment. Later refinements, not mentioned in the earlier chapter, include Kramer (1959), Maiden (1959) and Peiffer (1961), who derived equivalent expressions using slightly different postulates.

It was pointed out in section 1.5.1 that the missing parameter in these derivations was the grain size, and that both Russell *et al.* (1961) and Campbell and Marsh (1962) have shown the delay time t_d was markedly influenced by grain size, falling as the grain size increased. Russell *et al.* deduced from an extension of the Cottrell argument that

$$t_d = A/d^3 \exp\ [U/kT]$$

This may be rewritten using Yokobori's (1954) approximation to $\exp\ [U/Kt]$ as

$$t_d = (A/d^3)\ (\sigma/\sigma_{T_0})^{-1/nkT}$$

where, at room temperature, the exponent is ~ 9 (Campbell and Marsh, 1962). This gives

$$\sigma = kd^{-1/3}$$

which, while fitting the delayed yield curves well, cannot naturally reproduce the static upper or lower yield stress results.

If, instead of looking on $U(\sigma/\sigma_{T_0})$ as the activation energy for dislocations to break away from their atmospheres, one takes $U(\sigma)$ as the thermal activation stress for those dislocations which are **not** locked, then the theory of Russell *et al.* will give the same result. The variation of $U(\sigma)$ with temperature has been shown by Conrad (1961) to follow the analysis of static stresses at low temperatures, and the inverse cube grain size result also follows.

Alternatively, one may proceed as follows. Let us suppose that the first Luders band is nucleated after a fixed amount of microstrain, ϵ_s, which is sustained in a fixed fraction N of the crystals of the matrix. Following the analysis in Section 1.5.6.,

$$\epsilon_s = b \int_0^{t_d} (nv)\, dt$$

and substituting $n = Nd^3\rho$ where ρ is the dislocation density, and $v = (\sigma/\sigma_r)^{m*}$, we obtain

$$t_d = (A/d^3)\,(\sigma/\sigma_r)^{-m*}$$

Despite the assumption in this simple analysis, it is useful to compare it with Campbell and Marsh's (1962) solution given earlier. The value of $m*$ varies from ~ 15 for 0·05% C iron, to 5–6 for high purity iron which covers the observed experimental value of 9. Another theory, again using multiplication concepts, is given by Hahn *et al.* (1962b). By concentrating on nucleation conditions, one avoids the difficulty discussed by Moon and Vreeland (1969), namely that delay times are many orders of magnitude larger than the time taken for dislocations to cross grains.

All these grain-size relations predict t_d is zero in single crystals. Kramer and Madden (1952), Liu *et al.* (1956), and Kramer (1959, 1961) have both measured this quantity in single crystals of various metals. The delay times are always at least an order of magnitude smaller than in polycrystals, and in copper appears effectively zero. However, Kramer (1959) did find a slight effect in cold-worked copper, which may then be showing an unloading yield point. The effects of crystal orientation are indefinite. Kramer (1961) found an orientation dependence in aluminium, but not in zinc, and Liu *et al.*

(1956) found no orientation effects in zinc or β-brass. Orientation effects in iron have not been systematically studied, and the origin of the delay time remains obscure, but from Kramer (1961) we may see that the delay times are $\sim 10^{-4}$ times smaller than in polycrystals. One might, in fact, be measuring dislocation acceleration here and further extensive study is needed to cover the real meaning of these measurements.

2.4.4 *The lower yield point*

As with the case of the upper yield stress, the general characteristics of the Luders extension and the lower yield stress have been covered in broad outline in Chapter 1, using mild steel as the main example. It now remains to study in a little more detail the effects of the metallurgical variables.

(a) *Morphology.* The general morphology of the Luders bands has been seen, in Chapter 1, to depend both on specimen size as well as on the stress conditions. It is now well established from the work of Pomey *et al.* (1964) that in tensile tests the band nuclei are formed in the plane requiring minimum bending moment, i.e. across the front of a flat tensile specimen. As deformation proceeds, the bands, if single, turn to orient themselves at an angle of $\sim 45°$ across the front of the test piece.

It must also be remembered that as the grain size increases the Luders strain falls and the edge of Luders band becomes diffuse (see Hall, 1950, 1951a). This effect, which is difficult to allow for in theories of the lower yield stress, arises approximately at the point where Hutchison's curve for the upper yield stress intersects the lower yield (see Fig. 1.4). In other words, the stress differences are here so low that any small inhomogeneity in grain size will cause part of the band front to progress far in advance of the remainder.

(b) *Variations with grain size.* As will be recalled, the analysis of the lower yield stress in terms of grain size leads to the Hall–Petch equation. This was derived initially from considering the stresses ahead of a piled-up group of dislocations at the grain boundary. The objections to this explanation stem largely from the fact that pile-ups are not often seen; however, there may be good reason for this (Section 1.6.4). Nevertheless, others have revised the approach using alternative models to explain the grain-size relationship.

For example, Crussard (1963a, b, 1964) has developed a theory in terms of dislocation walls, instead of blocked slip bands. The stresses due to these walls are analysed, and a relation equivalent to the Hall–Petch equation may be derived.

Others (for example, Ball, 1959; Warrington, 1963 and Baird and McKenzie, 1964) have considered that the formation of sub-grains during deformation could lead to a reduction in effective grain size and even be a controlling factor in the Hall–Petch equation. X-ray microbeam and thin foil electron microscopy techniques allow reasonably accurate measures of sub-grain sizes, and a $t^{-1/2}$ relationship (where t is the sub-grain size) is readily determined. There were, however, discrepancies in magnitude between the former two and the latter papers, and a recent critique and some new results by Baird (1966a) show that the strengthening due to grain boundaries is greater than from sub-grain boundaries. Nevertheless, a clear effect exists, and any interpretation of k_y beyond the Luders strain should include the possibility of cell formation (see also Embury and Fisher, 1966).

It is obvious from this and the earlier treatments in Section 1.6.4 that a variety of models will lead to the same expression for the grain-size equation; the time may now be appropriate to study more closely the dislocation densities and configurations in fine-grained steel close to the edge of the Luders band. Until this is done in some detail, and for a variety of chemical compositions, it may not be possible to distinguish between these otherwise valuable theories.

(c) *Variations with strain rate.* As was again shown in Chapter 1, the lower yield stress variation with strain rate seems to be given accurately by the equation

$$\ln \dot{\epsilon} = m \ln \sigma_L + \text{constant}$$

over a wide range of strain rates, where $m > m^*$ as defined earlier (see Section 1.6.4).

The only uncertainty overlying this empirical relationship is at very high strain rates. Krafft (1962) has pointed out that under these conditions, 'delayed yield' conditions may apply behind the band front; a log–log relationship, however, still appears to hold between lower yield stress and band front velocity.

(d) *Variation with composition.* Most studies on the effects of composition have been concerned with the validity of the Hall–Petch relationship. Cracknell and Petch (1955) have shown that this holds over a variety of heat-treated, nitrided or carburized specimens, the important result being a variation in σ_0 and the relative constancy of k_y (Fig. 1.23) although minor variations in this parameter have occasionally been noted (see Anderson *et al.*, 1969).

The metallurgical condition must not vary too widely, otherwise apparent anomalies may result. For example, Morrison and Woodhead (1963) obtained an apparent failure of the grain size relation in Fe–Nb steels, but in a later paper Morrison (1963) showed that careful control of rolling conditions did result in the law being obeyed, provided the niobium existed in the same form.

(e) *Variation with temperature*. The discussion at this point will be restricted to temperatures below room temperature, since above this any results are rendered complex by the introduction of strain ageing.

TABLE 2.4. Values of σ_0 and k_y at low temperatures

Temperature K	σ_0 kg mm^{-2}	σ_0 MN m^{-2}	k_y kg mm$^{-3/2}$	k_y MN m$^{-3/2}$	Reference
291 77	7·1 43·3	70 425	2·3 5·5	0·71 1·7	Heslop and Petch (1956) 0·15% C
291 77	6·8 40·9	67 401	2·4 5·2	0·74 1·6	Heslop and Petch (1957) 0·04% C, 0·02% Mn
291 77	7·7 41·3	76 405	2·2 4·7	0·68 1·5	0·04% C, 0·47% Mn
291 77	14·0 53·1	137 531	1·9 3·0	0·59 0·93	0·05% C, 1·90% Mn
293 95 77	32·5 42·5 51·5	319 417 505	2·7 2·9 4·8	0·84 0·90 1·5	Hull (1961) 0·04% C, 3·25% Si
293	11·9	117	2·5	0·78	
173 123 98	24·6 42·2 53·4	241 414 524	2·4 2·2 2·6	0·74 0·68 0·81	Hahn *et al.* (1962a) 0·2% C
293 173 77	5·6 21·1 52·0	55 207 510	2·2 2·9 3·8	0·68 0·90 1·2	Hutchison (1963) 0·03% C
293 77	14·4 48·2	141 473	1·7 2·2	0·53 0·68	Jolley (1968) 0·002% C, 3·28% Ni

The major result of work by Petch and his co-workers (see Petch, 1958) is that again k_y is not markedly dependent on temperature, while σ_0 varies markedly. Since low-temperature results were excluded from Table 1.2 by reasons of space they are now given in Table 2.4.

The small increase in k_y at low temperatures is striking, and has some relevance to theories on the brittle fracture of mild steel (see Section 2.4.5 below). Any theory of the low-temperature strength must then consider the reasons for the variation in the single crystal strength, σ_0 (see also Section 2.7), which shows a characteristic and very rapid increase at low temperatures.

2.4.5 *Yielding, fatigue and fracture*

The mere existence of a yield point must necessarily affect mechanical properties other than the purely tensile ones described heretofore, and metallurgists are now called on to give attention to a wide range of associated variables, such as weldability, impact transition temperature and the like, when new steel products are proposed.

At high temperatures, the problem of creep studies will undoubtedly be complicated by strain ageing (see Section 2.5) but at room temperature and below, the influence of the yield point on two properties, namely fatigue and fracture, deserves further consideration here.

(a) *Fatigue.* Mild steel, in contrast to most other non-ferrous metals and alloys, shows a well-defined fatigue limit, i.e. a stress level (σ_c) below which fatigue failure apparently does not occur. This might be considered the result of a clearly defined yield stress, or alternatively might be associated with strain ageing during the running of the test. It is also interesting that this fatigue limit is likewise grain-size dependent, as Sinclair and Craig (1952) have found:

$$\sigma_c = C_1 + C_2 d^{-1/2}$$

where C_1 and C_2 are constants.

The fatigue limit is, of course, less than the lower yield stress (except possibly at coarse grain sizes), but the similar grain-size relationship is suggestive. Klesnil *et al.* (1965) confirmed this result, but showed that cyclic stressing led to a softening of the sample, as evidenced by an increase in strain amplitude. Thus, the formation of deformed regions, or persistent slip bands, as the case may be, is obviously allied to the microstrain problem, rather than to a gross yielding. The dislocation sub-structure is, as a result, clearly different, but it should also be expected that limited fatigue cycling will affect the subsequent tensile properties. For example, Holden (1959) shows a suppression of the yield point after cyclic stressing, and rightly considers the possibility of strain ageing as a contributing factor.

Some of the difficulties are seen from the work of Mintz and Wilson (1965) who noted indications of temperature rises within tensile fatigue specimens tested at room temperature and below. These temperature rises are quite adequate to increase markedly the ageing rate; other effects, possibly due to resolution of carbide, may also complicate the picture. A relation between tensile and fatigue properties is thus still lacking, particularly since Ferro and Montalenti (1963) have shown that decarburized specimens still possess a fatigue limit, although the value of σ_c is lowered.

(b) *Fracture.* The brittle fracture stress of mild steel at sub-zero temperatures has long been known to follow the grain-size relationship (Petch, 1953). This in itself has helped us to understand the brittle fracture problem, following the work of Stroh (1957) and later, Cottrell (1958). Cottrell has suggested, and this is currently the most acceptable theory, that slip dislocations in iron can nucleate numerous microcracks by a process of dislocation combination; these microcracks can grow until one of them reaches a critical size when catastrophic failure follows. The critical conditions were shown by Cottrell to lead to the Petch equation, and further that the condition satisfies

$$(\sigma_0 d^{1/2} + k_y)\, k_y = \beta\mu\gamma$$

where $\beta \sim 1$ for an unnotched specimen, μ is the shear modulus, and γ the surface energy of the crack. Factors which increase the parameters on the left of the equation tend to increase the propensity to brittle fracture, and vice versa.

In this respect, the increase in σ_0 and k_y at lower temperatures is vitally important, but unfortunately strain ageing, which can decrease k_y while raising σ_0, will lead to an increase in the impact transition temperature.

An alternative theory has been put forward by Petch (1958) giving the transition temperature directly as

$$\eta T_c = \sigma_0'' + C - (4\beta\mu\gamma/k - k)\, d^{-1/2}$$

where C and η are empirical constants, σ_0'' is the temperature independent part of the friction stress, and k the parameter in the grain-size equation applied to ductile failure while the other constants have been defined.

It is unfortunately not yet possible to give anything but qualitative answers from these expressions. This is because the surface energy term (γ) is not known precisely, while at low temperatures the possible incidence of twinning is a further complication. Twin intersections can nucleate cracks, and the conditions of twinning itself are again grain-size dependent. Likewise, twinning in mild steel may be suppressed by pre-strain at room temperature (Lindley, 1965; Jolley and Hull, 1964) but crack nucleation seems little affected.

The results of Hahn *et al.* (1962a) on two steels similar in all respects except manganese content show, as is well known, the higher Mn content attracts the lower transition temperature. Their values of k_y for the higher Mn steel are slightly lower; this in itself would lead to a drop in transition temperature (see Heslop and Petch, 1956 and Table 2.4), particularly if σ_0 is varying only slowly with temperature in this range. However, Hahn *et al.* were not convinced their values of k_y in the case of the two steels were markedly different, and in contrast to others, decrease at lower temperatures (see Table 2.4).

It should, however, be mentioned here that the Cottrell condition for fracture given earlier has been criticized because it fails to take into account any change in slip mode. Several alloys are known to show straight slip lines at low temperatures, and certain investigators, notably Johnston *et al.* (1965) and Jolley (1968) have suggested that low-temperature suppression of cross slip is responsible for increased brittleness. This will, in fact, increase the effective value of k_y by affecting the Taylor factor m (Section 1.6.4); thus, the Cottrell condition may be satisfied. This modification to the Cottrell theory deserves a more complete treatment than can be given here.

2.5 *Strain-ageing Kinetics*

As in the previous section, the basic introduction and theories lying behind the results which follow have been covered in Chapter 1 (Section 1.7). In this chapter, more detailed attention will be given to some of the results which might differentiate between the various theories, and following this, a study of dynamic strain ageing.

Since the origins of quench ageing and strain ageing are so closely related, no further account will be given of the compositional dependence, which was fully covered in Section 2.2.

2.5.1 *Effects of ageing time and temperature*

A review of early work in this field is found in the article by Kenyon and Burns (1939), and more recently in that of Baird (1963). In detail, most investigations have used the results listed in these reviews to correlate the theories of age hardening prevalent at the time, and there are indeed very few full analyses of the kinetics over a reasonable range of temperatures and times.

The Luders band which appears after strain-ageing treatment was termed the secondary Luders band in Section 1.7.1. At early stages of ageing it is diffuse (Hall, 1951a), but as ageing proceeds, it sharpens

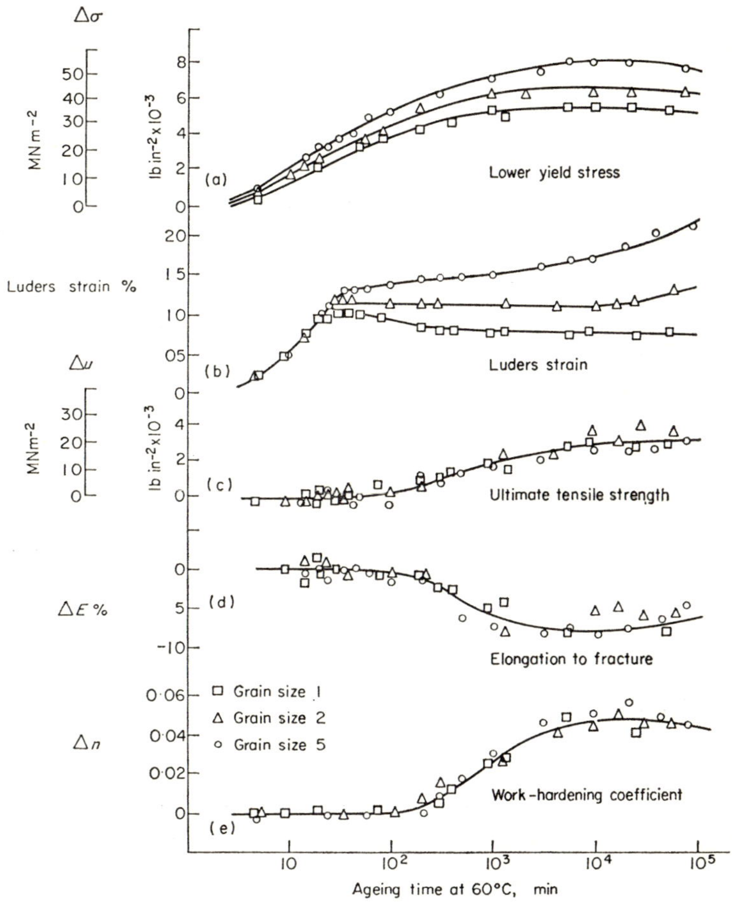

FIGURE 2.7. Variation of tensile test parameters during ageing (samples pre-strained 4%). Grain sizes are (1) 50, (2) 195 and (5) 1850 grains mm^{-2} (Wilson and Russell, 1960b)

up and becomes a typical Luders band in all respects. The growth in the lower yield stress of the secondary band has already been illustrated in Fig. 1.22(b).

Typical changes in the other properties of the aged sample are shown in Fig. 2.7, taken from Wilson and Russell (1960b). In the first of these sets of results, the increase in yield stress is shown, with over-ageing commencing after about 10^4 min at 60°C. The Luders strain initially shows a steady rise, then falls off, while the U.T.S. goes up and elongation falls. The elongation reaches a minimum at about the maximum in $\Delta\sigma$.

Similar results, at least for the increases in yield stress, were obtained by Sylwestrowicz and Hall (1951) using ageing at the edge of the Luders band. These have again been illustrated earlier as Fig. 1.23.

Together with the changes in mechanical properties, there is a change in hardness. This is more marked than in Fig. 2.6, which illustrated quench ageing; for example, in quench ageing at 80°C, a hardness increase of six points (Rockwell B) is noticed, and less at higher temperatures. In strain ageing, the hardness change in a comparable steel is twice that, and is achieved irrespective of the ageing temperature (Davenport and Bain, 1935).

As mentioned earlier, two other methods are available for studying the depletion of solute occurring during ageing. The first of these is the resistivity method, which has been employed by Cottrell and Churchman (1949a) and Cottrell and Leak (1952). The resistivity changes were roughly five times as fast in quench ageing as in strain ageing; nevertheless, in the latter case, depletion occurred at a rate proportional to $t^{2/3}$ up to about 50% of the total solute removed. The second, and more direct method, is to measure the solute remaining in solution using internal friction. As was explained in Section 2.3.2, there are again clear differences between quench ageing (i.e. ageing of quenched samples with and without pre-strain) and strain ageing in furnace-cooled samples. In the latter case, and despite the difficulties of determining smaller atom fractions, McLennan (1965) has shown a two-stage process in operation, with signs of resolution of precipitates.

A useful correlation of internal friction and mechanical properties has been carried out by Lautenschlager and Brittain (1962a, b). They show that the amount of strain ageing, as measured by internal friction, lags behind measurements of the yield point return (see Section 2.3.4), and, in the later paper, they also show that the return of the yield point will be suppressed if the pre-straining is carried out at a temperature lower than that used for testing after ageing. The explanation of this lies in the presence of work softening (Section 1.8.3), which, irrespective of any other locking mechanism, influences the yield stress at the higher temperature.

Within a temperature range of up to about 200°C, and with the limitations given above an increase in temperature will therefore merely accelerate the diffusion rate and hence the rate at which the hardness peak is reached. Beyond that temperature, there is the possibility that appreciable resolution of carbide and nitride will occur, and the structure will be altered by the ageing process.

2.5.2 *Effects of pre-strain and applied stress*

As might be expected, the ageing kinetics are comparatively insensitive to the amount of tensile pre-strain (see, for example, Edwards *et al.*, 1939). The density and distribution of the dislocations will of course change, but any effects on kinetics will be of secondary importance. Marked reduction in elongation to failure can, however, arise from repeated strain ageing tests on a single specimen, and under these conditions one is approximating to the conditions of an interrupted blue-brittle test (Section 2.5.5) (see Hundy and Boxall, 1957; Felbeck *et al.*, 1965).

These remarks do not, however, apply to prior deformation by rolling. Hundy (1956) shows that the secondary yield point elongation fell with increasing reduction followed by ageing for a fixed time; the secondary yield stress followed a similar pattern. The retained stresses are of course complex, but there also appears to be an effect on ageing kinetics, in that specimens cut transversely from the rolled sheet apparently age at a much slower rate (Tardif and Ball, 1956). However, it is now known that the result of rolling is even more complex than this (Butler and Wilson, 1963). Where the temper rolling reduction is less than the tensile Luders strain, a reduction in yield stress on ageing can be achieved. At higher reductions, ageing always increases the yield stress, and little difference is found between longitudinal and transverse samples (see also Section 2.5.4).

The question of the effect of an applied stress during ageing has led to further interesting results, since Holden and Kunz (1952) first noted that ageing under tensile stress apparently increases the rate of ageing. This has been confirmed by Brittain and Bronisz (1960), Mura and Brittain (1960) and Almond and Hull (1966). Three possible theories have been advanced. Paxton (1953) suggested that ageing under stress helped eliminate any stress concentrations and so increased the upper yield. There is also the possibility that vacancy-enhanced diffusion can occur, as happens in substitutional systems (see Chapter 5), and in this light Brittain and Bronisz (1960) found an unusually low activation energy for strain ageing in low carbon steel. Yet another theory (due to Haasen, (1959) and Mura and Brittain (1960)) suggests that if the dislocations are locked in a bowed-out position, this would produce an increased resistance to unpinning.

A comparison of these three theories was made by Almond and Hull (ibid.). They showed, using internal friction methods, that there was no change in the solute depletion rate, and that their activation energy determinations gave normal values. They also dismissed the effects of stress concentrations, saying that the upper and lower yield stresses should be increased on this theory. (This may not actually be so, since the stress concentrations at a band front are always high, and would, after yielding, swamp any machine misalignment.) With electron microscopy, however, they do show the dislocations are markedly bowed out, and this, in their view, will lead to an increased unpinning stress.

2.5.3 *The structure of strain-aged iron*

As an extension to the electron microscopy of quench-aged material, several investigators have endeavoured to find changes in the dislocation structure of strain-aged iron. If the initial yield point in the material is the result of a deficiency of free dislocations, because the majority of these already present are fully locked by ϵ-carbide, or α''-nitride, then it would seem equally plausible that the same mode of precipitation might occur, producing the same yield point effects, during strain ageing.

Hundy (1956) suggested that nitride precipitation could be responsible for strain ageing, and the possibility was further examined by Doremus (1958, 1960). Although high resolution electron microscopy was only then becoming available, Doremus and Koch (1960) found good evidence of precipitation on dislocations in strain-aged material using replica techniques.

More sensitive studies using thin foil methods have been carried out by Leslie and Keh (1962), Keh and Leslie (1963), and by Phillips (1963). Using a 0·03% C steel, Keh and Leslie could detect little change in a sample aged for over 100 hours at 60°C. Phillips likewise could find little change, although after prolonged times (two to three years) at 22°C, some evidence of platelet precipitation could be detected. This result, however, could not be repeated on ageing at higher temperatures. Certain contrast effects were noted around the dislocations, which in turn could be interpreted as segregation.

The levels of available solute naturally play an important role in the structures which might, and then only with difficulty, be resolved in the electron microscope. In quench-aged material, about 0·02% carbon can be available for precipitation both in the matrix and on dislocations. In furnace-cooled strain-aged material, the maximum available is only at least a tenth of this, while the dislocation density is up at least a hundred-fold, and although Phillips (ibid.) estimates there could be 30–40 atoms per dislocation plane, this does not appear sufficient to nucleate precipitate. Phillips did, however, ignore trapping of carbon atoms by vacancies produced in the deformation, which would reduce the estimate further. In any case, precipitates of this magnitude would be extremely difficult to detect, even with the latest techniques of electron microscopy.

2.5.4 *Correlation with theory*

From the introductory discussion on strain ageing (Section 1.7) and from the description of strain ageing in this chapter, it is apparent that the simple explanation of strain ageing is in terms of dislocation locking, either by an atmosphere of the Cottrell type, or from actual precipitation. It is probably fair to say that most investigators in the field would now agree that dislocations, once locked by either of these mechanisms, will remain locked, and that the yield point and subsequent secondary Luders extension should be described in terms of a dislocation multiplication theory. The difficulties arise in the detail, both of the theories, and also in terms in which experimental results may be correlated with these theories.

In the first instance it is apparent that the hardening in quenched and strain-aged material and in furnace-cooled and strain-aged material is of a different character. The former occurs more rapidly, and leads to a marked increase in the yield point at very short times. The latter is slower, and is not subject to the rapid over-ageing found in the former samples. Quench ageing is speeded up by pre-straining, strain ageing is little affected. Thus, if precipitates form, then the distribution must be different, and lead to quite different kinetics, particularly at short ageing times.

In fact, no precipitates can be found in strain-aged samples (Section 2.5.3). This, it may be argued, is because the precipitates, if formed, come from about 0·001% carbon and form on 10^{10} lines cm^{-2}, and the result may be beyond the resolution of the microscope. Thus, glossing over this difficulty, precipitation may still be present.

Precipitation would be preceded, on this basis, by the drift flow of solute to the dislocations, and from internal friction studies discussed above (Section 2.3.2) it is tempting to relate the two stages of solute depletion to atmosphere formation, followed by later precipitation. This conclusion is, however, at variance with simple theory, as the first stage varies as $(\text{time})^{3/2}$ rather than as the 2/3 power, and the later stage following approximately a $(\text{time})^{1/2}$ law (McLennan, 1965).

The difficulties of the original Cottrell theory, or Harper's (1951) modification of it, have led to a reappraisal of its basis. Baird (1963) points out that applications of the Cottrell theory give rise to high calculated densities of dislocations (see Table 2.3), higher in fact by an order of magnitude than is given, directly, by electron microscopy, Another way of expressing this statement is to say that the Cottrell theory, in terms of observed dislocation densities, is too rapid at short ageing times to agree with internal friction results. This discrepancy may be due to neglecting the concentration gradient build-up near the dislocation core, a fact which Bullough and Newman (1962) did include in their treatment. However, they were attempting, primarily, to produce a theory which was compatible with Harper's results (ibid.), and could not have been aware of those internal friction results on slow cooled strain-aged alloys which deviate from the $t^{2/3}$ law. Other attempts to modify the original Cottrell approach (for example, Mura *et al.*, 1961) have been severely criticized (Bullough and Newman, 1962c, Mura and Brittain 1962).

The theoretical difficulties of dealing with the kinetics of precipitation have been alluded to in an earlier section (1.5.5). Both Ham (1959) and Bullough and Newman (1962 a, b) have treated this problem in simple models, but although it is possible to correlate some of their cases with experimental results (McLennan and Hall, 1963), the basic physical conditions cannot be established by this means. Doremus (1960) also suggested a simple model, where, once precipitation has nucleated, the drift flow is controlled only by diffusion. However, the time exponents again could not give a $t^{3/2}$ relationship directly.

From this discussion it is seen that the theories put forward cannot be readily applied to quench-aged samples, where matrix precipitation is clearly associated with precipitation on dislocations. It is equally true that the theoretical models will be inadequate in strained material whether previously quenched or furnace cooled, since it is now well known that the dislocation distribution is not uniform, with roughly 80% of the dislocations distributed in cell walls and tangles. If these act as planar sinks (to a rough approximation) then their kinetics will be different from an isolated dislocation. If diffusion control is alone responsible, then isolated dislocations might act as sinks trapping material as $(Dt)^{3/2}$, while the planar arrays would have a time dependence of (Dt). This again does not agree with experiment.

Studies of solute depletion using electrical resistivity studies are similarly bedevilled with electron scattering arising from both discrete precipitates and from the dislocation cores. The only other approach to kinetic studies is that of mechanical properties. This, in turn, presents difficulties of correlation between the increase in yield stress $f\,(=\Delta\sigma/\sigma)$ and the solute depletion, as measured by either internal friction or resistivity methods. Here, as shown by Wilson and Russell (1960b) graphs of $\Delta\sigma/\sigma$ and $\Delta\rho/\rho$ (the fractional changes in yield stress and resistivity respectively), are of similar form when plotted against $(Dt/T)^{2/3}$ but the change in resistivity occurs at a slower rate. Part of the problem is relating f to the number of free dislocations, which in turn must be related to the fraction of solute removed. Some investigators have taken that $f \propto N_t/N_0$ (Blakemore and Hall, 1966). There is some justification for this if one considers ageing properties at the front of the Luders band (Fig. 1.23). Here, before ageing, the free dislocation density is ρ_2 and their velocity under the applied strain-rate is v_2. After ageing, and at the moment of stress release, the number is ρ_1, and velocity v_1. Thus, as before in discussing theories of the upper yield,

$$\dot{\epsilon} = \rho_1 v_1 b = \rho_2 v_2 b$$

$$\frac{\rho_2}{\rho_1} = \frac{v_1}{v_2} = \left(\frac{\sigma_U}{\sigma_L}\right)^{m^*}$$

$$\therefore \frac{\rho_2}{\rho_1} = \left(\frac{\sigma_L + \Delta\sigma}{\sigma_L}\right)^{m^*} \simeq 1 + m^* \frac{\Delta\sigma}{\sigma_L}$$

$$\therefore f = \frac{\Delta\sigma}{\sigma} = \frac{1}{m^*}\left(\frac{\rho_2 - \rho_1}{\rho_1}\right)$$

If it is assumed at this stage that a fixed number of solute atoms are sufficient to lock these dislocations, then

$$f = \frac{1}{m^*}\left(\frac{N_t}{N_0}\right)$$

where N_t is the number of carbon atoms removed from the initial concentration N_0. Provided that m^* does not change very rapidly with temperature, values of f may thus be used to determine the activation energy of solute diffusion.

Another useful, if less direct approach, is to consider the variation in parameters of the grain-size equation during ageing. This has been done by Wilson and Russell (1960a, b) whose results have already been illustrated in Fig. 2.7. It will be noticed there that the Luders extension increases fairly rapidly up to a value which is dependent on grain size, and then flattens out. This, according to Wilson and Russell, divided the ageing process into two stages. Stage I corresponded to atmosphere formation, stage II to solute clustering or precipitation. This theory was cross-checked with parallel changes occurring in the grain size relationship, and which can be seen from the following figure (2.8). After straining, k_y is much reduced, increasing up to the end of stage I, and thereafter remaining constant. Beyond this any increase in strength is associated with a change in σ_0.

The interpretation of these results was carried out in terms of the original theory of grain-size dependence as proposed by Hall (1951b). There are several reasons why this interpretation should be treated with reserve. In the first place, Hall (1951a) showed that at early stages of ageing, the secondary band was diffuse, or not even fully

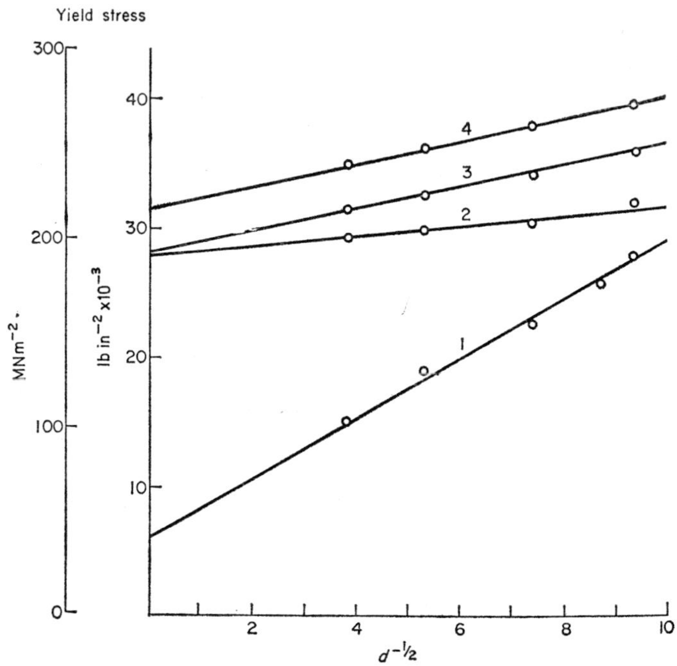

FIGURE 2.8. Variation of the lower yield stress with grain size. (1) annealed; (2) strained 4%; (3) strained 4% and aged to end of Stage I; (4) strained 4% and fully aged (10^4 min at 60°C) (Wilson and Russell, 1960a)

formed, and only when the sample had increased its yield stress by more than 44 MN m^{-2}, did the bands become sharp. (This applies to material with $d^{-1/2} \simeq 10$ $mm^{-1/2}$.) This is considerably after the end of stage I on Wilson and Russell's curve, and it becomes difficult to compare the modes of deformation if the band front is still becoming sharp. Secondly, the variation in k_y seems to parallel those found by Fisher (1961) (see Fig. 1.21) on quench-aged material, and it will be remembered there that the interpretation was based on the ability of new grain boundary sources to come into operation. Thus, an alternative theory could be developed in terms of the ability or otherwise of the aged material to nucleate new sources (probably from grain boundaries), or activate the old, and the division between 'stage I' and 'stage II' could be interpreted this way. Wilson (1968) has endeavoured to follow changes in k_y in these terms. Unfortunately, any such dividing line would be hypothetical at the moment. Careful dislocation density determinations, and a careful study of changes in the mode of deformation, could, perhaps, assist. But, at the moment, if we wish to interpret the return of the yield in terms of a deficiency of free dislocations, we are able to do so only in the most general terms, without any clear idea of the locking mechanisms present through the ageing range, and without knowing at what stage the original dislocations fail to move, and new dislocations are created.

Finally, in this discussion, there remains the equally difficult problem of interpreting the effects of complex stressing on the ageing kinetics. The kinetics of quench-ageing can, as has been stated, be speeded up by prior tensile straining, presumably because the number of nucleation (or condensation) sites has been increased.

In the case of normal strain ageing, as indicated earlier, the kinetics of ageing can be severely affected by straining after ageing in a different sense from the original pre-strain. This has been done in torsion by Wilson and Ogram (1968) and differences have also been noted in specimens cut from the longitudinal and transverse directions in rolled sheet (Tardif and Ball, 1956 and Butler and Wilson, 1963). The possible results of the Bauschinger effect, and the presence of locked-up stresses of varying distributions make it difficult to treat this subject adequately. Wilson and Ogram (ibid.) concluded that any effects due to the Bauschinger effect were obliterated at an early stage in the ageing process. In the case of rolling, the results are more complex. Butler and Wilson showed that in temper-rolled material, where the reduction was about 1%, the material in the deformed condition consisted of a regular mass of Luders band nuclei. Ageing in this state produced a lowering of the yield more marked in the longitudinal than the transverse direction. Temper rolling to 3% reduction, i.e. beyond the Luders extension, always produced an increase in yield, and the direction of rolling then had less effect. The kinetics of ageing may, however, still be affected (Tardif and Ball, 1956).

Part of the reason for these differences may well lie in the presence of locked-up stresses in the subsequent tensile testing direction, which would increase the apparent ageing rate (see previous section). However, Cottrell (1963) has suggested that one should not overlook the possible role of inclusions, since inclusions can activate dislocation sources on pre-straining and their stress fields consequently reduced. After ageing, the yield point should then return more rapidly if straining is in the same direction. In rolling, where the inclusions are elongated in the rolling direction, samples cut transversely should show a lower yield stress. This in accord with Butler and Wilson (ibid.) but only if the rolling reduction is small.

It is obvious from this section that we are still a long way from obtaining a clear physical picture of the mechanisms of strain ageing. In large measure this is due to the failure of electron microscopy to give a clear lead, as it has done in quench ageing, on changes in structure of the aged alloys. Failing direct observation, one might make some progress by tensile measurements on partially quench-aged samples, where the initial structure at least can be characterised, and determine the grain size effects and ageing rates. With enhanced solute concentrations, possibilities of useful electron microscopy are higher, and in any case, careful studies of the dislocation density may help to differentiate between unlocked 'old' and 'new' dislocations thus assisting the separation of ageing into a two-stage process. Theoretical studies of solute flow and precipitation should not be neglected, as there is an equal need for a lead from the predictions of sound theory.

2.5.5 *Blue-brittleness*

As the ageing temperature is increased, the diffusion rates of the interstitial carbon and nitrogen atoms increase, until the ageing rate is sufficiently high to cope with the creation of the new dislocations generated during deformation. This will lead to a serrated stress–strain curve, an inhomogeneously deformed specimen, and a decreased ductility. The decreased ductility, which occurs in the 'blue-heat' zone of the temper colour range (approximately 150–300°C) has led to the use of the term 'blue-brittleness', but it would be better described as dynamic strain ageing. The general background to this problem has been given in Section 1.7.4 and it now remains to examine the case of mild steel in some detail.

The diminished ductility in this temperature range is naturally not a truly brittle behaviour, but the lower elongation to fracture may cause complications if mild steel is worked around these temperatures. Consequently, blue-brittleness is no mere scientific curiosity, and substantial reviews of the problem already exist. Detailed discussion of the engineering aspects are given by Kenyon and Burns (1939), and both Baird (1963) and Keh and Leslie (1963) have reviewed the theoretical implications of these results.

The general results from these hot tensile tests may be seen in Fig. 2.9(a) and (b). The former shows the yield stress, and the latter the ductility of steels over this temperature range. The maximum

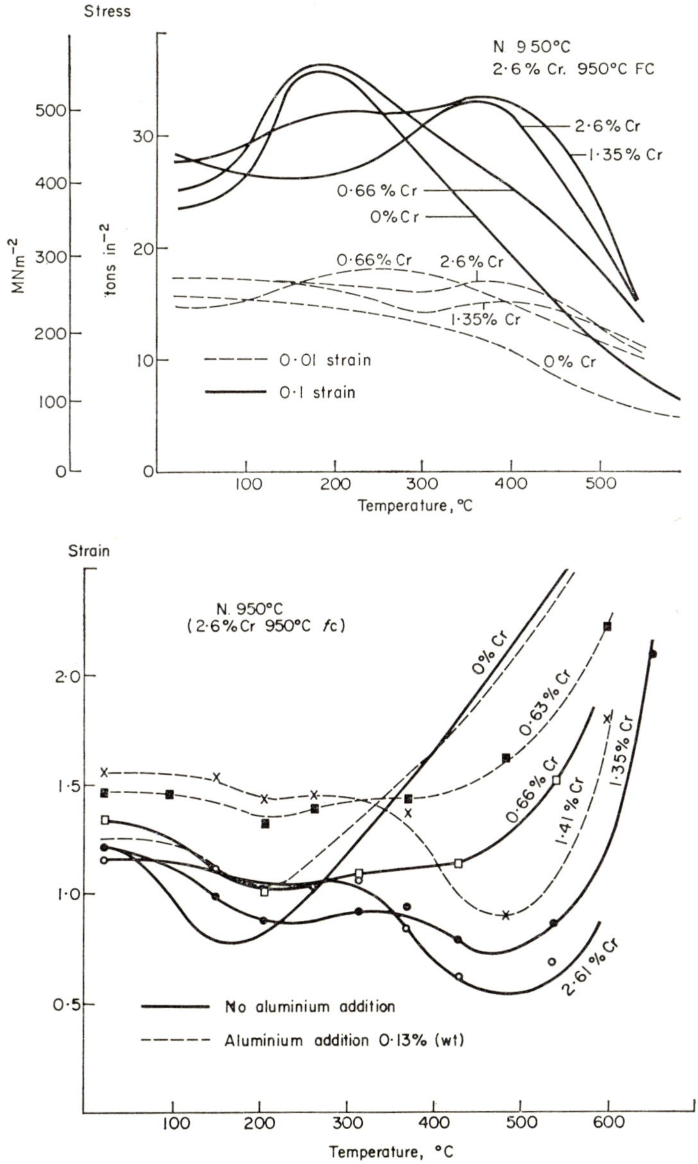

FIGURE 2.9. (a) Stress and (b) ductility variations with temperature for a series of Cr steels (Glen, 1957)

in the former does not occur at the same temperature as the minimum in ductility, for complex reasons which will be discussed later. The temperature of the maximum and minimum are, however, both sensitive to strain rate, and a plot of $\sigma_{y\,max}$ or ϵ_{fmin} against $1/T$ will lead to an activation energy which will be that of carbon or nitrogen in iron. Furthermore, the temperature range over which serrations can be noted can be extended to less than 50°C in very slow tests (Elam, 1938; Sleeswyk, 1958) to over 400°C in very high speed deformation (Kenyon and Burns, ibid.).

The variables which affect the nature and extent of dynamic yielding will now be described in turn before considering recent work on theoretical aspects of the problem.

(a) *Composition.* As in the case of quench ageing and static strain ageing, it is comparatively simple to show that the effect is conditional on having sufficient free carbon and/or nitrogen in the sample. Baird and Jamieson (1963) have carried this out with a series of Fe and Fe–Mn alloys. Neither pure iron nor pure Fe–Mn alloys showed any anomaly in the yield-stress/temperature curve (see Fig. 2.9). It was present in Fe–C and Fe–N alloys, and the peak was not so marked in Fe–Mn–C alloys, showing the weak affinity between these two elements. The Fe–Mn–N alloys, however, behaved in a similar way to the Fe–N samples.

The presence of carbide and nitride formers can restrict the appearance of blue-brittleness. Glen *et al.* (quoted in Baird, 1963) show that aluminium killed steels do not show a peak in yield stress in the 150–300°C range, while Epstein *et al.* (1950) demonstrated that serrated yielding could be suppressed by vanadium additions.

Glen (1957) has also examined in some detail the effects of additions of Al, Mn, Cr, Mo, W, V, Ti, Si, Ni and Cu. As might be now expected, the carbide and nitride formers could reduce or eliminate the maximum in the yield and the minimum in the ductility curve. However, with Cr, Mo, Mn, W or Cu the steels showed an additional peak in the yield stress curve at around 500°C. The reasons for this additional effect are obscure. Glen himself suggested precipitation was occurring at these higher temperatures, but Cottrell (1957a) showed that the diffusion rates were probably too low. On the other hand, the prospects of vacancy enhancement of diffusion cannot be ruled out. Yet, if this is so, why did the subsidiary maximum not occur in the steels containing Ti and V, which are both strong carbide formers ? Later studies by Baird and Jamieson (ibid.) on Fe–Mn steels do not completely explain this effect.

(b) *The nature of the stress–strain curve.* The form of the stress–strain curve has already been illustrated in Fig. 1.3(c) in Chapter 1. At a strain rate of about $10^{-4}\ s^{-1}$, testing at around 100°C will discover irregularities in the stress–strain curve beyond the primary Luders extension. At 138°C, there are clear signs of a secondary Luders extension, and then at slightly higher temperatures, the primary Luders strain becomes covered with strong serrations. At still higher temperatures, greater than 200°C, the serrations disappear firstly from the 'homogeneous' region beyond ϵ_L, and later from the primary band itself (Blakemore and Hall, 1966).

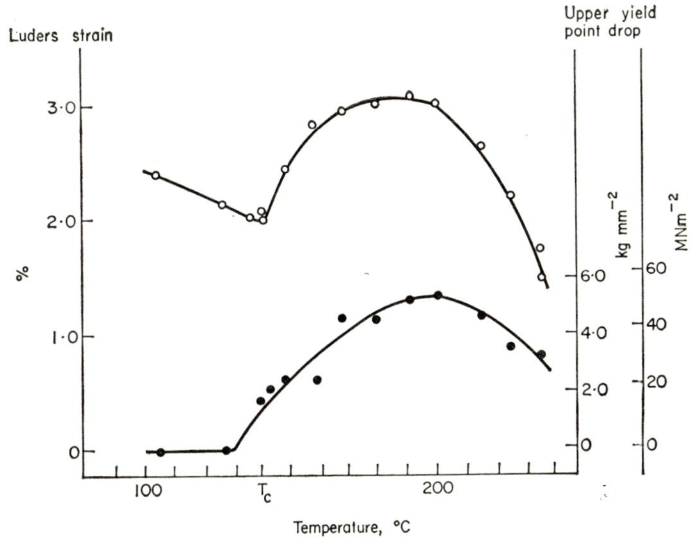

FIGURE 2.10. Variation of Luders strain (open circles) and serrations on the primary band (full circles) in mild steel ($\dot{\epsilon} = 0{\cdot}52 \times 10^{-4}\ s^{-1}$) (Blakemore and Hall, 1966)

Not so evident from these curves is a characteristic change in ϵ_L. Before serrations appear on the primary band, the Luders strain is falling, but as soon as the serrations appear, ϵ_L becomes notably larger, and only starts to fall again at much higher temperatures (Fig. 2.10). If T_c is the critical temperature at which these primary serrations first appear, then a plot of $\ln\dot{\epsilon}$ against $1/T_c$ will yield the activation energy for carbon in iron (Blakemore and Hall, ibid.). Sleeswyk (1960) obtained an identical result from a similar study.

(c) *Morphology of the Luders bands.* The serrations on the primary band have been related to irregular motion of the Luders front by

Hall (1952a). At each fall of stress, the front of the band moves forward rapidly, the stress falling meanwhile. The front then slows and becomes locked, and the stress starts to rise again.

This irregular motion is revealed by a series of parallel marks on the polished front face of a strip specimen (Plate 2.2 (a)). After the primary band has progressed through from one grip to the other, it is obvious that the last material to deform has been least strain aged, so that a secondary Luders band commences from that end and moves back through the specimen. It can also move irregularly; and leaves marks on the surface in the opposite sense to the primary. These are seen in Plate 2.2(b), and the markings which arise can conveniently be called Type A and Type B, a nomenclature which will also be applied to the markings found in aluminium alloys (see Chapter 5).

(d) *Effects of grain size.* Within the range of grain-sizes studied (50–170 gr mm^{-1}), Blakemore and Hall (1966) were unable to detect any major change in kinetics. The values of T_c as defined above appeared identical, irrespective of grain diameter, as would be expected if bulk diffusion is the controlling factor. With coarser grained material (5 gr mm^{-1}), however, Brindley and Barnby (1966) showed that the serrations, instead of disappearing at higher strains, appeared to increase, while at the yield point, the curve could be smooth. In this sense the curves correspond to Type B yielding in aluminium alloys (see Chapter 5). In a later paper (Brindley and Barnby, 1968) they also showed that this type of yielding could be associated with nitrogen depletion, caused by segregation to dislocation; the effect was absent in high nitrogen steels.

Wilson (1961) has studied the variation of σ_0 and k_y over a temperature range up to 200°C. The changes in σ_0 are small, with a slight rise at approximately 150°C, while k_y fell over the range 20–100°C, and 150–200°C, being roughly constant in the intermediate range. Similar results have been obtained by Brindley and Barnby (1966) and Dingley and McLean (1967).

(e) *Work hardening and dislocation density.* A further inspection of the stress–strain curves in Fig. 1.3(c) shows there is a systematic variation in the rate of work hardening for strains beyond the Luders extension. A careful study by Blakemore and Hall (1966) revealed that the maximum hardening rate occurred at the maximum in the Luders extension (Fig. 2.10), i.e. at around 200°C at this strain rate.

Coupled with this result is the observation by Keh and Leslie (1963) that the dislocation density is markedly increased in samples deformed over the blue-brittle range. Differences of an order of magnitude are found, and this result has been noted in other material showing serrated yielding (Section 4.5).

(f) *Theories of blue-brittle behaviour.* Obviously, the increased dislocation density and work-hardening rate are interconnected, as the problems of dislocation movement will be related directly. Of more importance is the mechanism whereby dislocations can become locked, and, moreover, locked permanently, as the new dislocations so produced will be largely responsible for the enhanced density values. Furthermore, values of the diffusion coefficient D are about 10^{-11} mm^2 s^{-1} at 100°C, so that the Cottrell equation

$$\dot{\epsilon} = D\rho$$

is satisfied with $\dot{\epsilon} = 10^{-4}$ s^{-1} and $\rho = 10^7$ lines mm^{-2}.

First attempts to understand the nature of blue-brittle behaviour were performed by Cottrell and Jaswon (1949). This theory assumed that at low dislocation velocities, the dislocation moved dragging its atmosphere behind it. As the stress level of the sample is increased, a velocity value is reached when the dislocations break free of their atmospheres. They can then propagate at a lower stress, and a serration arises. At higher temperatures, the atmospheres evaporate, and the serrations disappear.

This theory, though basically attractive, does not explain the marked increase in dislocation density in the blue-brittle region. Baird (1963) has likewise suggested that the degree of locking predicted is too low, of the order of 5 MNm^{-2}. It is probably more attractive to consider the upper stress of the serrations as a multiplication stress, the lower stress as a locking stress. Although the difference in stress levels may be approximately 60 MNm^{-2}, on a locking stress of 180 MNm^{-2}, this represents a velocity ratio for the dislocations of $(1.33)^{15}$, or approximately 80 : 1, and even larger in silicon iron. Thus at the lower stress, conditions may well be right for these newly created dislocations to be locked, and once locked, they remain so.

The only question in doubt here is again the locking mechanism. Blakemore and Hall (1966) show that the kinetics of dynamic strain-ageing agree with static tests; beyond this is the conjecture given in the earlier section.

2.6 *Effects of radiation damage*

The general outlines of the effect of radiation damage on metals has been covered in Section 1.4.5. Briefly, the interaction of ionizing radiation or fast elementary particles with a crystal lattice is to produce an enhanced population of point defects, both interstitials and vacancies. Under appropriate conditions of temperature and diffusion rates, these may drift to dislocations – to form jogs, and so pin them at points along the dislocation line. On straining, a small yield point may be developed as these pinned dislocations multiply, and this type of argument is of value in discussing austenitic steels (see below). At the same time, the yield stress and rate of work-hardening are raised, and the ductility lowered.

The situation in the case of mild steel will be more complex, since the yield point itself is already present and is due to the presence of interstitial impurities, while the number of free dislocations is low.

An extensive literature exists on radiation damage in metals and alloys, but studies on mild steel are comparatively sketchy. The sections below deal in turn with the more scientific aspects – there is a review of the technological aspects by Harries (1960). Unless otherwise specified, remarks below apply to neutron irradiation.

2.6.1 *The yield point*

The increase in yield stress following irradiation has received quite wide attention. At low doses (approximately 10^{18} nvt (total)) the characteristic yield drop is still present, but at higher doses (greater than 10^{19} nvt (total)) there is some evidence that a normal rounded yield is observed. (Wilson and Billington, 1956, Berggren and Wilson, 1957; Hall, 1962) (see Fig. 2.11). There is some conflict

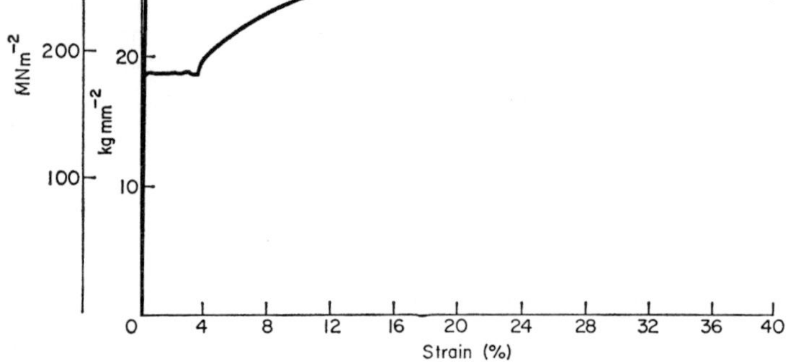

FIGURE 2.11. Stress–strain curves of irradiated mild steel (Hall, 1962)

however, with the results of Chow *et al.* (1962) and others, but the suppression of the normal yield drop may be composition dependent, and its detection a question of specimen dimensions and shape, as well as dose.

The mechanical properties of the irradiated material do appear to follow the normal Hall–Petch grain-size relation. Churchman *et al.* (1957), Hull and Mogford (1958), Nichols and Harries (1962) and Mogford and Hull (1963) claim that k_y is virtually unaffected by irradiation and that the whole of the hardening arises from an increase in σ_0. (Table 1.2.) Chow *et al.* (1962) confirm this at room temperature and at fluxes up to 2×10^{17} ϕ (total); at higher doses k_y appears to fall. Conflicting results were also obtained by Stefanovic (1966). Nichols and Harries (ibid.) determined σ_0 and k_y by the extrapolation method, but it is more likely that composition may have an unknown effect on this result.

Assuming for the moment that the increase is in σ_0 (see also Harries *et al.* (1964)), Nichols and Harries (ibid.) show that this increase is proportional to the square root of the flux, i.e.

$$\Delta\sigma_0 = K\phi^{1/2}$$

This power law appears preferable to the $\phi^{1/3}$ law proposed by Blewitt *et al.* (1960) for copper.

The increase in σ_0 critically affects other properties of the steel, and in particular the impact transition temperature is markedly raised. Experimentally, this was found in EN2 steels by Churchman *et al.* (1957) and subsequently in most ferritic steels, and is interpreted on Cottrell's (1958) theory. In Section 2.4.5 the brittle fracture condition for an unnotched specimen was given by

$$(\sigma_0 d^{1/2} + k_y)\, k_y = \beta\mu\gamma$$

and an increase in σ_0 through irradiation will result in this equation being satisfied at higher temperatures. If, however, as Chow *et al.* (1962) maintain, k_y falls then it is difficult to see how this increase in I.T.T. can arise. Fine-grained, irradiated samples also show a lower transition temperature, as this theory would predict.

High resolution electron microscopy of irradiated samples has now revealed some of the basic processes. Hull and Mogford (1961) have shown that after irradiation evidence of clustering of solute is seen, and after a suitable heat treatment these clusters appear to have developed into precipitates lying on (100) planes. This precipitate, probably ϵ-carbide, should be compared with that arising from quench ageing. However, this will probably not be the fate of the majority of the defects, which will either trap the carbon or nitrogen interstitials (see below) or else associate with dislocations. On the other hand, Harries *et al.* (1964) associate the major portion of hardening with the dispersion of these clusters of solute.

The effects of irradiation may be reduced or removed either by irradiation at higher temperatures or by annealing samples after irradiation. Barton *et al.* (1965) have carried out the former experiment, and show that the effects of damage start to fall off rapidly above about 150°C; however, significant damage can be retained at 350°C, the highest temperature employed. Stefanovic (1966) claimed that on irradiating above 350°C, the Hall–Petch relation no longer holds.

While it is possible to measure the recovery of the yield directly, many prefer to examine such properties as internal friction, or resistivity, where the irradiation needed is lower, and the hazards consequently reduced. Internal friction studies by Wagenblast and Damask (1962) and McLennan and Hall (1963) both show that the free carbon level is lowered by irradiation, presumably by vacancy-solute interaction, and that the levels may be largely recovered by low temperature annealing. Resistivity studies by Fujita and Damask (1964) bear this out, and using isochronal annealing, suggest there are at least five stages in the recovery process. Trapping of the interstitials also largely prevents the formation of ϵ-carbide in quenched samples.

Three direct measurements of the yield point recovery have been made by Kunz and Holden (1954), Nichols and Harries (1962) and Hall (1964). Nichols and Harries, using a 0·24% C steel, obtained some evidence for a singly activated process with $E = 3{\cdot}2$ eV, which is close to the value for self-diffusion in iron. The other two papers are on low carbon irons. The latter is the more extensive, and shows, as might be expected, that the process cannot be singly activated.

The disappearance of the yield point may be adequately explained in terms of the multiplication theory of the yield. Instead of having relatively few long lengths of dislocation line, the jogs from the point defects thus create numerous short lengths of dislocation, as well as increasing the friction stress σ_0. As their density increases, the yield drop decreases. It is also likely, although there is no experimental proof, that the Luders bands will become progressively more diffuse as irradiation proceeds, until the yield point effect disappears entirely.

The case of austenitic alloys will be very different from mild steel. Yield points can be induced in these alloys by irradiation at and slightly above room temperature – see Fig. 1.9. This result on Type 347 stainless steel was obtained by Wilson and Berggren (1955); Broomfield *et al.* (1965) have obtained similar results on Type 316 and on 25 Ni/20 Cr–Nb stabilized austenitic steels. Much of the effect of irradiation, and in particular this yield point, may be removed by annealing at 500°C or above, or even carrying out the irradiation in this temperature range; the ductility, however, is still reduced. Pfeil and Harries (1965) and Broomfield *et al.* (ibid.) both show that this reduced ductility results from enhanced precipitation of NbC in the irradiated sample. However, this is not the sole reason, as it is also shown this reduced high temperature ductility is a function of thermal (slow) neutron doses, a surprising result, which can only be attributed to reactions with boron or lithium as minor impurities which would result in an artificial increase in contained helium.

2.6.2 *Ageing behaviour in irradiated material*

In contrast to the yielding behaviour and yield stresses in irradiated steels, the effects of the treatment on ageing have been little studied. Hall (1962) showed that the incidence of strain-ageing was effectively

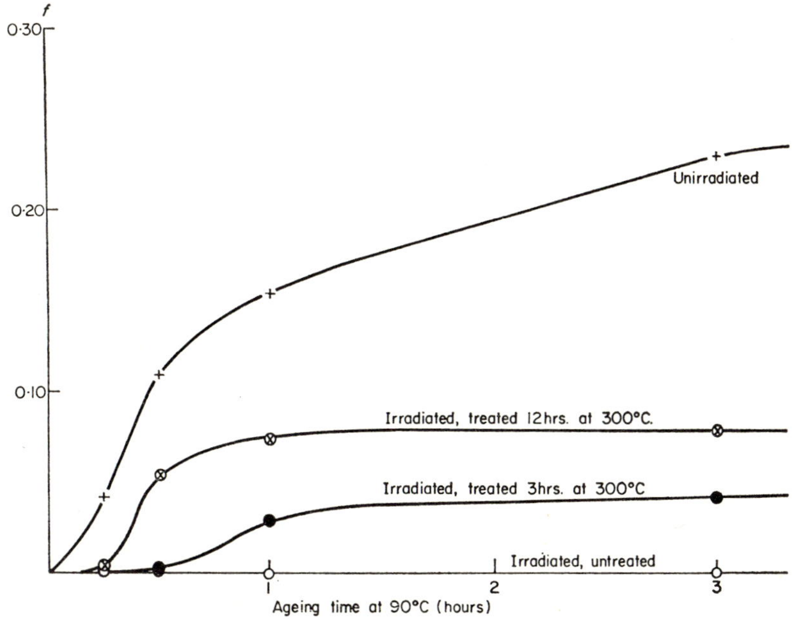

FIGURE 2.12. Ageing curves for irradiated mild steel (Hall, 1962)

zero in a mild steel irradiated to 10^{19} ϕ (total). Ageing would return if specimens were annealed at 200°C or above. Figure 2.12 shows the f values derived from sets of ageing curves; even after 12 hours at 300°C the ageing rates are nowhere near back to normal values.

A simple explanation of this effect lies in the trapping of carbon or nitrogen by the vacancies. The effective carbon levels can be measured by internal friction techniques, and both Wagenblast and Damask (1962) and McLennan and Hall (1963) showed these were reduced by up to a factor of 4, following irradiation. There is consequently insufficient free carbon to contribute to ageing. The free carbon level recovers following low temperature annealing, and so too does the ageing index *f*.

2.7 *Single crystals*

2.7.1 *Yielding in single crystals of iron*

The old grain boundary film theories of the yield point effect (see Section 1.5.1) would necessarily have predicted that single crystals of iron should show no yield point or strain-ageing effect. Indeed, even the more recent measurements by Hutchison (see Fig. 1.4) would also predict that an upper yield point drop would be absent in single crystal and coarse polycrystalline samples. Thus the yield point would appear to be of a smaller order of magnitude.

Certainly, earlier investigators (see Edwards and Pfeil, 1925; Elam and Taylor, 1926; Elam 1936; Holden and Hollomon, 1949) found the yield point difficult to detect, but in the latter year Schwartzbart and Low (1949) and Cottrell and Churchman (1949b, 1950) – see also Paxton and Churchman (1953) – conclusively demonstrated that single crystals of carburized or nitrided iron did show both yield point and strain ageing phenomena.

Some inconsistencies still exist between these papers. For example, Holden and Kunz (1953) found marked yield points in crystals near [111] orientations; elsewhere the yield drops seem variable and independent of orientation. [111] oriented crystals also formed marked kink bands. Paxton and Bear (1955), however, could find no orientation effects, and noted typical single crystal Luders bands spreading in parallel mode from one end of the test piece. Strain-ageing was also detected by many of these investigators, and both yielding and ageing were absent in crystals that had been treated in wet hydrogen.

Later papers have also failed to show much significant change in yield drop with orientation. Allen *et al.* (1956), and more recently Keh (1965) both confirm this, using a 0·0027% C iron, and Ferrovac–E, respectively. Some account has now also been taken of carbon content. Stein *et al.* (1963), have developed a method of producing iron of very low interstitial content and have shown that yield points may be obtained with 40 p.p.m. of carbon, but below that the yield is smooth and rounded (Stein and Low, 1966). Good reviews of the deformation and fracture of iron may be found in the papers of Low

(1963) and Allen (1963), and, on Fe–N crystals, by Nakada and Keh (1968).

The strain rate sensitivity of single crystals is of importance in discussing theories of the low temperature yield strength of the metal. This factor has been studied by Mordike and Haasen (1962), Stein (1966) and Nakada and Keh (ibid.) (see also Sections 2.7.2. and 3.9).

The kinetics of ageing in single crystals of iron have been little studied, but Evers (1961) has shown ageing curves from tensile tests, which, as in polycrystalline material, appear to show two-stage hardening. Evers interprets this, on indirect evidence, as precipitation on dislocations.

Orientation of the crystals is important in two other respects. Firstly, it would appear that the critical resolved shear stress law does not accurately hold for these crystals. Stein and Low (1966), for example, show that while it holds near room temperature, deviations are noted at subzero temperatures, especially for crystals of [110] orientation. Similar discrepancies, although not necessarily of the same magnitude, have been noted by Allen *et al.* (1956) and by Takeuchi *et al.* (1967). Cox *et al.* (1957), on the other hand, found no anomaly, nor did Hull (1963). The reasons for this discrepancy are obscure.

The second noteworthy effect is the appearance of three-stage hardening, similar to that found in face-centred cubic single crystals. This can be inferred from the work of Paxton and Bear (1955) mentioned above, but it is in more codified form in the paper by Keh (1965). Here it is shown that slip on a single system, e.g. (110) $[1\bar{1}1]$ gives rise to the characteristics of easy glide, but if other systems are active, parabolic hardening is recorded. These results, however, are more relevant to strain hardening studies, and do not appear to be directly related to the presence or absence of a yield point drop.

2.7.2 *The low temperature yield strength of iron*

It has been known for some years that the yield stress of iron increases markedly at low temperatures. This was first noted by McAdam and Mebs (1943), but has since been confirmed and reviewed by many others, including Heslop and Petch (1956), Conrad (1960), Low (1963) and Allen (1963).

It will be recalled that in Section 1.5.1 this variation with temperature was used by Cottrell and Bilby (1949) as evidence of thermal unpinning of dislocations. In this respect, it is of interest that single crystals and polycrystals all show a similar trend in carburized material. However, it is now known that decarburized samples, both polycrystals and single, now exhibit the same trend, as can be seen in the work of Allen *et al.* (1956), Jaoul and Gonzales (1961) and Burbach *et al.* (1966).

This discovery has led to considerable study of the temperature variation of yield stress of all the b.c.c. metals, and the various theories which have been put forward will be discussed together in the next chapter (Section 3.9). It is of some value to summarize the results for iron at this stage and to discuss them later.

The variation of single crystal yield stress can be adequately studied from the variation of the parameter σ_0 in the Hall–Petch equation. As shown by Heslop and Petch (1956) this parameter may be represented by

$$\sigma_0 = \sigma_0' + \sigma_0''$$

where σ_0' is composition independent and temperature dependent, while σ_0'' is dependent on composition only (Section 1.6.4 and Figs. 1.19 and 1.20). σ_0' can then be compared with other work on decarburized single crystals, such as Allen *et al.* (1956), and Jaoul and Gonzales (1961) and, allowing for some spread through orientation, reasonable correlation is found (Dingley and McLean, 1967). The influence of other alloying elements, which has been subject of further considerable study (see Allen, 1963) may then be separately included in the grain size equation through their effect on σ_0'', since k_y is again virtually insensitive to composition.

Measurements of the yield stress at sub-zero temperatures is complicated by the appearance of twinning. The formation of twins is quite sensitively a function of orientation, as has been shown by Allen *et al.* (1956), and Jaoul and Gonzales (1961). However, Biggs and Pratt (1958) showed that twinning may be at least partially suppressed by pre-straining at room temperature, prior to low temperature testing. Crystals whose orientations lie close to [100] tend to cleave at 77 K, while those close to [111] tend to slip, so that by suitable choice of orientation and the application of the critical shear stress law, twinning and/or cleavage may be avoided at the yield. Twins form in iron on {112} planes carrying the maximum shear stress, but whether a critical shear stress law applies to twinning as well as for slip is in doubt due to the wide scatter of results (see, for example, Allen *et al.*, 1956, Cox *et al.*, 1957, Hull, 1963).

The occurrence of twinning or cleavage will therefore terminate the yield-stress/temperature curve at some temperature below this; nevertheless, careful study can differentiate between yield, twinning and fracture stresses (McRickard and Chow, 1965) and it will be found the curve is rising steeply with decreasing temperature. Since this is common to all the b.c.c. refractory metals, discussion will then be postponed to the following chapter (Section 3.9).

2.8 *Steels*

Most of the basic studies on yielding and strain ageing in steels have been carried out on material of the so-called mild steel category, containing less than about 0·2% C and varying amounts of nitrogen. Comparatively little, other than engineering studies, has been done on samples of higher carbon content, but what results exist are nevertheless interesting and will now be briefly reviewed.

2.8.1 *Plain carbon steels*

As the carbon level of a steel is increased, excess carbon will normally be precipitated as cementite (Fe_3C) and in a normalized steel will appear as lamellar pearlite. The proportion of pearlite then increases up to the eutectoid composition (0·8% C). In pearlite, the ferrite

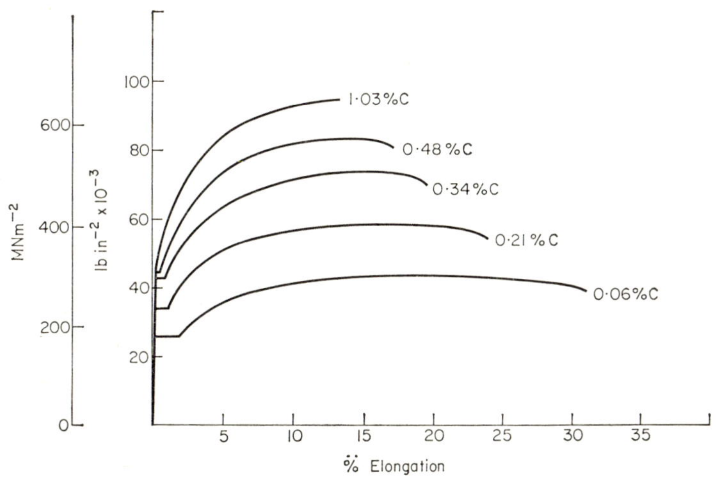

FIGURE 2.13. Stress–strain curves of plain carbon steels at room temperature (Winlock, 1953)

lamellae are thin, and it then appears that the free ferrite path length is too short to allow excessive dislocation multiplication. The Luders extension thus diminishes and finally the yield point itself disappears. This is shown in Fig. 2.13 (Winlock, 1953). Gensamer *et al.* (1942) also proved that a plain carbon eutectoid steel showed a yield point again if the pearlite was spheroidized, in line with the above observations. In this area, an early microstrain study has been carried out by Roberts *et al.* (1952) showing that the elastic limit (i.e. in the microstrain region) varied as the logarithm of the mean ferrite path, while Liu and Gurland (1968) show that the Hall–Petch relation applied to spheroidized steels, provided d was interpreted as the distance between particles.

Strain-ageing likewise has been little studied in these higher carbon ranges. Peterson (1963), using 0·09, 0·46, and 0·77% C steels, with varying nitrogen and manganese levels, did show there was a minimum in ductility and a maximum in yield and ultimate in these alloys, a fact he attributed to blue-brittle behaviour. Only a single strain rate was used, however, and no activation energy determined.

Another and more basic study of ageing in these plain carbon steels is that of Nishino and Takahashi (1962) and Nishino (1962) using steels up to 0·9% C. Their results may be interpreted as a two-stage ageing process (see Section 2.5.1), the first and more rapid being normal ageing in the ferrite, the second (in the range 150–350°C) due to a redistribution of cementite.

Quenching these high carbon steels from the austenitic field leads, as is well known, to the formation of martensite. This structure is normally considered too brittle to work, and is tempered to precipitate carbides to compromise between reasonable strength and ductility. However, it may be shown that the martensite itself does have limited ductility, especially in compression, and some interesting yield point and strain–ageing effects. Stephenson and Cohen (1961) showed that Type 4340 steel, quenched and tempered at 480 K developed a yield point. Straining such a sample say, 3%, and re-tempering in the range up to 530 K caused the yield point to reappear with an increase in yield stress, analogous to strain ageing. This work was extended by Breyer and Polakowski (1962) who showed that using a similar steel, a yield point could be induced at room temperature by cold drawing. A minimum reduction in area of 3% was needed. The rapidity of the ageing at room temperature is remarkable; this point was re-examined by Breyer (1966) who held deformed samples of the steel under strain in a hard tensile machine; in this way a sequence of yield drops could be obtained on one sample. Yield drops of 55 MNm^{-2} could be obtained in 30 s, 110 MNm^{-2} in 30 min in ageing at 297 K. This is more rapid and much more marked than in mild steel, and a rough attempt at an activation energy gave the very low figure of 1·8 kcal $mole^{-1}$ (7·6 kJ $mole^{-1}$). Bowen *et al.* (1967) also illustrate the rapidity of ageing, although in their steels the yield was smooth. The only locking process which might apply here is the Schoeck–Seeger local ordering theory.

2.8.2 *Ferritic alloy steels*

The boundary between 'plain' carbon steels and their low alloy neighbours is often tenuous, but in this section reference will be made to certain alloying additions which have led to an extension of our understanding of yield behaviour.

The addition of certain alloying elements to mild steel has, in several ways, been discussed in Section 2.2. Here it may be seen that the additions of carbide and nitride formers could suppress yield point and strain-ageing effects. In silicon steels, for example, the effect is limited, but at higher silicon content twinning and brittle fracture can occur at room temperature. Chromium additions increase the yield stress even more (Horne *et al.*, 1963), but twinning is not observed, even down to 78 K. Stress–strain curves from these crystals do show some instabilities; these may, however, be orientation effects for these single crystals.

Another interesting example of changes in mechanical properties is found in the paper by Tekin and Kelly (1965) on vanadium steels. High purity Fe–V–C alloys were quenched and tempered and secondary hardening studied using electron microscopy. At peak hardness, V_4C_3† was found to have nucleated on dislocations, and on tensile testing, the samples were found to have developed a yield point. This presumably arises from the absence of free dislocations. Such a yield point is not found in commercial steels of the same type.

Rather similar results were obtained by Raynor *et al.* (1964) on a Fe–4 Mo–0·2% C steel. Here the as-quenched carbide is Fe_3C, but on tempering at 820 K, Mo_2C grows on dislocations, martensite needle boundaries and at the ferrite/cementite interface. If tensile testing had been carried out at peak hardness, a yield point might have been found and for a similar reason. The appearance of a small yield point has even been noticed in a very dilute Fe–C–Nb alloy, rapidly quenched to produce lath martensite, and then aged to cause, again, precipitation on dislocations (Whiteman, private communication).

Some evidence of carbon diffusion control is seen in the paper by Steigerwald and Hanna (1963). Using a fully tempered ferritic steel (SAE 4340) (with an analysis 0·40 C – 0·90 Cr – 1·85 Ni – 0·40 Mo – 1·75 Si) and testing at different strain rates, minima were found in the notch tensile strength curve against test temperature which showed

† V_4C_3 is normally cube oriented with respect to the matrix (Baker and Nutting, 1959).

strain rate sensitivity. The usual plot of $\ln\dot{\epsilon}$ against $1/T$ yielded a straight line with roughly the correct slope for carbon diffusion in ferrite.

The same authors also cursorily investigated the high-temperature properties of a maraging steel containing 18 Ni–4·7 Mo–7·64 Co. These steels develop their martensite during quenching to tempering temperature, but the properties of martensite subsequently change with time. Again, the elevated temperature properties showed strain rate dependent effects.

Strain ageing in several high chromium steels has been investigated by Garofalo *et al.* (1957). Whereas Edwards *et al.* (1940) (Section 2.2.2) found 6% Cr could suppress strain ageing, Garofalo *et al.* found that susceptibility to strain ageing increased beyond 12% Cr, a result confirmed by hot hardness and room temperature tensile testing. However, it is not clear if this is due to carbide, or to the chromium atoms themselves, as no activation energy determination was made.

2.8.3 *Austenitic steels*

It was shown in Section 2.1 that the interstitial sites in austenite (γ-iron) are considerably larger than in ferrite. Thus, it would be expected that the binding energy between carbon or nitrogen and dislocations is here very small, and that if Cottrell locking occurs, it will be from substitutional atoms. Table 2.5 lists the common elements usually found in austenitic alloys. From this, it can be seen that the elements commonly added to form austenite steels, i.e. Ni, Co, and Cr, all have radii close to that of γ-iron, and Cottrell locking, if it exists at all, could come from the other elements in the table, which are usually present in much smaller quantities.

TABLE 2.5. Atomic radii of alloying elements in austenitic steel

Element	γ–Fe	Ti	V	Cr	Co	Ni	Nb	Mo	W
Atomic radii (Å)	1·26	1·47	1·36	1·28	1·25	1·25	1·47	1·40	1·41

There is, however, another possible mode of locking. Certain grades of stainless steel have a low stacking fault energy, and the possibility of some form of Suzuki locking does arise. This, as will be seen, might be considered in terms of diffusion of these elements to the faults, or by the precipitation of a phase such as carbide upon them.

Metallurgically, the problem is even more complicated, since the stacking fault energy is strongly composition dependent, and at the same time the transformation characteristics can be equally drastically changed. There are, however, two broad fields of interest which could with relevance be discussed here. The first of these has applications in ausforming, where the mechanical properties of the austenite are influenced by strain ageing. The second is in the field of high-temperature creep-resistant alloys, where in niobium stabilized grades some marked precipitation occurs in dislocations. There is, indeed, some overlap between these two, but the essential features will be covered by taking the latter first, leading then into the former which covers a wider range of steels, and finally some discussion will be given to ausforming and maraging types.

The development of austenitic creep-resistant steels such as Type AISI 347 (18 Cr–11 Ni–0·8 Nb–0·08 C max.) have proved of considerable use in turbines and heat exchangers, and similar steels with higher chrome and nickel contents proposed as canning materials in nuclear reactors. The material without the niobium is, however, difficult to weld (see, for example, Younger and Baker, 1960), because of precipitation of $Cr_{23}C_6$ in grain boundaries, and this fact in turn engendered a notable amount of research into minor alloying additions and the subsequent properties of the steel related to its microstructure. In this Type 347 steel stabilized with Nb, Younger and Baker (ibid.) did in fact show that strain induced precipitation can occur, while in the same year, Irvine *et al.* (1960) demonstrated a minimum in ductility at approximately 650°C, and, using replicas, suggested Nb was precipitating on dislocations. The first direct observation of precipitation on dislocations was by Haddrill *et al.* (1961), who showed that the particles of NbC were cube related with the matrix, although there was a large mismatch. The dislocations observed by Haddrill *et al.* were undissociated, but in 1962 van Aswegen and Honeycombe (1962) showed that after ageing the solution treated steel for five hours at 700°C, dissociation of the dislocations commenced, and precipitation then occurred on the stacking faults (see Plate 2.3). This appeared at first to be support for the form of locking of dislocations proposed by Suzuki (see Section 1.5.2), and in this paper it was pointed out that the dispersion of precipitates was thereby rendered finer, with a subsequent improvement in creep resistance.

A study of the dissociation of dislocations as a function of ageing temperature was carried out by Silcock (1963) and in a subsequent paper (Silcock and Tunstall, 1964) she was able to show that the dissociation reaction was not that of Shockley partials, i.e.

$$a/2\,[1\bar{1}0] \rightarrow a/6\,[1\bar{2}1] + a/6\,[2\bar{1}\bar{1}]$$

but rather the formation of a Frank partial, i.e.

$$a/2\,[1\bar{1}0] \rightarrow a/3\,[1\bar{1}1] + a/6\,[1\bar{1}\bar{2}]$$

and that the stacking fault grew by climb. The microstructure was also carefully examined by Honeycombe *et al.* (1963) and van Aswegen *et al.* (1964), who concluded that Suzuki segregation is not responsible for the nucleation of the precipitate. This was inferred from the regular arrangement of precipitates within the fault, and from the fact that, edge on, the precipitate has formed on only one side of the fault. Rather, they suggest that the stress field of the partial is responsible. This would explain why undissociated dislocations are always sites for nucleation, and that in the partial, coherency strains repel the dislocation from the growing particle. A nucleation, followed by a repulsion, leads to the regular spacing of the particles. Thus, we still lack an unequivocal example of Suzuki locking.

The presence of material with such a high proportion of dislocations locked in situ could be expected to lead to some interesting yield point phenomena. Naybour (1965) noted that the stress–strain curves of a steel similar to Type 347 (the actual analysis was 17·84 Cr–12·05 Ni–1·52 Mn–0·82 Nb–0·075 C) and tested at approximately 650°C were remarkably sensitive to strain rate. This would in fact be so if simultaneous straining and ageing were occurring. Serrations could be induced at a late stage of deformation, or at an earlier stage if the material were strained 1–2% first at a much slower rate. This as will be seen in Chapter 5, may be the result of enhanced diffusion caused by vacancy generation in the pre-straining. The distribution of precipitate between undissociated and extended dislocations is not unimportant – in a later paper Naybour (1966) showed stacking-fault precipitation was the more effective hardening agent, and that conditions for this mode depended, inter alia, on quench rate – another indication of the role of vacancies.

The alloy containing 20 Cr–25 Ni–0·7 Nb was originally developed as a possible nuclear fuel can material, and it is interesting to compare it with the alloy just discussed above. The additional nickel will raise the austenite stacking-fault energy, and so both Sumerling and Nutting (1965) and Dewey *et al.* (1965) found no evidence of stacking faults. The dislocations, on ageing at 650°C, became clearly decorated with NbC, but other sites such as austenite twin and grain boundaries were also nucleation sites. In the latter case, some of the precipitate was M_6C, and the proportion of this phase grew as ageing proceeded. In the overaged condition, sigma phase can be detected. Cold working prior to ageing increased the number of nucleation sites, and precipitation on dislocations competed for solute, thus decreasing the proportion of grain-boundary material. Sumerling and Nutting (ibid.) also showed that tensile testing at 600°C led to serrated yielding (Fig. 2.14), the serrations being more evident in the freshly quenched or lightly aged alloys than in the fully aged. The activation energy is believed to be that of niobium.

Finally in this series, it should be remembered that certain other carbides are ismorphous with NbC. Harding and Honeycombe

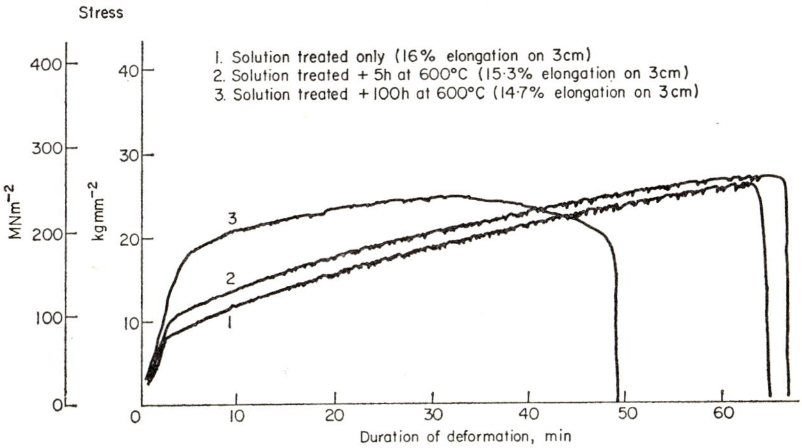

FIGURE 2.14. Serrated yielding in a 20 Cr–25 Ni–0·73 Nb–0.03 C austenitic steel (Sumerling and Nutting, 1965)

(1966) have carried out a comparative series of tests on 18 Cr–10 Ni–1 Nb and 18 Cr–10 Ni–1 Ti. Both showed similar results, and TiC could be identified forming on stacking faults. Again, the stress–strain curves serrated yielding (see Fig. 1.3(d)), but the strain rates had to be very slow ($2{\cdot}5 \times 10^{-8}\,s^{-1}$). A similar series of alloys containing 35 Ni–15 Cr–1 C with niobium or titanium also showed similar effects, and, although the fault energies were not measured in these alloys, faults, and precipitation on them, were again observed. The matrix precipitation in this case was probably $\gamma'(Ni_3Ti)$, and the morphology somewhat different, but considerable age hardening occurred. A further study of precipitation in these Ti-stabilized steels has been made by Singhal and Martin (1967).

Harding and Honeycombe (ibid.) also showed in this paper the beneficial effects of strain ageing in these alloys. Straining samples 3% after quenching caused a higher yield stress on ageing at 700°C, and peak properties were reached in shorter times (cf. the case of quench-aged and strain-aged mild steel). This was naturally at some cost in ductility, but the elongation values are still in excess of 30%.

The realization that austenite, although possessing an f.c.c. structure, is a good medium of study for both quench ageing and strain ageing has led to a further series of papers covering widely different alloying additions. For example, Irani and Weiner (1965) showed that dislocations could become decorated with V_4C_3 in a 25 Ni–20 Cr–5 V alloy, and the ageing was studied by Harding and Honeycombe (1966), using 18 Cr–10 Ni–1 V and 35 Ni–15 Cr–1 V alloys. Stacking faults were rare, and matrix precipitation more marked. In this case, strain-aged properties were inferior to quench-aged.

The possibility of strain-ageing leading to serrated yielding in austenite was in fact reported as long ago as 1950 in a footnote to a paper by Boulanger (1950). Powell *et al.* (1958) reported serrated yielding in Types 301 and 304 austenitic steels at 10 K, but this is probably a case of yield instability due to local heating (see Section 1.8.5). In Type 316 steels Garofalo *et al.* (1961) showed that $M_{23}C_6$ carbides could form on dislocations in a 18 Cr–8 Ni–2 Mo–0·06 C steel, prestrained at room temperature and aged between 500 and 700°C. This improved the high-temperature creep properties. Using a similar alloy, Barnby (1965) showed serrated yielding could take place in solution treated alloys subsequently tested in the range 300–700°C. Room temperature tensile testing of samples aged at 350°C also showed a yield point, thus demonstrating a true ageing effect. In a later paper, Barnby (1966) showed again the improvement in high-temperature creep properties resulting from locking these dislocations. However, the identification of the element responsible is difficult, as the diffusion rates are not accurately known, but using the Cottrell condition that

$$\dot{\epsilon} \simeq 10^9 D$$

Barnby (1965) suggests diffusion of chromium, assisted by vacancies, is responsible. This is surprising, for as shown in Table 2.5, the size factor differences are small.

A similar conclusion was, however, reached by Tamhankar *et al.* (1958) on a study of the tensile properties of a 10 Cr–35 Ni–0·39 Mn–0·03 C alloy. This shows serrated yielding which was studied as a function of grain size, strain rate and testing temperature. Since the stress–strain curves show a smooth initial yield, enhanced diffusion is probably present. The activation energy was determined as 50–80 kcal mole^{-1} (210–340 kJ mole^{-1}) and ascribed to chromium or nickel diffusion, associated with Suzuki locking. As was noted above, it is probably inadvisable to do this without a careful electron microscope study, which in the case of this particular alloy, has not yet been done. Nishino (1963) has also carried out static strain–ageing tests on 18 Cr–8 Ni and 25 Cr–20 Ni austenitic steels and claims an activation energy of 26 kcal $mole^{-1}$ (110 kJ $mole^{-1}$). This marked variation in observed activation energy has been partly resolved by the work of Jenkins and Smith (1969). Their activation energy at the lower end of the range (~200°C) is 34 kcal $mole^{-1}$, and 90 kcal $mole^{-1}$ at the upper temperature limit for various Fe–Ni–Cr alloys. The lower energy is put down to vacancy-carbon 'pairs' interacting with dislocations, the higher to chromium diffusion.

Hot working of austenite as discussed above has important technical aspects in the process known as ausforming. Here the isothermal transformation properties are such that there is an appreciable time interval before the austenite decomposes, and forging or other hot working processes may be carried out in this interval. The subsequent transformation to martensite or bainite further strengthens the steel. The alloying contents are generally lower and carbon content higher than in the cases discussed above, but could still be sufficient to cause dislocation locking by carbide. For example, Gerbereich *et al.* (1965) and Raymond *et al.* (1965) have examined a series of steels in this category (with type numbers D11, H11, 5M21, UHS-260, D6, Hy-tuf and 300M), and suggest a minimum of 1% (Cr + Mn + V) is necessary. An SAE Type 4340 (see previous section) did not show serrations but decomposed during testing. The ageing of the austenite would appear to influence the strength of the martensite product, and dispersion of carbide, as was seen above, appears to arise when serrated yielding occurs. Further complementary evidence has been supplied by Steigerwald and Hanna (1963) to underline the role of carbide formation. A study of the high temperature tensile properties of AM 355 (with composition 15 Cr—4·25 Ni—2·75 Mo—0·15 C) revealed a minimum in the notch tensile strength around 200°C which was strain rate dependent, and could be associated with carbon diffusion.

McEvily *et al.* (1963) clearly established serrated yielding in two other ausforming steels, 3 Ni–3 Mo–0·2 C, and 24 Ni–0·38 C. Serrations were noted in the vicinity of 500°C, but associating these effects with specific dislocation configurations is difficult, since the martensite product which forms during final cooling obscures most of the detail. However, from the mechanical properties, there is no doubt the ausforming markedly improves the yield strength of the final product.

The Fe–Ni–C alloys mentioned in the last paragraph have been recently examined by Rose and Glover (1966). A higher nickel content (32–42% Ni) ensured that the M_s temperature was well below room temperature, and conventional straining and ageing studies could be carried out. The most important result achieved with these alloys is that ageing is extremely rapid; only seconds are needed to produce an observable yield at room temperature, and serrated yielding occurs above 100°C. No ageing or serrated yielding occurred in an alloy free from carbon. If ϵ_0 is the strain to the first serration, then a plot of $\ln\epsilon_0$ against $1/T$ can give an activation energy (see Chapter 6). Alternatively, a plot of ln (strain rate) $\dot{\epsilon}$ against the reciprocal of the temperature at which jerky flow begins can also give an activation energy. In both cases, these are well below that of carbon in austenite (33 kcal mole^{-1} or 140 kJ mole^{-1}). Rose and Glover are forced to reject Cottrell and Suzuki locking as possible mechanisms, and to plump for Schoeck–Seeger locking (Section 1.5.3). This theory cannot as such be applied to austenite, as the carbon occupies octahedral sites and so produces a symmetric distortion, but it is possible that pairs of carbon atoms, or even better, carbon vacancy pairs, will explain the rapid room temperature ageing, the low activation energy and certain internal friction results from other research.

While Rose and Glover did appreciate the possibility of an unloading yield point (see Chapter 5), it is conceivable that they have underestimated it and it might be advantageous if this could be done with lower carbon contents, i.e. in the range 0·02–0·10%. It is also interesting that the two alloys discussed in this section with anomalously low activation energies, or very rapid ageing effects, are either martensitic (Section 2.8.1) or on the point of becoming martensite at room temperature. There may be some pre-nucleation effects in the iron-nickel system which increase the rate of material transport.

Finally, there is an isolated result on a cobalt/chrome/nickel alloy, in that Arrowsmith (1959) has found evidence of precipitation on dislocations in G18B steel. This again should thus show high-temperature ageing properties.

3

The Group Va and VIa Metals

3.1 *Introduction*

The groups Va and VIa metals form a series of high melting point elements which have collectively become known as the refractory metals. Like iron, they are all body-centred cubic in structure, and like iron, their mechanical properties show a marked dependence on the presence of small amounts of impurity, particularly interstitial elements.

The particular property which precluded any extensive investigation of these elements for many years was the existence of a ductile-brittle transition temperature; that is, below a certain temperature, the material did not deform, but broke by cleavage, usually along the {001} planes, although {110} cleavage is sometimes found. The same transition occurs in iron, and by careful control of composition, the transition temperature can be kept below 0°C; but in these newer metals in their relatively impure state, this temperature may be several hundred degrees above zero.

With improved purification techniques, such as zone refining and electron beam melting, the purity of these refractory metals may now be raised to such a level that the transition temperature is well below room temperature, and yet enough impurities remain to cause interesting yield point effects.

Failing any more logical order, these metals will now be examined in turn, in order of increasing atomic weight.

3.2 *Vanadium*

3.2.1 *Polycrystalline material*

In the 1950's, a large effort was put into the production of ductile vanadium, usually by bomb reduction of the oxide, and it was not long before acceptable mechanical properties were obtained with reasonably low transition temperatures. A deliberate effort was made to keep values of all the possible interstitial impurities as low as possible, C, O_2, N_2 and H_2 (particularly) were at such levels that the metallic contents could be regularly given as 99·8 + % V.

First mention of a yield point effect is in a paper by Lacy and Beck (1956) and in the discussion to this paper by Pugh (1956a) who had used calcium reduced oxide and arc-melted powder respectively. Pugh also mentions serrated stress–strain curves.

In later years, the role of the interstitial elements listed above have been examined in turn. Hydrogen is known to be primarily responsible for excessive increases in the transition temperature (Eustice and

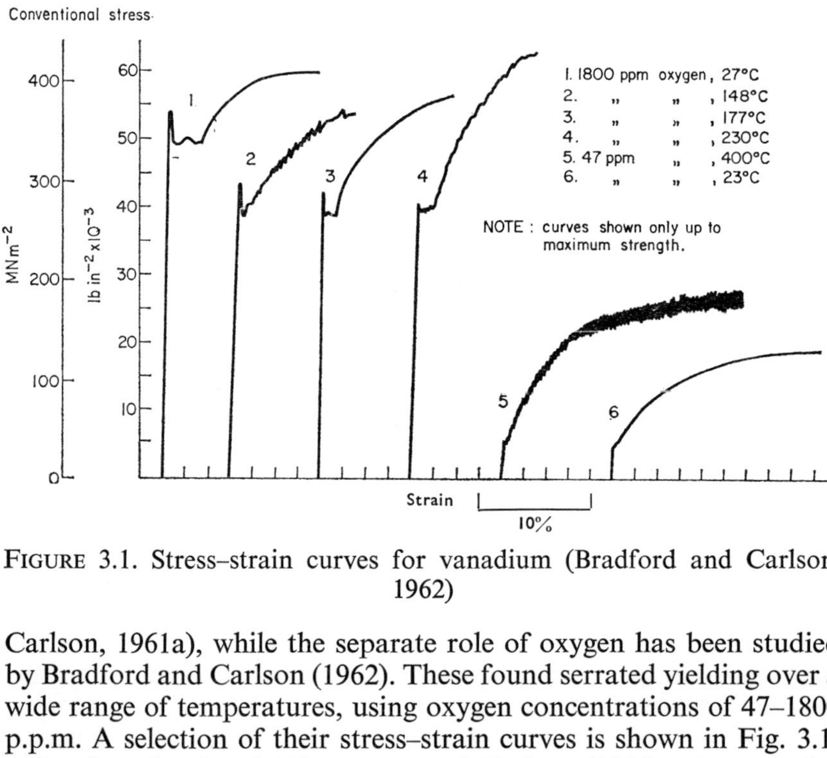

FIGURE 3.1. Stress–strain curves for vanadium (Bradford and Carlson, 1962)

Carlson, 1961a), while the separate role of oxygen has been studied by Bradford and Carlson (1962). These found serrated yielding over a wide range of temperatures, using oxygen concentrations of 47–1800 p.p.m. A selection of their stress–strain curves is shown in Fig. 3.1.

On the other hand, Thompson and Carlson (1964) using samples with varying nitrogen contents, showed that likewise the level of this interstitial was vital in controlling the level of both yield and U.T.S.

In view of the uncertainty of the separate role of these elements, however, it is not surprising that conflicting reports of the existence of a yield point effect and/or strain ageing did exist. For example, Clough and Pavlovic (1960), amongst others, found no yield point present, while Farrell (1961) reported yield points in all room temperature specimens of vacuum arc-melted material, although a 'soft' machine was used, rendering careful observation of yield points difficult. Curves of yield strength against temperature showed little change from 200 to 400°C, a fact attributed to strain ageing.

The kinetics of the strain-ageing reactions have been studied by Bradford and Carlson (ibid.), and Edington *et al.* (1964). The latter found that, at early stages of ageing a $t^{2/3}$ time law held for the return of the yield point, and an activation energy of 25·45 kcal mole^{-1} (106·5 kJ mole^{-1}). This is in good agreement with Bradford and

TABLE 3.1. Diffusion coefficients of interstitials in vanadium.

Element	D_0 mm^2 s^{-1}	Q kcal mole^{-1}	Q kJ mole^{-1}
O_2	0·3	28·0	117
	1·9	29.3	123
	0·25	27·5	115
C	0·47	27·3	114
N_2	1·8	35·1	147

Carlson's result of 27·7 kcal mole^{-1}, (115·9 kJ mole^{-1}). Unfortunately, it is not possible uniquely to identify the element concerned, for as seen in Table 3.1 (from Edington *et al.* (1964)), the diffusion coefficients of oxygen and carbon are very similar.

Bradford and Carlson (ibid.) measured the height of serrations on the stress–strain curve, and found two peaks from the results, one at 350–450°C, (confirmed by Edington *et al.*, 1964) and one at ~150°C. Using the Cottrell (1953) equation for the appearance of serrations

$$\dot{\epsilon} = 10^9 D$$

they obtained reasonable agreement with O_2 (or C) for the 400°C effect, but were unable to explain the results at 150°C.

Thomson and Carlson (ibid.) were able to prove their contention that nitrogen is another interstitial capable of producing strain ageing. This was done in two ways. Firstly, the peak in the U.T.S. temperature curve at 300–450°C, which occurs in the 'blue brittle' range, is sensitive to strain rate, and its shift can be used to give an activation energy of 37·3 ± 3·7 kcal mole^{-1} (156 ± 15 kJ mole^{-1}). The more usual 'return of yield point' method gave 36·2 ± 4·5 kcal mole^{-1} (151 ± 19 kJ mole^{-1}), and obeyed a $t^{2/3}$ law. Both of these figures are in reasonable agreement with the value for nitrogen in Table 3.1.

Lindley and Smallman (1963a) showed that the yield stress of the metal was governed by the usual Hall–Petch equation.

$$\sigma_y = \sigma_0 + k_y d^{-1/2}$$

and that the twinning stresses (apparent at temperatures less than 77 K) also obeyed the same law, although with much larger k_y.

Indeed, the only apparent difference between the material and α-iron of equivalent composition appears in the dislocation sub-structure – whereas in iron dislocations form a marked cellular structure after deformation (Plate 1.7), vanadium does not appear to do so (Edington and Smallman, 1964).

3.2.2 *Vanadium single crystals*

Smallman and his co-workers have extended their studies on polycrystalline specimens into the single crystal region (Edington and Smallman, 1965). The crystals, being prepared by electron beam melting, were much purer than the polycrystalline samples discussed above, impurity concentrations being C 100, O_2 300, and N_2 700 p.p.m. The crystal orientations were such that the specimen compression axis was close to [111] and at 77 K, the specimens deformed by twinning rather than slip, in contrast to polycrystals. A single crystal at 20°C showed a distinct yield point with a single crystal Luders type extension of about 4%. However, strain ageing a twinned sample does result in a lowering of the yield stress at short ageing times on re-straining; this effect is probably due to relief of stresses around the first batch of twins.

3.3 *Chromium*

3.3.1 *Introduction*

Information on yield behaviour in chromium is, it must be admitted, the most unsatisfactory of all this group of metals. In some respects this is surprising, for it is to be assumed that a marked yield exists in all these metals, and Wain *et al.* (1957) have used this concept to explain changes in the transition temperature with heat treatment.

Most interest has, however, centred around the ductile–brittle transition temperature. This has been a perennial problem with the production and application of chromium metal. A discussion of the problem up to 1957 is given in the book *Ductile Chromium* (published by A.S.M., Cleveland), and more recent papers by Allen *et al.* (1963), Solie and Carlson (1964) and Wood *et al.* (1964). It is now generally agreed that among the interstitial atoms, carbon and nitrogen are the most deleterious additions, with oxygen a very poor third; substitutional elements also in general are undesirable, although with deliberate and careful alloying successful high-strength and high-temperature creep-resistant alloys have been made for possible gas-turbine applications. The situation in chromium is more complex than in other metals in this group; not only do some of the impurities promote intercrystalline failure by grain boundary precipitates (Hook and Adair, 1963), so giving single crystals different transition temperatures from the polycrystalline samples, but there is also a transition at 40°C which may be the result of an antiferromagnetic–paramagnetic transition and which leads to Young's modulus and other anomalies in the physical constants of the metal. Consequently, most investigators doing transition temperature determinations have used bend tests, and this mode of deformation may well mask a yield point effect unless it is very pronounced.

3.3.2 *Yielding*

The number of investigators who have used extensive tensile testing is therefore limited, and first notation of a yield point is apparently by Johansen *et al.* (1954) in a U.S. Bureau of Mines report. Other

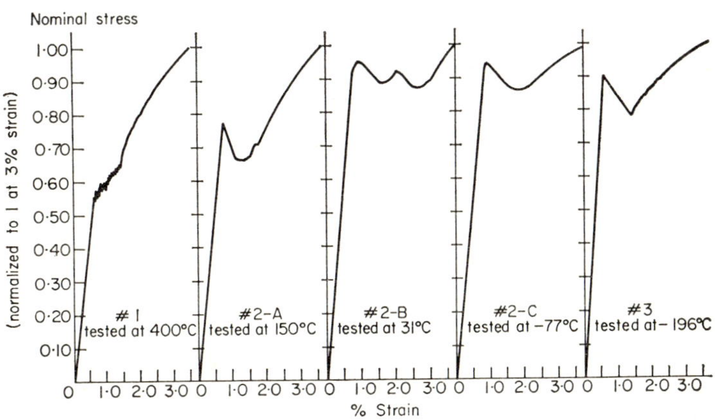

FIGURE 3.2. Stress–strain curves (in compression) for chromium (Marcinkowski and Lipsitt, 1962)

incidental investigators who noted a similar effect are Smith and Seybolt (1956, 1957) and Weaver (1957). The 1956 paper does not mention yield points. However, their 1957 paper in *Ductile Chromium* does. Pugh (1958) found no yield point, but suggested the existence of strain ageing, which Weaver (ibid.) had in any case conclusively established.

The most extensive study of the mechanical properties has been by Marcinkowski and Lipsitt (1962) using compression testing on a hard Instron machine at a very slow strain rate. The analyses are not given. A selection of their stress–strain curves at various temperatures is shown in Fig. 3.2. Strain ageing was apparent at temperatures above

~ 150°C, leading to serrated yielding at still higher temperatures. Studies of the lower yield stress against grain size again indicated that the usual Hall–Petch equation

$$\sigma_y = \sigma_0 + k_y d^{-1/2}$$

was followed with k_y virtually independent of temperature. At 78 K deformation was principally by twinning, with a similar effect of variation with grain size, as in vanadium. Values of the upper yield stress, whether for twinning or slip, again gave a similar result for $d^{-1/2}$; however in twinning the lower yield stress measurement gave k_y as about 5 times that for slip, and σ_0 lower. This result is interpreted as meaning slip dislocation movement is necessary for creation of a twin dislocation and once formed, this dislocation is more difficult to move.

Another paper to note is that by Bullen *et al.* (1964b) who repeated results of pressurization on iron (see Section 1.4.4) on recrystallized chromium. This material was not only rendered ductile under pressure, it remained ductile after removal of pressure, and whereas the original material fractured in the elastic region, the yield stress of the pressurized specimens was about $\frac{1}{2}$ the fracture stress. The interpretation is similar in this case, namely, that free dislocations are engendered around inclusions and other imperfections and which then multiply at the upper yield. The pre-yield flow is quite marked in these specimens, and implies that the original dislocations are heavily locked in position. Similar results have been obtained by Yoshida *et al.* (1960) using rolling to induce prior deformation, free the dislocations and so improve the ductility.

There is some evidence for this locking; as Garrod and Wain (1965) have noted the appearance of defects associated with loops of dislocation (Plate 3.1). Whether these precipitates are responsible for locking is a moot point – not enough examples were found to analyse their nature properly, and further research is needed.

Single crystal studies have been carried out by Gilbert *et al.* (1963) and by Reid *et al.* (1967). The larger crystals were used to discuss the variation of the yield point drops with orientation, the largest drop being found for orientations close to [110], which is similar to the case of tungsten (see Section 3.7).

3.4 *Niobium*

3.4.1 *Introduction*

This metal possesses a low capture cross section for thermal neutrons, and consequently has applications in canning nuclear fuel elements.

Thus, a considerable effort has gone into the examination of its physical properties, and it is the best documented of all the six metals in this group. For general references, the reader may consult the review volumes *Columbium and Tantalum* edited by E. T. Sisco and E. Eprennan (Wiley, N.Y., 1963), *Columbium Metallurgy* by D. L. Douglass and F. W. Kunz (Interscience, N.Y., 1961), and *Tantalum and Niobium* by G. L. Miller (Butterworth, London, 1960). (There is still a tendency to call the metal Columbium in the United States.)

Inherently, pure niobium possesses a very low transition temperature, making it workable at room temperature, and the mechanical properties have not been so dogged by premature failure as other metals in this group.

3.4.2 *Mechanical properties*

First reports of yield point phenomena in niobium are contained in papers by Kochs and Maddin (1956) and Wessel (1957b), after many engineering studies on the mechanical properties had apparently not reported the effect. Wessel's samples showed brittle failure at 77 K, but later papers (e.g. Dyson *et al.*, 1958) indicated that with still higher purity, the transition temperature was even lower. Churchman (1959) showed that coarse-grained material had a higher transition temperature, and a detailed study of the metal by Adams *et al.* (1960) appeared in 1960. Both Adams *et al.* and Churchman showed that yield and fracture stresses followed the usual grain-size relationship, but the value of k_y in the Hall–Petch equation was very small. Churchman (ibid.), Johnson (1960a), Evans (1962) and Fourdeux and Wronski (1964), all take k_y as effectively zero, but Adams *et al.* give k_y as $2{\cdot}76 \times 10^6$ c.g.s. units ($1{\cdot}3$ N mm$^{-3/2}$). It is clear that k_y cannot be precisely zero, for otherwise yield point effects could not arise, and this is clearly not the case.

From the discussion in Section 1.5.6, the strain rate exponent m in the equation

$$\sigma = K\dot{\epsilon}^m$$

is simply and inversely related to the dislocation velocity parameter m^*, if extrapolation to zero strain is carried out. Values of m have been given by Mincher and Sheely (1961), and were lower for electron-beam melted material; generally, however, other engineering studies have given higher values (Tankins and Maddin, 1961).

A welcome study on Luders bands in material other than steel has been carried out on niobium by Conrad and Stone (1964). Using $\frac{1}{16}$ in. diameter wire, coated with 'Stresscoat', the velocity could be determined. Evidence of rotation of the band front was obtained, something which cannot happen in strip specimens. In agreement with Butler (1962) on mild steel, it was confirmed that the velocity is uniquely dependent on stress, being independent of specimen length, crosshead velocity and number of Luders fronts, However, room temperature tests only were used. Results were used to determine the activation volume v^* and to compare it with work on thermal dependence of yield stress (see Section 3.9.1). The Luders strain is of the order of approximately 4 per cent for a grain size of 15μ.

Shaw and Sargent (1964) carried out a microcreep study on the metal, and suggested that the dislocation multiplication theory could account for their pre-yield results. The analogy between niobium and mild steel seems complete.

3.4.3 *Single crystal studies*

The stress–strain curves of niobium single crystals have been studied by Mitchell *et al.* (1963). In contrast to early work on other crystals

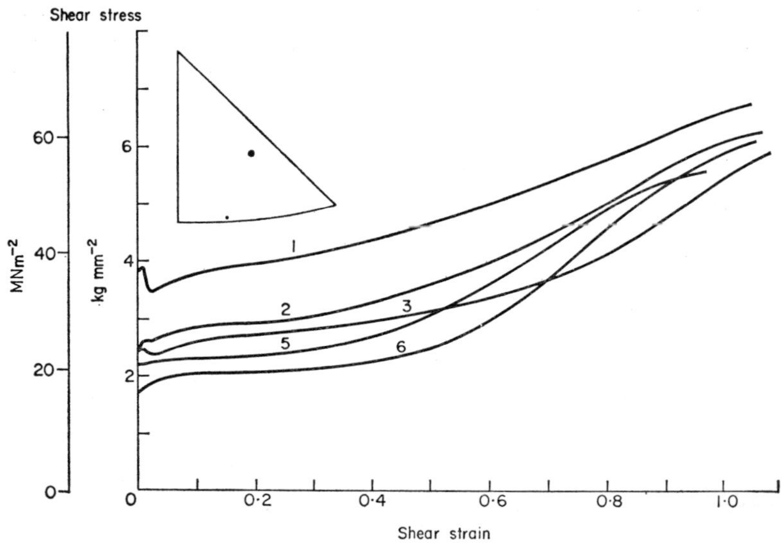

FIGURE 3.3. Shear stress–shear strain curves for zone-refined niobium. (The numerals indicate the number of zone-refining passes) (Mitchell *et al.*, 1963)

in this group, zone refined single crystals showed curves superficially similar to the f.c.c. metals, with an ‘easy glide’ region and three-stage hardening. Yield points were only observed if crystals with only one zone pass were used – as refinement proceeded, the yield point disappeared (Fig. 3.3).

Some sign of yield points were noticed in these pure crystals below room temperature, but the ductility rapidly decreased. The unexpected appearance of easy glide in these b.c.c. crystals has led to further investigation on the deformation mechanisms, notably by Christian and his co-workers (see, for example, Christian and Masters, 1964a) but because these are mostly electron beam purified or zone refined crystals, yield point effects are missing. High-temperature vacuum annealing is another powerful method of removing unwanted interstitials to lower still further the transition temperature (Taylor and Christian, 1965).

Later studies by Reid *et al.* (1966), Taylor and Christian (1967) and Votava (1968) have now clarified the slip and twinning systems, as well as the mode of hardening, in these crystals. As in the case of iron, the critical resolved shear stress law does not apply to twinning.

3.4.4 *Strain ageing phenomena*

Dyson *et al.* (1958) first noted the existence of serrated yielding and the other usual strain-ageing characteristics during tensile tests, and these effects can also be inferred, as has been done by Mincher and Sheely (1961), from a broad maximum in the curve of UTS against temperature and a corresponding minimum in the elongation, corresponding to conditions prevailing in the blue-brittle region of mild steel.

Dyson *et al.*, using the Cottrell equation for serrated yielding,

$$\dot{\epsilon} \simeq 10^9 D$$

suggested from the temperature range of serrated yielding that oxygen was responsible. Mincher and Sheely came to the same conclusion and so too have Enrietto, Sinclair and Wert (1961).

TABLE 3.2. Diffusion Coefficients of Interstitials in Niobium (from Fountain and McKinsey (1963))

Element	D_0(mm^2 s^{-1})	Q (kcal mole^{-1})	Q (kJ mole^{-1})
O_2	1·47	27·6	116
N_2	0·72–6·1	34·8–38·6	146–162
C	0·46	33·3	139
H	2·15	9·37	39·2

Leadbetter and Argent (1961) have shown that the yield stress of niobium is strongly dependent on oxygen content, and kinetic studies using resistivity measurements by Köthe (1968) have shown the activation energies agree with oxygen diffusion (see Table 3.2). A useful study of strain ageing on the mechanical properties of niobium have been made by Szkopiak (1968). He followed the changes in σ_0 and k_y during ageing, and, as in the case of mild steel, found that k_y increased during ageing to a maximum value, while longer term ageing simply increased σ_0. The effects were more marked, the higher the oxygen content.

The views given above have been challenged by Wilcox and Huggins (1961). Considering the time taken to reach a small arbitrary value of load drop (1.5 MN mm^{-2}) at various temperatures, over the range 121–260°C, Wilcox and Huggins (1961) obtained a value of $E = 10{\cdot}5$ kcal $mole^{-1}$ (43 kJ $mole^{-1}$) in reasonable agreement with the value for hydrogen given above. This study will be quoted again in the following chapter. Evans (1962) and Taylor and Christian (1967) have shown that nitrogen is also a potent solid solution strengthener in this metal, and cannot be discounted as another active interstitial, while Votava (1965) discovered a yield point in niobium after swaging, which was put down to possible carbon pick-up. Steigler *et al.* (1964) suggest carbon can precipitate on dislocations in niobium, and Dollins and Wert (1969) show evidence of continuous precipitation of nitride on dislocations. Thus, all four solute atoms may act to induce yield point and strain-ageing effects.

Electron microscopic studies of dislocation patterns in deformed niobium have been published by van Torne and Thomas (1963), Wronski and Fourdeux (1964), and others. Normal cellular structures and dislocation tangles are seen.

Finally, the effects of radiation damage. Makin and Minter (1959) showed that after extensive neutron bombardment (10^{20} nvt) the yield point in polycrystalline material was suppressed. Evans *et al.* (1963) showed that k_y is unaffected by irradiation and σ_0 is raised. These effects are similar to mild steel. However, Makin and Minter (ibid.) proved that after ageing at 200°C, there was a further **increase** in the yield stress, and the yield point returned. They considered that since the activation energy for vacancy movement is high, $\sim 1{\cdot}3$ eV, they migrate on annealing to the dislocations and pin them. In f.c.c. metals, the activation energy is much lower, and pinning can occur at room temperature. The damage has been observed in electron microscopy by Tucker and Ohr (1967).

3.5 *Molybdenum*

The metal molybdenum was the first material in this group where a yield point was noticed on deformation. In 1936 Túry and Krausz

(1936, 1937) noticed that a yield point could be induced in molybdenum wires heated in an atmosphere of nitrogen, and this result later stimulated work on nitrogen gas dissolved in cadmium and zinc (see Chapter 7).

Following the developments of powder metallurgy and arc-melting and casting techniques, massive ingots of the metal became available, which could be hot worked and so lower the transition temperature. Bechtold and Scott (1951) reported the existence of yield points in the metal, and later, Bechtold (1953, 1954), and Pugh (1955) examined the mechanical properties over a wide range of temperatures and strain rates. These samples all normally contained about 0·01–0·05% C, 0·001–0·006% N_2, 0·002–0·003% O_2, and still less H_2, giving transition temperatures at around room temperature in the as-cast state, and this could be lowered to well below room temperature by suitable heat treatment and grain refinement. All these writers showed the existence of yield points, and there is little doubt their existence is associated with some or all of these interstitial impurities. Alers *et al.* (1958) for example, showed that vacuum annealing at high temperatures would remove the yield point, and it also appears that the yield point is very sensitive to grain size. For example, Johnson (1959) showed that coarse-grained samples did not show a yield point, while fine-grained samples did. The samples did, however, obey the relation

$$\sigma_y = \sigma_0 + k_y d^{-1/2}$$

and Wronski and Johnson (1962) point out that the value of k_y is higher than iron and at least 10 times higher than niobium (Table 1.2).

The Luders extensions are usually small, ~ 1%, and no studies of Luders bands appear to have been made, although Johnson and Peacock (1959) note irregularities which could have been caused by the small slenderness ratio of their Hounsfield testing machine specimens.

Molybdenum is strongly strain–rate dependent: the value of the exponent m in the strain rate equation

$$\sigma = K\dot{\epsilon}^m$$

is ~ 0·05, and this leads to a marked variation of lower yield stress with strain rate (see, for example, the curves in Carreker and Guard (1956)). This has been the basis of considerable speculation, see for example Orava (1964). As discussed above, the effect is related to the dislocation velocity exponent term m^*, and Gilbert *et al.* (1965) have checked this by taking dislocation counts in foils which have been strained very slowly, and very rapidly. The dislocation densities were essentially the same, and they conclude the increased strain rate is merely accommodated by the increase in dislocation velocity. Lawley and Gaigher (1964) showed as in other metals of this group, that the dislocation density is approximately proportional to strain.

The existence of delayed yielding has also been established by Hendrickson *et al.* (1956b). Here differences were found between arc-cast and sintered metal, perhaps due to differing nitrogen contents, but the results (including those on microstrain) are similar to those on steel (see Section 1.4.2).

Hendrickson *et al.* (ibid.) also showed the existence of pre-yield microstrain, which was later examined in more detail by Shaw and Sargent (1964), Sargent and Shaw (1966) and Kossowsky (1967). These studies suggest the yield point may be satisfactorily explained in terms of the dislocation multiplication theory. In single crystal studies, Davies and Gilbert (1967) claim that the critical resolved shear stress law is obeyed for microstrain, but does not hold in macroscopic yielding.

The effect of strain ageing has been little studied. In hot tensile tests Carreker and Guard (1956) showed any drop in load is virtually eliminated by 200°C, and King and Spretnak (1964) using etch pitting techniques, suggested that below 260°C dislocations were locked and multiplied at the yield, while above that temperature any discontinuous yielding which may occur arises without any significant increase in dislocation density. On the other hand, Pugh (1955) reports serrated yielding over a range from 540 to 1095°C, a result which has not always been confirmed by others (see Carreker and Guard, ibid.). Brock (1961) has made a reasonably intensive study of strain ageing kinetics, and has obtained serrated yielding in the range 760–860°C. This range is not associated with a minimum in ductility, as might be expected from the case of mild steel and other metals in these groups; the serrations seem small and are perhaps not always picked up unless the testing machine is very sensitive. By plotting ln t (where t is the time for a yield point to return) against $1/T$ he obtains an activation energy of 36·3 kcal mole^{-1} (152 kJ mole^{-1}), which agrees reasonably well with the diffusion of carbon in molybdenum (38 kcal mole^{-1} or 159 kJ mole^{-1}), and the temperature range for serrations to appear also agrees with the Cottrell equation

$$\dot{\epsilon} \simeq 10^9 D$$

The small strain ageing effects are put down to the low solid solubilities of interstitials in the metal.

On the other hand, Hartley and Wilson (1963) found an activation energy for the return of yield as 25·1 kcal mole^{-1} (103 kJ mole^{-1}), which corresponds with the activation energy of nitrogen diffusion. (Values for oxygen were not then available.) From the values of the parameters, they concluded that precipitation was probably occurring on dislocations.

It should be noted in this assessment that the temperatures used here are high enough for recovery of the cold-worked matrix to occur; Lytton and Tietz (1964b) have shown, for example, that the yield point diminishes as recovery proceeds – a curious effect because if there is a limited and low solute content there should be a higher yield noted if the dislocation density is reducing. The temperatures used (725–901°C) were below those for recrystallization and there was an activation energy 'spectrum' for the recovery process, so the interpretation of this effect is very difficult. It does, however, suggest that at these temperatures bonding between dislocations and atmospheres is weak and this is in line with the small serrations involved during straining. This is, on the other hand, not in accord with large k_y values noted in this metal.

Since coarse-grained material does not always show a yield point, single crystal studies (e.g. Lawley and Gaigher, 1964) are of little value, except in so far as they elucidate the dislocation picture which, as mentioned above, is more similar to iron with cell wall tangles than to chromium.

Finally, even if these samples of the metal show no yield point, one can be induced, as might be expected, by neutron irradiation (Makin and Gillies, 1957; Johnson, 1960b). The irradiated samples do not appear, however, to obey the Hall–Petch grain-size relationship.

3.6 *Tantalum*

The metals tantalum and tungsten remaining to be discussed are those with the highest melting points of the group, and in so far as effects of yielding might be determined by a homologous (T/T_m) temperature factor, the effects noted in other metals in this group should be sought at higher temperatures.

Yield points were noted first by Bechtold (1955) using metal containing 0·01% of both carbon and nitrogen; oxygen and hydrogen were not determined. Yield points were noted in samples tested at low and room temperatures and serrated yielding, although small, was noted above about 400°C. The transition temperature was low, possibly because of the higher solubility of this metal for oxygen. Similar results were obtained by Pugh (1956b) who noted, in addition, the appearance of strain markings on his specimens (Plate 3.2), true strain ageing return of the yield point and the usual maxima in UTS and minima in elongation indicative of 'blue-brittle' behaviour. Pugh also measured the variation of strain rate sensitivity (m) with temperature, but Schussler and Brunhouse (1960), in a later paper, suggest this parameter seems dependent on grain size, as might now be expected.

Koo (1962) and Gilbert *et al.* (1962) showed that the usual grain-size relationship applied to the yield stress, and Hull *et al.* (1963) showed that in the microstructure, in common with all other metals of this group except chromium, the dislocations group into well-defined cell walls. Twinning is also observed at low temperatures

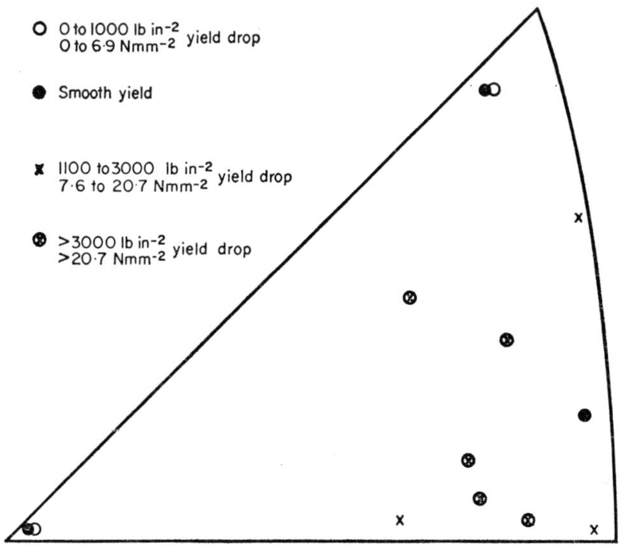

FIGURE 3.4. Yield point effects in tantalum single crystals at 300 K (Ferris *et al.*, 1962)

(Barrett and Bakish, 1958; Anderson and Bronisz, 1959), and Koo (1962) also showed that k_y is high for twinning (see Table 1.2).

Electron-beam melted single crystals have been examined by Ferriss *et al.* (1962). No yield point was observed in crystals whose orientations lay near the (001)–(111) boundary in the stereographic triangle (Fig. 3.4), i.e. in crystals where slip on several systems could circumvent those dislocations locked by interstitials. This has been already noted in chromium (Section 3.3.2) and is also seen in tungsten (Section 3.7). More extensive studies by Mitchell and Spitzig (1965) showed that three-stage hardening, already noted in niobium above, could occur with tantalum crystals provided the temperature is in the range $0{\cdot}10$–$0{\cdot}18T_m$, and the orientation effect on the yield point was not observed. In this latter work none of the single crystals showed a yield point, perhaps because the purity was higher, although a 'hump' was observed between 147 and 157 K.

Strain ageing in single crystals has been studied by Rosenfield and Owen (1963) following the initial investigations of Ferriss *et al.* mentioned above. Ferriss *et al.* have shown that pre-straining at 300 K did not remove the upper yield point at 77 K, and explained this result in terms of exhaustion of free dislocations at the pre-straining temperature. Rosenfield and Owen (1963) strained samples beyond the Luders extension, and then determined the changes in the parameters σ_0 and k_y in the grain size relationship, following the method used by Wilson and Russell in mild steel (Section 2.5). In this case ageing is sluggish, and while their results are interpreted as showing a two-stage process, electron microscopy showed no evidence of precipitation on dislocations, and physical processes could not be assigned to the two effects. This work does, however, show the difficulties of working with strains $> \epsilon_L$, and interpretative problems are increased.

The element responsible for strain ageing has been studied by Formby and Owen (1965), who, using samples with 48 p.p.m. O_2, 10–20 p.p.m. N_2 and 67 p.p.m. C, followed further the technique outlined above by Rosenfield and Owen (ibid.). σ_0 was unaffected by quenching and/or ageing but k_y was changed, and an activation energy of 25·4 kcal mole^{-1} (106 kJ mole^{-1}) was determined for the process. This was identical with the activation energy for oxygen diffusion, which clearly differs from carbon or nitrogen diffusion (respectively 38·7 and 37·7 kcal mole^{-1} or 162 and 158 kJ mole^{-1}). No precipitate occurs because of the high solubility of oxygen in this metal.

A more recent study of strain ageing is by Hartley (1966) who, using the Cottrell–Bilby approach, showed that, at least at early stages of ageing, the $t^{2/3}$ law was obeyed. There was notable scatter, which did not stop Hartley determining the activation energy as 23·6 kcal mole^{-1} (99·8 kJ mole^{-1}), and relating the process to oxygen diffusion. The $t^{2/3}$ law is only obeyed for a very limited time, when some other, slower process takes over.

There is, however, little doubt that oxygen is responsible for these effects, because of the high affinity between the gas and tantalum, and internal friction studies by Carpenter and Baker (1965b), and resistivity studies by Köthe (1968) have confirmed the activation energy. Formby and Owen (1966) have considered the variation of k_y mentioned above, and claim it is consistent with the formation of Cottrell atmospheres.

3.7 *Tungsten*

Tungsten has been manufactured in a variety of forms, particularly for the electric lamp and vacuum tube industries, for many years, and the variation of engineering properties with temperature has been known for a corresponding period (see, for example, Jeffries, 1919). It is only in recent years that the yield stress behaviour, and also the modes of single crystal deformation, have been studied extensively.

Because of the high melting point of tungsten, the yield stress is rising very rapidly near room temperature $T/T_m \sim 0{\cdot}08$) and consequently any grain-boundary segregation will at least tend to bring about a transition temperature well above room temperature in recrystallized material. Further, the high rates of cooling which may be present in cooling from the correspondingly higher recrystallization temperatures may leave the interstitials responsible for yield point effects largely in solution. Hence it is hardly surprising that yield points in polycrystalline material have been difficult to detect, and only cursorily examined.

Bechtold (1956) noted a small yield point effect, but only at temperatures **below** 250°C, and at strain rates of $10^{-3}\ s^{-1}$ or greater. Pugh (1957) has also examined tensile and creep properties over a wide range.

Values of σ_0 and k_y for tungsten are listed in Table 1.2, but no studies of Luders band effects have been found, and few papers on ageing properties. Baird and Hartley (1963) examined the return of the yield point on ageing deformed cylinders of tungsten at 950–1100°C. The activation energy was 96 kcal mole^{-1} (402 kJ mole^{-1}), a value much higher than usual for these metals. In view of the high ageing temperatures, they may in fact have been observing some form of recovery. Using internal friction studies, Carpenter and Baker (1965a) found the usual $t^{2/3}$ law applied, and an activation energy of 47·5 kcal mole^{-1} (100 kJ mole^{-1}), tentatively identifying the impurity with carbon.

Single crystals, prepared by electron beam melting, have of late become available and relatively more complete data exists on these. In fact 'Bamboo' type single crystals have been used in tungsten filament lamps for many years, and first data on the slip system is by Goucher (1924). Schadler (1960) also examined the slip systems, and confirmed that $\{112\}\ [11\bar{1}]$ slip was present, as well as $\{110\}\ [1\bar{1}1]$ not found by Goucher at room temperature. At lower temperatures, twinning replaced slip on $\{112\}$. At high temperatures, Berlec (1963) claims $\{110\}$ slip is absent.

None of these investigators noted any yield point effects, and it was Rose *et al.* (1962) who noted, first, a strong dependence of yield stress with orientation, and secondly the fact that crystals whose axis lay near the [110] corner in the stereographic triangle showed yield points, whereas others did not. Figure 3.5 shows this effect clearly, and is in some respects similar to tantalum. Note also the rapid work-hardening rate in other samples. This unexpected result was confirmed by Garlick and Probst (1964), Beardmore and Hull (1965) and Argon and Maloof (1966), although Koo (1963b) and Schnitzel

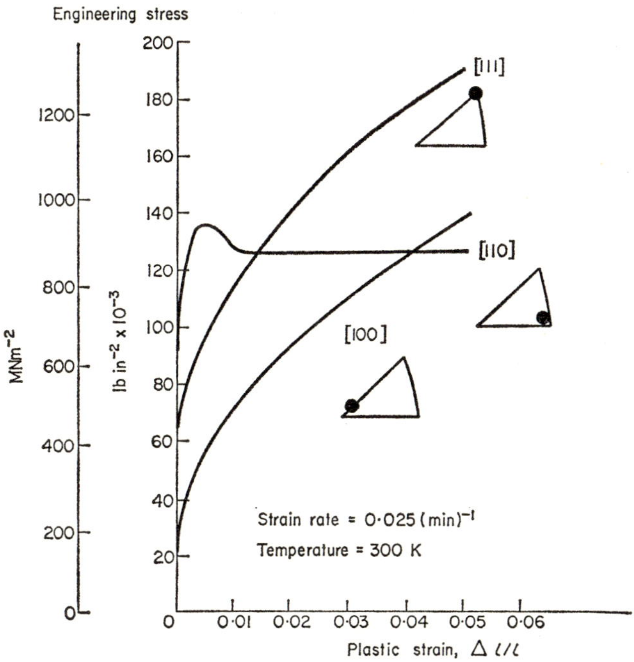

FIGURE 3.5. Yield point effects in tungsten single crystals at 300 K (Rose *et al.*, 1962)

(1965b) failed to find the effect. It should also be noted that the critical resolved shear-stress law does not hold with these crystals (Beardmore and Hull, ibid.). Rose *et al.* point out that in a [110] orientation, only two slip directions are operative, [111] and [11$\bar{1}$], and these lie in the common plane {110}. The resolved stress on this plane is zero. This explains the high yield stress for this orientation, but does not explain the existence of the yield point, unless it is a form of geometric softening, for it should be noted that this yield does not return on strain ageing.

3.8 *Alloys of these metals*

The six metals which have just been discussed are interesting in yet another way – unlike iron, they retain their body-centred cubic structure up to the melting point, and when alloyed with their neighbours in the periodic table often form a complete range of solid solutions. Thus, in metals where the rapid rise in yield stress at low temperatures is considered significant for theory (Section 3.9 below) then the additional advantage of alloying will be of considerable value.

Studies on single crystals of several of the alloy systems have now been carried out, and as mentioned by Van Torne and Thomas (1966) a curious effect seems to exist, namely, at about equiatomic composition, the alloys of Group Va and VIa within themselves are ductile, but in alloying across the columns of the Periodic Table the alloys are relatively brittle. This proposition cannot be validly extended to include vanadium and chromium, as Cr-Nb forms an intermetallic compound $NbCr_2$, and data is lacking on the Nb–V and Cr–Mo systems.

The alloys crystals which have received any extensive study mostly contain tantalum as one component, but numerous studies have been made over a limited range of compositions in other systems. The notable feature is the virtual absence of any reference to yield points in these alloys. Van Torne and Thomas (1966) for example, illustrate the stress–strain curves for Ta–Mo single crystals; the addition of Mo to the system now seems to make it possible for crystals whose axis is close to the [100]–[111] boundary to exhibit a yield point. In addition, the three-stage hardening seen in pure molybdenum is suppressed and a yield point appears with the addition of 7% Ta. In the centre of the binary field, alloys for 19–58% Ta are brittle. No long range order exists in this region; but because of local variations in composition and the large difference in elastic constants between the two metals, local cracking occurs and the specimens cleave on {100}.

A different system was covered by Peters and Hendrickson (1966) who examined single crystals of Nb–Ta alloys. Here the elastic constants are roughly equal, and the critical resolved shear stress falls smoothly from pure tantalum to pure niobium. Unfortunately, no comments are given on the change in the three-stage hardening curve, or the appearance of any yield point effects. Rudolph and Mordike (1967) note little change in adding 4·4% Mo or 2·6% Nb to their tantalum base.

The properties of the complete range of Nb–Mo alloy single

crystals has been investigated by Milne and Smallman (1968). Again, crystals of roughly equiatomic composition were the strongest and relatively brittle but the elongation to fracture was dependent on orientation. In most cases where limited ductility existed, a yield point was seen.

Amongst other systems which have received incomplete study is the tantalum–tungsten system. Tedmon and Ferriss (1962) have examined a polycrystalline Ta–10% W alloy, for comparison with pure Ta, and measured the grain-size dependence of yield. After tests on pure Ta at 77, 195 and 300 K, and on Ta–10% W at 300 K, the value of the parameter k_y was effectively zero but the scatter was high. A yield drop was still obtained in each case so that a small but finite value is still necessary. At the other end of the phase diagram, Schnitzel (1965a) has examined the effects of carbon additions to both pure tungsten and W–0·35% Ta single crystals. However at this low level of addition, there is little difference between this alloy and pure tungsten. The addition of carbon gave rise to an increased strain-ageing rate at 300°C, and to serrated yielding at 600°C, not present in the as-grown crystals. The effects are put down to carbide precipitation. Kieffer *et al.* (1959), in the course of a survey of tungsten-base alloys, shows that maximum hardness is again attained with equiatomic alloys of W–Mo, W–Nb and W–Ta alloys, but no mention is made of yield point effects.

It should be noted that the brittleness discussed above between certain elements in this group only applies to alloys of roughly equiatomic percentages. At each end of the phase diagrams useful increases in strength can very often be achieved by minor alloying. For example, Wilms (1964) has examined chromium base alloys, containing 2% Ta, 0·5% V and 0·05% B. These alloys produced serrated yielding at 800°C and have useful creep rupture lives at still higher temperatures. Many alloying additions seem to push the temperature range for serrated yielding to higher temperatures, as can be seen from the work of Sheely (1962a) who quotes results on niobium, with 5 at.% additions of V, Zr and/or Ti. Strain ageing at about 250°C was suppressed and serrations now appeared in the 700–850°C range, which in turn would imply improved high-temperature properties. Arsenault (1966) and Arsenault and Lawley (1967) have likewise noted that the addition of 9·1% W or 9·2% Nb to tantalum tends to suppress the yield point. Finally, in this field where little systematic work has been done, the quaternary alloy 60 Nb–20 Ta–15 W–5 Mo (wt%) has been studied by Harris and Peacock (1965). This alloy shows yield point and strain ageing phenomena, and has an activation energy for the yield point return of 29·5 ± 2 kcal mole^{-1} (123 ± 8 kJ mole^{-1}). Thus, again, the ageing effects in niobium are identified with oxygen.

3.9 *Discussion*

3.9.1 *Temperature dependence of the yield stress*

As is apparent from the earlier sections of this chapter, these metals form a group with remarkably similar properties. All have a body-centred cubic structure, all display yield point and strain-ageing effects, all obey the Hall–Petch equation, and yet in detail there are

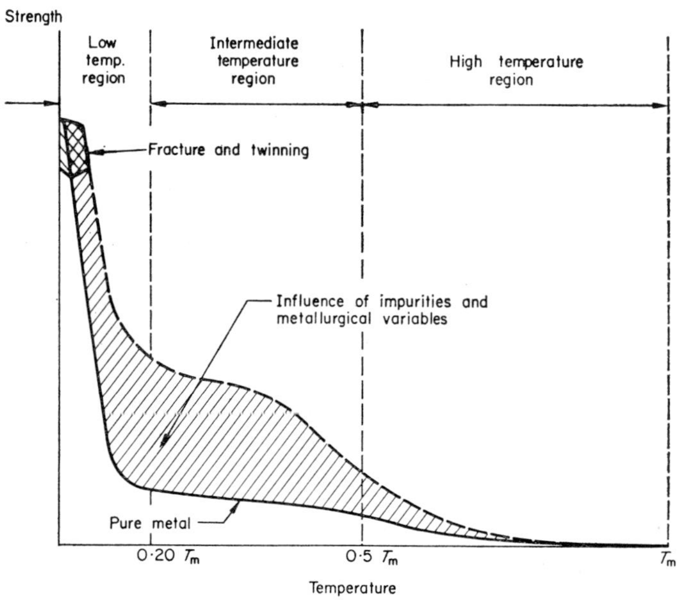

FIGURE 3.6. Variation of yield stress with temperature for the b.c.c. refractory metals (Bechtold *et al.*, 1961)

some equally remarkable differences. The purpose of this section is to discuss both the similarities and the differences, and to see how far current theories can account for these. The first point of note is the remarkable variation in yield stress with temperature, much larger than for h.c.p. or f.c.c. metals. Within limits, the variation may be described by a single curve by plotting yield stress against homologous (or reduced) temperature T/T_m. This, as shown in Fig. 3.6, shows a steep rise at $T/T_m < 0·1$ (Bechtold *et al.*, 1961). Within the 'intermediate' temperature region, the curve flattens out, but in this region there are maxima in U.T.S. coupled with minima in elongation, characteristic of blue-brittle behaviour. The shift of these maxima with strain rate can be used to determine the activation energy for the ageing process, as has already been noted. Above about $0{\cdot}5T_m$ ageing effects have largely disappeared, and the strength falls steadily to the melting point.

Considerable interest has been shown in the low-temperature region in recent years, and in the remarkably rapid rise in low-temperature strength. Numerous theories have been put forward, and these may conveniently be classed in four groups:

(a) Variations on the lattice friction stress;
(b) Interaction of dislocations with impurities;
(c) Theories using jog densities, including dislocation intersection;
(d) Changes in slip mechanism.

In many studies on the variation of yield stress with temperature, it has been found that the parameter k_y is only weakly temperature dependent. Increases in the yield stress are accounted for by the change in σ_0, and it will be recalled in Section 1.6.4 that Heslop and Petch (1956) showed σ_0 may be made up of two components:

$$\sigma_0 = \sigma_0' + \sigma''$$

where σ'' is independent of temperature but dependent on purity. Heslop and Petch then suggested that the low-temperature rise in σ_0' was due to changes in the so-called Peierls–Nabarro forces, which is the name given to the interatomic forces controlling the basic properties of dislocations, including the yield stress.

It was originally thought by Heslop and Petch (ibid.) that it would be the **width** of the dislocations which would be basically affected by temperature, but it was later realized that the Peierls–Nabarro forces would also control the height of the potential barrier which the dislocations would have to overcome. If we leave for the moment the nature of this barrier and the detailed manner in which it is overcome, it is possible to develop a simple theory whereby the dislocations are thermally activated over this barrier. The basic equations have been developed by Conrad and his co-workers. (See Conrad and Schoeck, 1960; Conrad and Wiedersich, 1960; Conrad and Frederick, 1962; Conrad and Hayes, 1963a, b; Conrad, 1960, 1961, 1963b.) The theory starts with the Heslop and Petch hypothesis, put in terms of shear stress, viz.

$$\tau = \tau^* + \tau_\mu + k_y' d^{-1/2}$$

where τ^* is the thermal and τ_μ the athermal component, the latter depending on purity and morphology, and its only temperature variation being in the shear modulus μ. Measurements of τ^* can then be made at varying temperatures and strain rates.

To overcome the potential barrier presented by the Peierls–Nabarro forces, energy H has to be supplied to the dislocation by thermal means. Thus, the strain rate is given by

$$\dot{\epsilon} = \rho b s \nu^* \exp\left(-\frac{H}{kT}\right) \tag{3.1}$$

where ρ is the dislocation density, b the Burgers vector, s is the product of the number of places where thermal activation can occur and the area swept out per thermal fluctuation, and ν^* is the frequency of vibration of a dislocation segment.

Writing $bs\nu^* = \nu$, and assuming that the activation energy H is a decreasing function of the effective shear stress τ^* (the difference between the applied stress τ and the internal stress field τ_μ) and the temperature T, the following equations are obtained (for derivation, see Appendix I):

$$H = -k\frac{\partial\,(\ln \dot{\epsilon}/\nu)}{\partial\,(1/T)} \tag{3.2}$$

or

$$H = -kT^2\frac{(\partial\tau^*/\partial T)\,\dot{\epsilon}}{\partial\tau^*/[\partial\,(\ln \dot{\epsilon}/\nu)]_T} \tag{3.3}$$

From these results, an activation volume v^* for the deformation process may be defined by the equation

$$v^* = kT\left(\frac{\partial \ln \dot{\epsilon}/\nu}{\partial\tau^*}\right)_T \tag{3.4}$$

and H and v^* are related by the equation

$$H = -v^*T\left(\frac{\partial\tau^*}{\partial \mathrm{T}}\right)_{\dot{\epsilon}} \tag{3.5}$$

H is an increasing and v^* a decreasing function of the temperature.

Conrad's theory has found wide support. It allows a unified treatment of iron and the group Va and VIa metals, illustrating, as shown in Fig. 3.7, that the low-temperature deformation may be described by a single, thermally activated process. It may be applied to theories of delayed yielding (Conrad, 1961), and to the Luders deformation of iron (Conrad, 1963c; Violan *et al.*, 1967) and niobium (Conrad and Stone, 1964). Since radiation damage affects σ_0 more than k_y, it may be used on material so treated (Arsenault, 1967b; McRickard, 1968). It has been used to cover the yield stress of iron whiskers (Conte and

Groh, 1968) and sapphire (Conrad *et al.*, 1965) (see Chapter 7) and has been applied to the deformation of f.c.c. metals at high temperatures (Gallagher, 1967) and to potassium, which is another b.c.c. metal (Bernstein and Gensamer, 1968). (See also the paper by Arsenault and Koppenaal (1965) on neutron irradiated copper.) It has been modified to include varying dislocation density (i.e. changes in ρ) by Bailey (1967) and Bailey and Flanagan (1967, 1969), while a series of dilute alloys have also been investigated with generally satisfactory

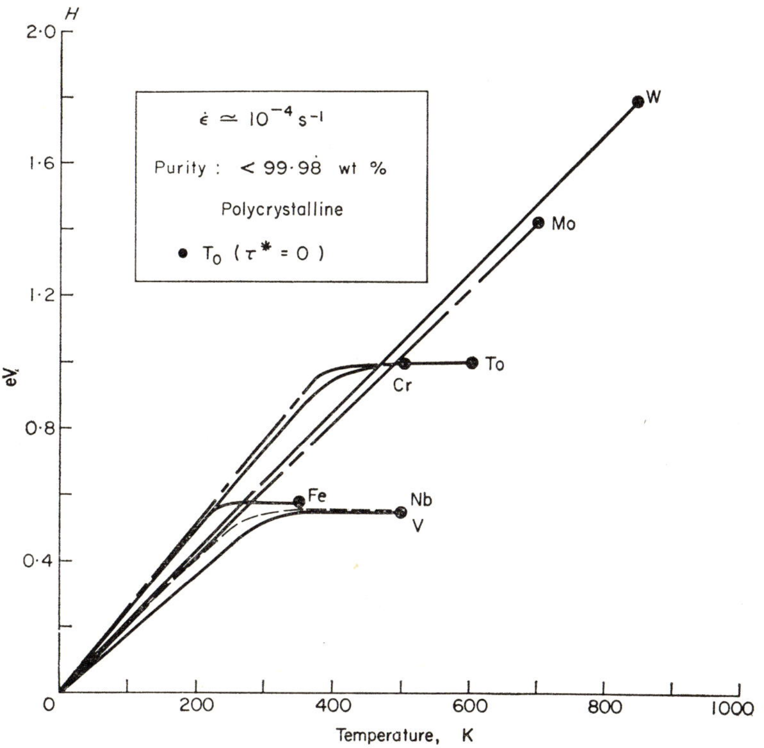

FIGURE 3.7. Variation of the activation energy with temperature for b.c.c. metals (Conrad, 1963b)

results. Among alloys studied using this theory are Fe–Mn alloys (Wynblatt *et al.*, 1965; Wynblatt and Dorn, 1966; Rawlings and Newey, 1967) and Ta-base alloys (Arsenault, 1966). In the latter case, alloying may affect the temperature at which athermal effects take over, but in general, the analysis is not very sensitive to impurity level, and this holds for the very important case of decarburized iron, where, as Conrad (1967) himself has shown, the results are virtually identical to the mild steel case.

It is now necessary to look in more detail at the process whereby the potential energy barrier is overcome. A dislocation will spread over the barrier forming a small loop (Fig. 3.8), bounded by a double kink. Under these circumstances, it may collapse or grow, and the configuration is often known as a Seeger oscillator, after Seeger (1956) who examined the case under zero stress. The conditions for growth of the double kink have been extended by Dorn and Rajnak (1964), Evans and Flanagan (1966), Lau (1967), Arsenault (1967a), Prekel and Conrad (1967) and Prekel *et al.* (1968). While there are some differences in detail amongst these, it is worth noting that in principle there are many advantages in pressing for an acceptable

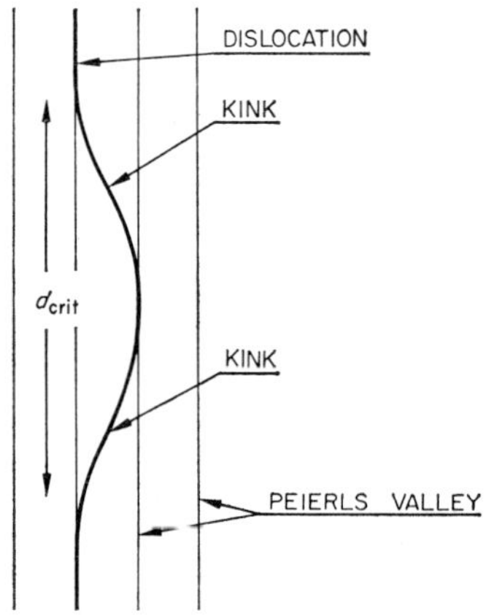

FIGURE 3.8. Dislocation loop spreading on the double kink model (after Arsenault, 1967a)

theoretical basis. For example, Prekel *et al.* (ibid.) show, following Seeger (1956), that in terms of the variables defined above

$$H = H_k \left[1 + 1/4 \ln \frac{16\,\tau_P}{\pi\tau}\right]$$

where H_k is the kink energy, and τ_P the Peierls force, so that the stress exponent of the dislocation velocity m^* follows as

$$m^* = H_k/4kT$$

Thus, as well as suggesting m^* is inversely proportional to temperature, it does lead the way to a marriage of the thermal activation theory of yielding, and the dislocation multiplication theory. This union has much to recommend it, and has been urged recently by Krausz (1968) and by Christian (1967) in commenting on a paper by Yada (1967) on the determination of m^*.

Leaving aside for the moment the objections which have been raised to this theory, the next group which deserve consideration are those which consider the impurity level as of paramount importance. It will be recalled (Section 1.5.1) that the original Cottrell–Bilby theory comes in this category and that the breakaway stress could be related to the yield stress at absolute zero. Reid and Owen (1962), working with a low carbon iron, found reasonable agreement with Cottrell's formula down to 160 K, while Lawley *et al.* (1962) using zone-refined molybdenum, suggested that the relationship held to liquid helium temperatures. Now that concepts of breakaway have largely been discounted, other theories of interaction with solute atoms have been proposed. Of these, the most successful is due to Fleischer (1960, 1961, 1962a, b). This, in many ways, is a development from the tetragonal distortions around dislocations considered by Cochardt *et al.* (1955), Schoeck and Seeger (1959) and others, but in this case the important defects are considered to be those lying within one atom distance of the slip plane. This gives a force–distance curve with a maximum at $x = 0$, in contrast to the Peierls–Nabarro theory which predicts the maximum at some fraction of the lattice spacing along the slip direction. Yielding then takes place by thermally activated jumps through the distortions. The variation of yield stress is then calculated by Fleischer (1962b) as

$$\left(\frac{\tau}{\tau_0}\right)^{1/2} = 1 - \left(\frac{T}{T_0}\right)^{1/2}$$

where T_0 is a characteristic temperature, and τ_0 the yield stress at absolute zero. Further, Fleischer showed that the variation of yield stress at absolute zero varies with composition, c as

$$\tau_0 \propto c^{1/2}$$

The theoretical basis of the theory has been extended by Kumar (1968), while experimental results in support of the theory have been obtained by Stein (1966) and Stein and Low (1966) for iron, by Conrad (1966) and Jones and Conrad (1967, 1968) for α-titanium, by Rudolph and Mordike (1967), and by Raffo and Mitchell (1968) on tantalum alloys, where it is found that the addition of rhenium can lead to alloy softening in a certain temperature range. Koppenaal (1968) has also used the theory to study hardening in copper single crystals after irradiation.

A basically different theory part-way between the Conrad and Fleischer approaches, is that put forward by Mordike and Haasen (1962). Here thermal activation over **clusters** of solute atoms or precipitates was considered to be the rate-controlling step; but in a later paper Burbach *et al.* (1966) showed that decarburized samples of iron displayed the same temperature sensitivity and this theory must be abandoned. However, at higher temperatures, the barrier concentration effect still has a place, particularly in discussing precipitation hardening, and Wilcox and Gilbert (1967) have shown that the yield stress of a Mo–0·48 Ti–0·11 Zn alloy varies with (barrier spacing)$^{-1/2}$.

The third group of theories, that involving jog densities, has been put forward in somewhat different terms by Schoeck (1961), Rose *et al.* (1962), Mordike (1962), and Gregory (1963). As is well known, for jogged dislocations to move, point defects have to be created, and the ability to do this will be impaired at low temperatures. A rise in yield stress will result.

The fourth and final group of theories have, as their basis, the fact that microstrain effects and particularly the anelastic limit, are not temperature sensitive (see Section 1.4.3 and also Lawley and Gaigher, 1964 and Stoloff *et al.*, 1965). On this fact, it is necessary to base a multiplication mechanism which is strongly temperature sensitive. Various cross-slip models (see, for example, Low and Guard, 1959), have been proposed and Brown and Ekvall (1962) have used this theory to explain their microstrain results, since thermal activation may well be necessary to activate cross slip. Again, if cross slip is necessary to overcome obstacles, then thermal activation may help the dislocations over these. Ferriss *et al.* (1962) have used this argument to discuss the deformation of tantalum.

Each of these theories has had its active group of protagonists and it could not yet be said that any one of these theories is definitive. There are however, clear unanswered arguments against the third group, in that Conrad (1961, 1963b) has shown that v^* and H are independent of structure, which would not be the case if jogs were controlling the deformation. Furthermore, the mobility of edge and screw dislocations is identical, which would not be expected since jogs largely control the free movement of screw dislocations.

Later work on microstrain studies (Kossowsky and Brown, 1966) has also shown that the elastic limit σ_E is, to some extent, temperature sensitive, and interpretations of σ_E and σ_A (the anelastic limit) do now call upon Peierls–Nabarro forces, as well as the dislocation geometry, to explain the results. Bacon (1968) further calls on an orientation dependent friction stress to cover some of the experimental results. σ_A is now known to be dependent on pre-strain, and the temperature dependence becomes more marked, the higher the pre-strain (Kossowsky and Brown, ibid.). One is thus forced to consider double kink models which are easier to move than those shown in

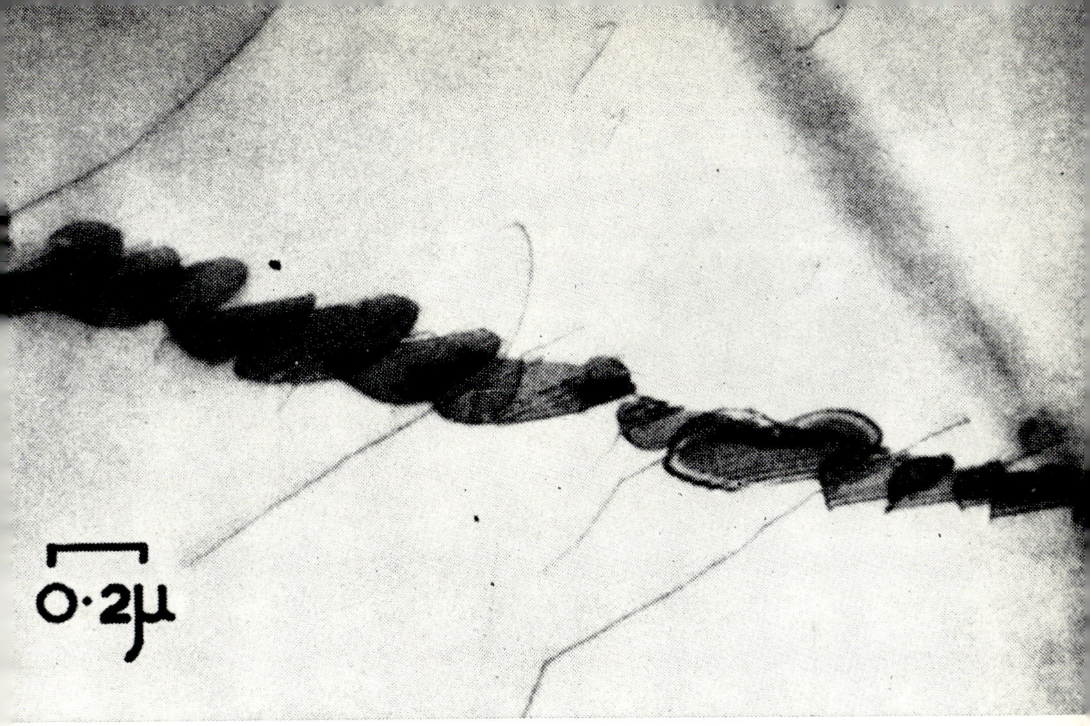

PLATE 3.1 Defects associated with a dislocation in chromium (Garrod and Wain, 1965)

PLATE 3.2 Strain markings on a deformed tantalum sample tested at 600 K. (Pugh, 1956b)

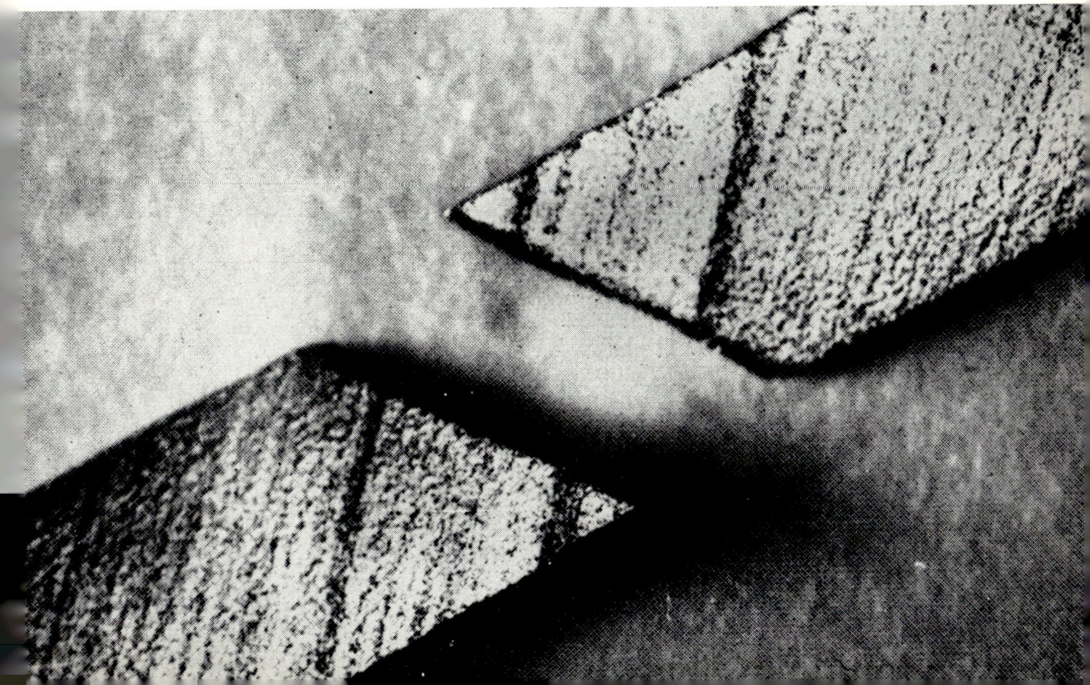

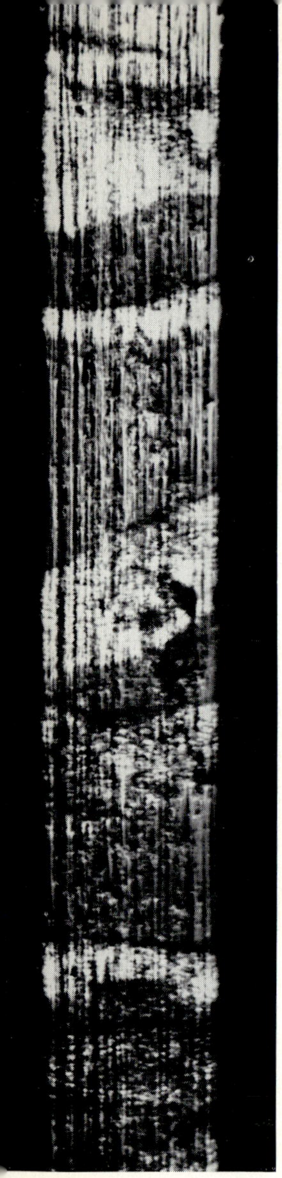

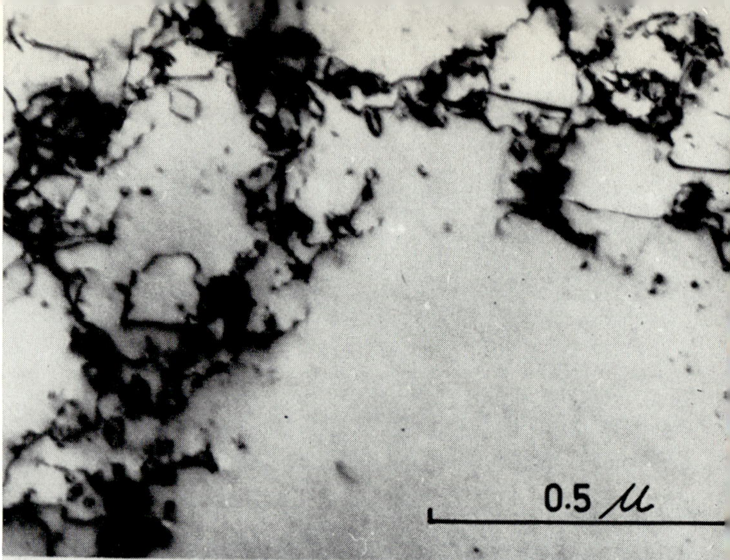

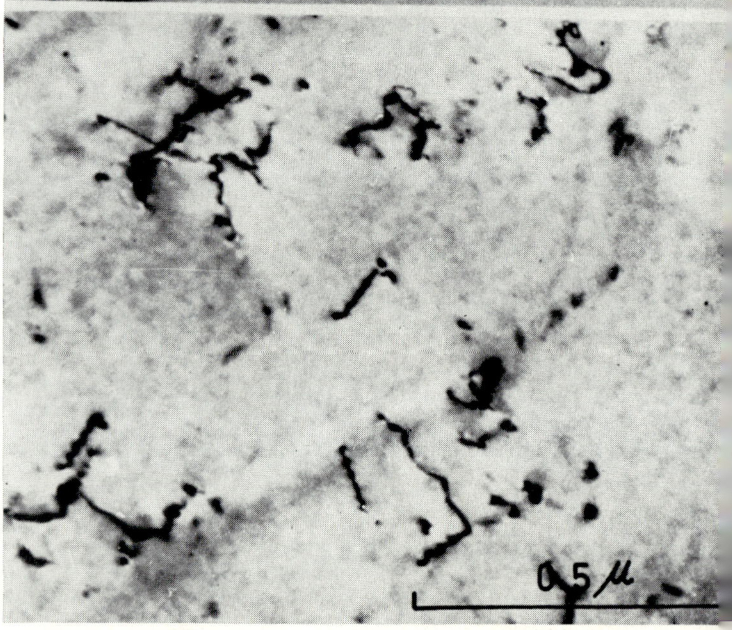

PLATE 4.1 (*above left*) Complex Luders bands in thin mild steel strip, hydrogen charged. (*By courtesy of* Mr. J. Grahame)

PLATE 4.2 (*top right*) Dislocation patterns in (a) hydrogen charged and (b) (*above right*) hydrogen-free nickel, strained 9%. (Blakemore, 1968)

Fig. 3.8, but the detailed discussion of these models is beyond the scope of this review.

For the temperature dependence of macroscopic yielding, then, the two main theories are those of Conrad and of Fleischer. In 1960, Bazinski and Christian (1960) studied the flow stress of iron and suggested from thermal softening data, that theories based on Peierls–Nabarro forces, or on a uniform distribution of impurities would explain the results. This is still the position (Baird, 1967).

For the Conrad type theories, it follows from the equations given earlier that v^* and H are independent of strain, and further that these parameters are independent of interstitial content. Those who raise objections to this theory are in two categories, (a) where their experimental results do not satisfy the internal checks of the Conrad theory and (b) those who object that the results are dependent on interstitial content.

In the first of these come papers by Mordike and Haasen (1962), Davidson *et al.* (1966), or Carnahan *et al.* (1967); the latter case showing clearly that the activation volume v^* for microstrain is both large and very strain sensitive.

The more basic, and perhaps the more important arguments at this stage depend on the sensitivity of impurity level. Stein (1966) using very high purity iron (0·005 p.p.m. C) showed that, in his view, the yield stress was a function of interstitial content at sub-zero temperatures, and that the stress–strain results on this, and less pure, material, could not be adequately represented by a singly activated thermal process. Conrad (1967) contended that the results could be so interpreted, that the values of v^* and τ^* were within the scatter band obtained by others for less pure iron, and that near absolute zero the yield stress of this pure iron would be the same as less pure material. In the reply to this note, Stein (1967) maintains that this latter point is not correct and that changes in pre-exponential factors in equation (3.1) could well occur which would make the results dependent on structure.

A similar difference of opinion has arisen on a series of results on tantalum by Arsenault (1964b, 1966). These were originally interpreted in terms of Conrad's theory but have been criticized in detail by Fleischer (1967). It is contended that, as in Stein's results just quoted, the yield stress of tantalum is controlled by impurities and not by the Dorn and Rajnak double kink model. In reply, Arsenault (1968) claimed that the impurity level had been wrongly estimated and that his data could not be made to fit Fleischer's formulae. Arsenault's (1966) results on Ta–9% W alloy also did not fit. In a further reply Fleischer (1968) would still maintain his argument is correct, and further draws on results by Evans (1962) on niobium–nitrogen alloys which conform to his (concentration)$^{1/2}$ formula.

There is obviously little advantage here in exploring this controversy further. The proof may well have to await the accumulation of careful data on a variety of interstitial and substitutional systems, or the chance discovery of a system where the yield stress is markedly more sensitive to impurities. Perhaps the b.c.c. intermetallic compounds may provide an example. As it is, the general trend, even in 'gettered' material (Leslie and Sober, 1967) or in the high-purity material of Stein mentioned above, a rapid increase in yield stress occurs at low temperatures. If, however, this is to be explained on the Conrad theory, then the relative insensitivity of the elastic and anelastic limit to temperature changes must also be included. If microstrain is to be included, it will remain the most difficult part of the Dorn-Rajnak analysis until the dislocation kink configurations can be adequately detailed.

3.9.2 *Comparison of the Va and VIa metals*

Having stressed in the previous section the similarities between these seven metals, it is now necessary to discuss their differences.

The first difference that can be noticed is the range of brittle-ductile transition temperature. This variation must be tied in turn to the variations in solubility of interstitials which these metals possess. The atomic radii of the elements are given in Table 3.3.

TABLE 3.3. (Goldschmidt) Atomic Radii of the Refractory Metals

Metal	Radius (Å)	Metal	Radius (Å)
V	1·36	Cr	1·28
Nb	1·47	Mo	1·40
Ta	1·47	W	1·41

Thus, where interstitial atom size is important, one would expect an increased solubility in the Group Va metals, and this is true of oxygen in Ta–W (~ 4 as against 0·06 at.% at 1700°C), Nb–Mo (5·5 as against 0·027 at.% at 1110°C), and V–Cr (4 as against 0.043 at.% at 1500°C). Similar results seem to apply to other interstitials where data is available. However, the possibility of segregation to grain boundaries, larger degrees of supersaturation, and possible heterogeneous precipitation are not the only criteria to determine the transition temperature; as covered in Section 2.4.5., Cottrell (1958) has predicted the criterion for brittle failure in terms of the equation

$$(\sigma_0 d^{1/2} + k_y)\, k_y = \beta \mu \gamma$$

where the symbols were earlier defined. The most important variable in this equation is k_y; this has been derived for all metals in these groups and is listed in Table 1.2 (Chapter 1).

In this table the group Va metals appear to have notably lower values of k_y: niobium has been given the values of Adams *et al.* (1960) rather than a zero value which has no meaning; thus Nb and Ta are expected to be more readily ductile. It is probably true to say that the problems of producing ductile Group VIa metals were more intractible than the other, and it is certainly true that Cr, Mo and W are more influenced by small amounts of impurities than the V, Nb, Ta group.

Values of the shear modulus, μ, also vary between the two groups, and are given in the next table (Table 3.4).

TABLE 3.4. Values of Shear Modulus for the Refractory Metals

Metal	Modulus ($\times 10^4$ N mm^{-2})	Metal	Modulus ($\times 10^4$ N mm^{-2})
V	4·67	Cr	11·53
Nb	3·75	Mo	12·56
Ta	6·92	W	16·06

This difference has also been called on to explain why alloying Group Va to VIa metals leads to embrittlement, while within their own group ductile alloys are formed.

Despite these differences, it would seem that all these six metals display an almost uniform pattern when it comes to their deformation behaviour. All display yield point and strain-ageing effects, and while the solute elements responsible for this have not always been identified, it is highly likely that carbon, nitrogen and oxygen can act together as effective dislocation locking solutes. This must undoubtedly complicate kinetic studies, and close control of impurities is essential.

Cases are reported where precipitates have been noted on dislocations, but this is only to be expected where the solute solubility is very low. For example, no oxide precipitate would be expected, at the impurity levels quoted above, in niobium or tantalum, and yet strain ageing is present. One must therefore conclude that Cottrell–Bilby locking is evident, but again from dislocation studies we must also assume these dislocations are permanently locked, and the upper yield point associated with dislocation multiplication. In developing a theory of strain ageing, there are as many unknowns here as in the case of mild steel.

The uniformity of properties should lead to a unified treatment of the low-temperature yield strength as has been outlined in the previous section. Here again there are several unknowns, particularly the effects of very low controlled amounts of interstitial impurities, but undoubtedly there will soon be agreement on the reasons for the very rapid increase in the low-temperature strength of these metals.

In single crystal studies, we are at last starting to see systematic and careful examination of these metals. As well as being important in any study of dislocation dynamics, deformation behaviour of these crystals is basic to our knowledge of yield propagation. The existence of 'easy glide', and its relation to alloying, may well lead to an understanding of the anomalous yield points noted in tantalum and tungsten single crystals, and further studies on twinning and its grain-size dependence may be of value in assessing grain-size effects generally. While the experimental difficulties are greater and the possibilities of unwanted contamination higher than in iron, no theory of yielding and strain ageing in b.c.c. metals could be considered complete without correlation within this remarkably homogeneous group of metals.

4

Hydrogen in Metals

4.1 *Hydrogen embrittlement*

The decreased ductility which results when hydrogen is introduced into many metals has led to the term 'hydrogen embrittlement', and to a massive literature on the subject. Technically, it is of considerable importance because it can occur readily in iron with disastrous cracking as a result, but it is not the purpose of this exercise to consider in detail the mechanism of hydrogen embrittlement. We will be concerned mainly with the introduction of hydrogen as an interstitial element which can produce yield points and strain-ageing effects; because of the smaller size of the atom, it can act interstitially in face-centred cubic and hexagonal (both close-packed lattices) as well as in the body-centred cubic systems. However, for the sake of completeness, a very brief review of the embrittlement process will be made, and then the effects of hydrogen on the yield point and strain ageing of metals will be considered. Reviews of this very large subject will be found in Somalowski (1962) and Cotterill (1961). The metals will be taken in the same order as they have been dealt with in this book: iron, the group Va–VIa metals, face-centred cubic and finally the hexagonals.

4.2 *Solubility of hydrogen in metals*

The actual amount of hydrogen taken up by these metals will depend on many electrochemical factors, but for the purposes of this discussion, the metals under discussion may be divided into two groups:

(a) The so-called endothermic occluders, where the solubility increases with increasing temperature. In this group are the Group VIa metals, the transition metals (Fe, Co, Ni) and also Cu, Ag and Au. These metals form only solid solutions and/or hydrides which are very unstable.

(b) The exothermic occluders, which form pseudo-metallic hydrides, and where the solubility limits decrease with increasing temperature. Since the hydrides which form are invariably brittle, these hydrides platelets can act as crack nuclei or as crack paths. Metals of interest in this group are Ti, Zr and Hf, and the Group Va elements.

There are cases where other chemisorption effects may complicate the issue, as for example, in the case of nickel, but these divisions will suffice. There are two simple methods of introducing the gas into bulk samples of these metals:

(a) cathodically charging, using a suitable electrolyte. In the case of mild steel, for example, this can be 10% sulphuric acid, with As_2O_3, carbon bisulphide, or yellow phosphorus added as a poison.
(b) by heating to a suitable temperature range in a dry hydrogen atmosphere, if necessary under increased pressure in an autoclave.

Smallman *et al.* (1953) bubbled steam through molten aluminium, and ascribed the small ageing yields which resulted to hydrogen locking. This interpretation is doubtful as these could be unloading yield points or due to some other impurity so introduced.

Concentrations of gas are usually expressed in ml/100 g of metal.

4.3 *Mild steel*

In brief, three theories have been proposed from time to time to explain hydrogen embrittlement of mild steel. These are

(a) *The pressure theory.* Here it is assumed that the hydrogen exists in the metal as molecular hydrogen under pressure, and concentrated in voids or formed in blisters at the surface. The pressure within these crack nuclei causes deformation and hence hardening at the crack tips, which then fail more readily under stress.

(b) *The adsorption theory.* The stress needed to propagate a crack can be given by the usual Griffiths formula $\sigma_f = \sqrt{(2E\gamma/\pi C)}$, where E is the modulus and C the crack length. Absorption of hydrogen on the surface of a crack will reduce the surface energy (γ) and hence make a previously static crack now grow.

(c) *The triaxial stress theory.* Here the hydrogen is assumed to diffuse to regions under triaxial tension, i.e. those regions ahead of a crack, and build up high hydrogen concentrations there. The lattice is here weakened and the crack can further propagate.

Each of these theories has its protagonists, and it is not within the scope of this work to examine them in any detail. However, it can be said most recent papers support the side of the pressure theory. In this monograph we are most concerned with hydrogen-induced yield points and strain-ageing effects, and any effects on the ductility will be of minor importance.

One of the general features of hydrogen embrittlement is that the loss of ductility is considerably more marked at slow strain rates. This fact, coupled with the generally low activation energy for diffusion of hydrogen in steel and other metals, has led to many considering that hydrogen is acting as does carbon and nitrogen in the blue-brittle range, i.e. decreasing the ductility by simultaneous straining and ageing.

Rogers (1954) first discovered an ageing effect which he ascribed to hydrogen. Strain-age testing an SAE 1020 steel at −150°C, he found that hydrogen charged specimens could exhibit a yield point, whereas uncharged strained specimens did not. In the following year, Cracknell and Petch (1955b) found that hydrogen charging eliminated the yield point extension at room temperature, but (rightly) suggested that this was due to the local stress from hydrogen in voids. (See also Schuetz and Robertson, 1957.) Nevertheless, in a later paper, Rogers (1956) confirmed this result, and showed that the yield point reappeared at low temperatures (Fig. 4.1). He considered that there were two possible explanations of this effect; either

(a) hydrogen is more tightly bound to dislocations than carbon or nitrogen. At room temperature the hydrogen moves with the dislocations, and the yield point is eliminated, while at lower temperatures the yield point can appear; or
(b) hydrogen is less tightly bound to dislocations than carbon or nitrogen, and would attach itself to loops freed by thermal stresses, thus lowering the energy and hence the stress to form a loop of critical size. Lowering the temperature raises the yield stress and the yield point effect thus reappears.

Rogers accepted the second alternative and in a further paper (1957) undertook a series of low-temperature tests on both Armco and SAE–1020 iron to verify it. However, Vaughan and de Morton (1956a, b, 1957) used the first alternative to explain the hydrogen embrittlement of their samples.

The critical experiment was to measure the activation energy for the process, and this was not done until 1959 by Rogers himself.

Using yield point **elongation** as a measuring parameter, he determined the activation energy as 23·6 kcal mole^{-1} (98·7 kJ mole^{-1}) almost identical with carbon or nitrogen diffusion within experimental error, and very much larger than the value, for hydrogen diffusion. Rogers therefore concluded that localized plastic deformation occurred during charging, and that this acted as stress nuclei which suppressed the yield point at room temperature.

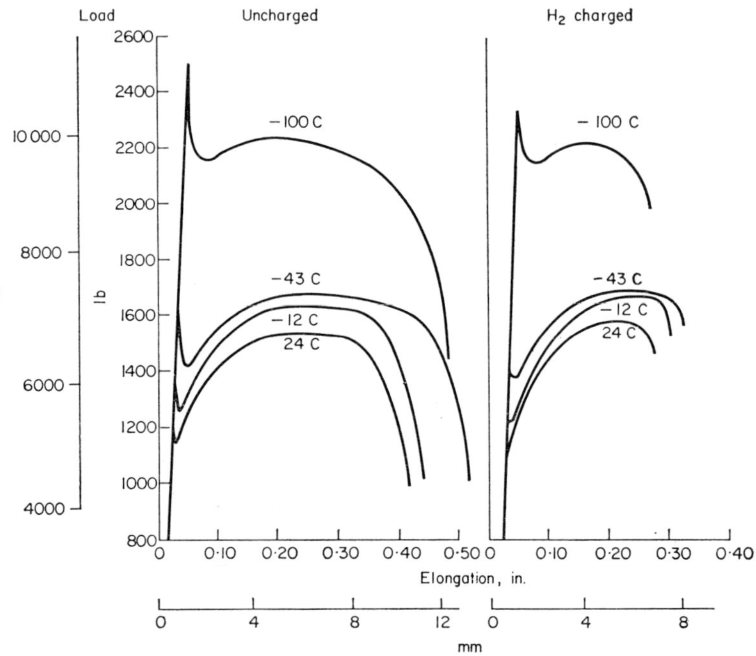

FIGURE 4.1. Yield point suppression in hydrogen-charged Armco iron (Rogers, 1956)

The Luders deformation has received some attention, and following the discussion above, de Kazinczy (1959b) examined the velocity of Luders bands in charged and uncharged samples. Using the Fisher and Rogers (1956) relation that $\log v_L = C - D(\sigma T)^{-1}$ (see Chapter 1) he managed to show that the Luders band velocity was unchanged in hydrogen charged specimens. However, the flow stress was slightly raised. de Kazinczy suggested this could be the result of drag from a hydrogen atmosphere, while the suppression of the yield point would result from pressure in voids (de Kazinczy, 1959a).

In work on Luders bands, it is important that the surface of the steel is not blistered during cathodic charging. Grahame (unpublished work) has shown that in steel strip where a single Luders band is expected, the pattern of Luders bands becomes complex when straining after cathodic charging (Plate 4.1). This is probably due to nucleation of new band fronts in or near blisters, or from pressure in voids.

One other small, but possibly important point remains for discussion, in that the initial slope of the homogeneous deformation curve (after the Luders extension) shows a significant rise in work-hardening rate in hydrogen charged material (see Fig. 4.1). Vaughan and de Morton (1956b) associate this with a restrictive influence of the hydrogen atmospheres on dislocation movement (cf. blue-brittle behaviour). Farrell (1965), however, points out that since the Luders extension is wholly or partly suppressed, the whole homogeneous deformation section of the curve is moved to the left which gives the appearance of a higher work-hardening rate.

The question of strain-ageing kinetics is still raised from time to time. Adair and Hook (1962, 1963) repeated Rogers' experiments and showed, in their view, that hydrogen does interact with dislocations and produce its own yield. However, as they point out in their later (1963) paper, an unloading yield point could well be present, which contributes to the effect, and makes any study of the kinetics difficult. There is still scope for careful controlled work in this field.

4.4 *Group Va and VIa Metals*

In the discussion of the effects of hydrogen on these two groups of metals, we notice another difference between them, over and above those brought out at the end of the previous chapter. Group Va metals are exothermic occluders, and capable of forming stable hydrides, whereas Group VIa metals are endothermic occluders and the possibility of hydride formation is very unlikely.

It is not therefore surprising that an extensive study has been done on the first group, and little on the second, and that a great deal of it is concerned with the effects of hydrogen on the ductile-brittle transition temperature, with comparatively little on the effects on yield point and strain ageing.

In many respects these metals should behave in a similar manner to iron charged with hydrogen, and show strain rate embrittlement as another manifestation of the transition temperature. An extensive investigation on vanadium, for example, is that by van Fossen *et al.* (1965), some of whose results on the ductility surface are shown in Fig. 4.2. The results are in many respects similar to mild steel, and the slow strain rate embrittlement shows up clearly, even with hydrogen levels as low as 6 p.p.m. No hydride forms at these levels, and van Fossen *et al.* found some difficulties in applying the 'void pressure' theory in this case.

Earlier, Eustice and Carlson (1961a) had noticed that 10 p.p.m. of hydrogen produced a minimum in ductility at −100°C, this being a function of strain rate. At the same time the yield stress showed a marked increase over the vacuum annealed yield stresses, as shown in Fig. 4.3. In contrast to mild steel, yield points were present in all cases, and the authors decided that the low-temperature embrittlement was the result of interaction of dislocations with a hydrogen Cottrell atmosphere. Eustice and Carlson (1964b) also showed the effect is confined to vanadium rich V–Nb alloys. However, van Fossen *et al.* (ibid.) objected to this as precipitate formation – no serrated

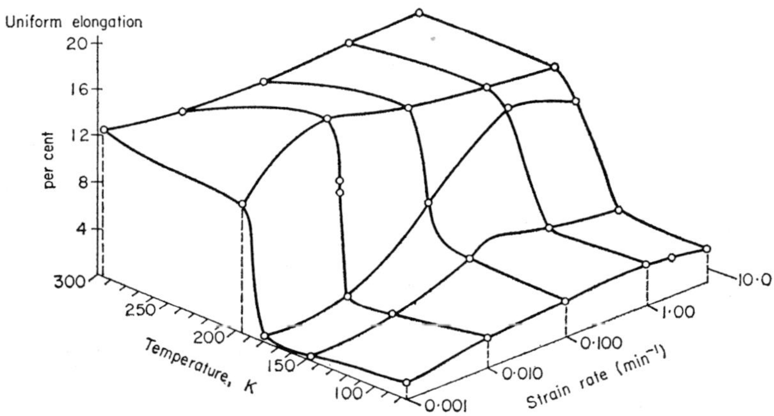

FIGURE 4.2. Ductility surface of hydrogen-charged vanadium (van Fossen *et al.*, 1965)

yielding resulting from hydrogen has been observed. It may be that there are many effects combining here, void and incipient hydride formation, as well as the possibility of strain ageing. Some activation energy results in this temperature range would be of immense value.

Fewer doubts can remain about the metal niobium, which has been subject to much more searching investigation. Even the structure of the hydride is known, which can form at high gas concentrations (Wainwright *et al.*, 1964), and which had earlier been inferred from metallographic observations by Paxton *et al.* (1959). At the lower hydrogen concentrations we are indebted to Wilcox and his co-workers who, in a series of papers (Wilcox and Huggins, 1960, 1961; Wilcox *et al.*, 1962) examined the possibility of hydrogen as an interstitial causing strain ageing.

Using hydrogen charged samples of niobium, activation energies for return of yield, and also for changes in dynamic modulus, were calculated for their samples. The activation energies all lay in the range 8·08–9·92 kcal mole^{-1} (33·8–41·5 kJ mole^{-1}) all very different from the activation energies for diffusion of C, N or O (Chapter 3). From the modulus data, the ageing results at short ageing times could be fitted to a $t^{2/3}$ law but some over-ageing was apparent and this was ascribed by Wilcox and Huggins (1960) to the possibility of precipitation on the dislocations. With 89 p.p.m. of H_2 present, Wilcox *et al.*

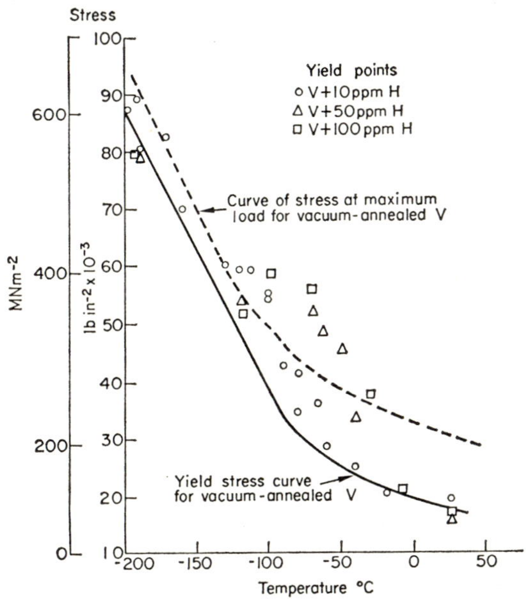

FIGURE 4.3. Effect of hydrogen on the yield stress of vanadium (Eustice and Carlson, 1961a)

(1962) found serrated yielding at 25°C (Fig. 4.4), a sign that strain ageing is present at much lower temperatures than for other interstitials. Westlake (1969) has shown that the tensile properties depend critically on the distribution of hydride.

The more commonly performed tests, i.e. a plot of ductility (or R. of A.) against temperature has been carried out on vacuum annealed and hydrogenated niobium by Wood and Daniels (1965). As in the case of vanadium, a minimum is found at about −100°C, and bearing in mind the evidence above, this can be interpreted as a hydrogen strain-ageing effect.

Similar results have been quoted in one of the few papers on tantalum (Ingram *et al.*, 1963). This, and later work by Pink (1968), shows clearly that hydrogen can induce strain ageing in all three metals.

The metals of Group VIa (Cr, Mo and W) are all endothermic occluders. Compared with iron and the Group Va metals, little information exists on the effects of hydrogen on the yield point, but for molybdenum, one paper is of some importance, that of Lawley *et al.* (1961b).

Molybdenum wires were charged with hydrogen by heating in the gas under one atmosphere pressure at temperatures up to 2550°C. This would imply, from solubility data, a hydrogen content of

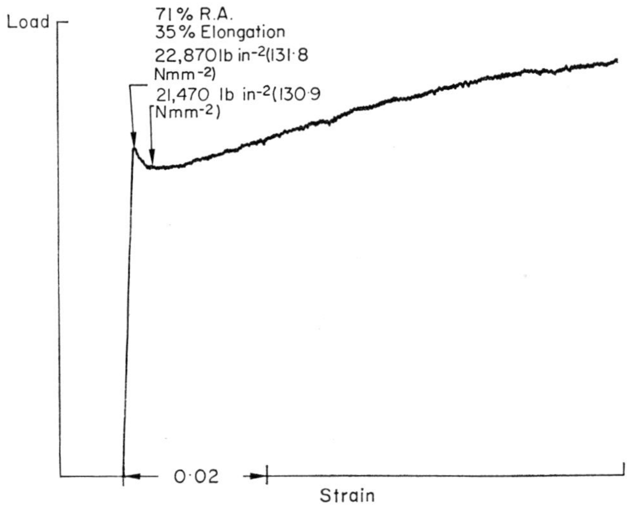

FIGURE 4.4. Serrated yielding in niobium containing 89 p.p.m. H_2, tested at 25°C (Wilcox *et al.*, 1962)

approximately 250 p.p.m. On quenching, no yield drop is present, although there is a notable increase in the yield stress from quench hardening. On ageing, a rounded yield point appears, and at increased ageing times the stress falls as the yield point disappears and the hydrogen is rejected by the lattice.

The activation energy was found from the time taken to reach maximum hardness, and gave 0·4 eV (9·25 kcal mole^{-1} or 38·7 kJ mole^{-1}) for the process. This is in only fair agreement with the activation energy for diffusion of hydrogen (0·63 eV).

Although Lawley *et al.* (ibid.) explain their results in terms of dislocation locking by hydrogen in much the same way as Rogers did for mild steel (Section 4.3), the critical and distinguishing value of the activation energy is, in this case, derived for what is almost certainly not a singly activated process. The relatively high concentration of point defects from the 2550°C quench might cause pinning, giving the rounded yield which is observed and which is slowly removed by pressure generating new dislocations at the hydrogen/gas nuclei. Further work on this interesting ageing phenomenon should include variation in quenching temperature as well as a closer examination of the over-ageing process.

4.5 *Nickel*

Nickel is normally classed as an endothermic occluder of hydrogen (Cotterill, 1961), and as such is not supposed to form hydrides, yet known chemisorption effects occur and from time to time papers appear on the crystal structure of nickel hydride (see, for example, Bonizewski and Smith, 1961). Hydrogen embrittlement of nickel (often called 'steam' embrittlement) has been known for many years – for a later paper see Blanchard and Troiano (1960) – but it was not until a series of papers by Smith and his co-workers appeared that it was realized the dissolved hydrogen could produce many intriguing interactions with dislocations (see also Macherauch and Vöhringer, 1963).

4.5.1 *Polycrystalline material*

Using cathodically charged samples of nickel, Boniszewski and Smith (1963) examined polycrystalline samples of nickel over a range of temperature. Care must be taken to avoid formation of surface hydride in this treatment, otherwise grain-boundary cracking will occur. At room temperature, a slight increase in yield stress and a marked increase in the rate of work hardening are apparent, as the temperature is lowered, the Portevin–le Chatelier effect is noted (Fig. 4.5), and this disappears at still lower temperatures. The analogy with the case of mild steel is apparent, and Boniszewski and Smith give the upper and lower critical temperature for the appearance of serration as a function of strain rate. These are listed in Table 4.1.

TABLE 4.1.

Strain rate (s^{-1})	Lower critical temperature (*K*)	Upper critical temperature (*K*)
$3{\cdot}33 \times 10^{-4}$	153	228
$8{\cdot}33 \times 10^{-4}$	157	243
$1{\cdot}67 \times 10^{-3}$	163	251
$2{\cdot}50 \times 10^{-3}$	—	256

Assuming, as in the case of mild steel, that these phenomena are thermally activated, then the activation energy for the process is calculated as 8·2 kcal mole^{-1} (34 kJ mole^{-1}). The binding energy of the hydrogen atoms to dislocations can also be calculated from Cottrell's (1956) equation:

$$T_0 = U/k \ln(1/C_0)$$

where T_0 is taken as the lower critical temperature at the lowest strain rate and C_0 is the average atomic concentration of solute. This gives $U = 0{\cdot}08$ eV, a figure confirmed by Nechay and Popov (1965).

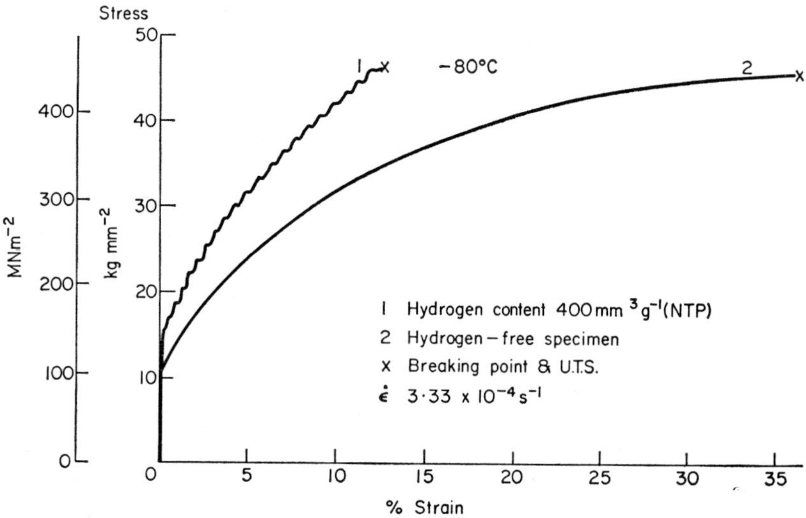

FIGURE 4.5. Stress–strain curves of hydrogen-free and hydrogen-charged nickel ($\dot{\epsilon} = 3{\cdot}3 \times 10^{-4}$ s^{-1}, $T = -80$°C) (Boniszewski and Smith, 1963)

Boniszewski and Smith also examined the kinetics of strain ageing in the usual manner, taking the time for the return of yield at various temperatures. This method gave a value of 9·3 kcal mole^{-1} (34 kJ mole^{-1}), and this and the figure quoted above, are well within the range of experimentally determined activation energies for the diffusivity of hydrogen in nickel.

It is also interesting to note that the curves of ductility against temperature show a marked minimum in the region of −50° to −100°C, where serrated yielding is present, showing that as in the case of niobium, the embrittlement is very similar to the mild steel situation and the result of straining and ageing. The ductility in this case can be recovered by vacuum annealing.

This work was later extended by Wilcox and Smith (1964) using hydrogenation by annealing in a dry hydrogen atmosphere; the method gives a lower hydrogen concentration but a more uniform one. Closer attention to the mode of yielding showed that the phenomena was indeed more similar to that of aluminium–magnesium alloys (Chapter 5), the yielding being comparable with Type B. However, like mild steel, the dislocation density was notably higher in samples which had undergone serrated yielding than in hydrogen free nickel (Plate 4.2). Wilcox and Smith interpreted their results in terms of dislocation multiplication and localized yielding, and Windle (1966) has noted Luders type markings on polycrystalline material.

4.5.2 *Single crystals*

Windle (1966) and Lantanision and Staehle (1968) have further extended studies on polycrystals given above to the case of single crystals of nickel containing hydrogen.

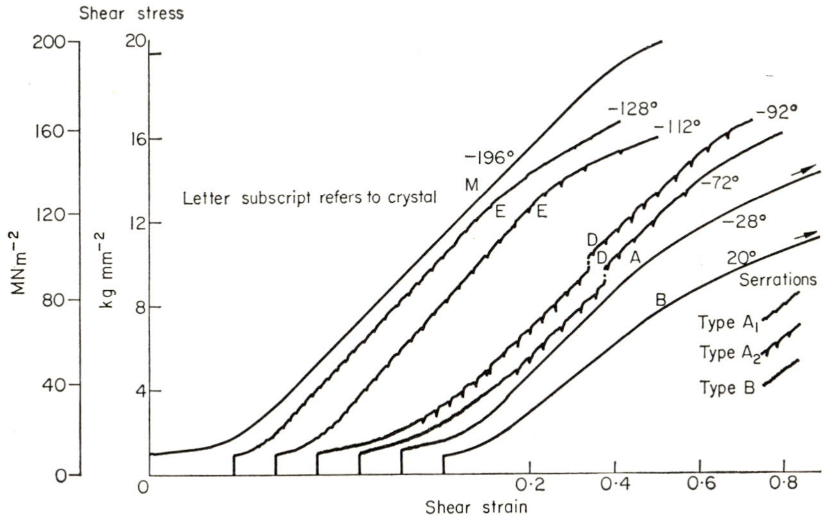

FIGURE 4.6. Serrated yielding in hydrogen-charged nickel single crystals. $\dot{\epsilon} = 2 \times 10^{-4}s^{-1}$ (Windle and Smith, 1968)

Most of the face-centred cubic crystals, under favourable conditions of purity, orientation (close to [110]), and temperature, illustrate the phenomena of three-stage hardening; namely an easy glide region (Stage I), a linear hardening region (Stage II), and a parabolic hardening region (Stage III). This is true of nickel, but in the presence of hydrogen the deformation is complicated by the appearance of serrations if tested in the correct temperature range, in this case 0 to

−130°C. (Fig. 4.6). Three types of serration were noted, and designated Types A1, A2 and B. Type B are noted during Stage I, whereas both A1 and A2 appear during Stages II and III. However, the presence of hydrogen does not affect the critical resolved shear stress in Stage I, but it does increase the rate of work hardening in Stage II. This is interpreted as a decrease of the stacking-fault energy which allows the easier formation of dislocation barriers to cross slip. In the easy glide region, prominent slip lines are formed which approximate to the usual single crystal Luders band.

It should be noted in passing that the fracture of hydrogen-charged nickel single crystals is completely ductile, whereas considerable grain-boundary intercrystalline failure occurs in polycrystalline material. Windle (ibid.) considers this the result of a hydride layer in the boundary, although Wilcox and Smith (1965) favoured the 'pressure-void' theory in microcracks which opened at grain boundaries.

4.5.3 *Nickel-base alloys*

An interesting extension of the case of polycrystalline nickel has been made by Blakemore and Hall (1968). By adding copper to nickel, the structure is unaffected, but the stacking-fault energy is rapidly

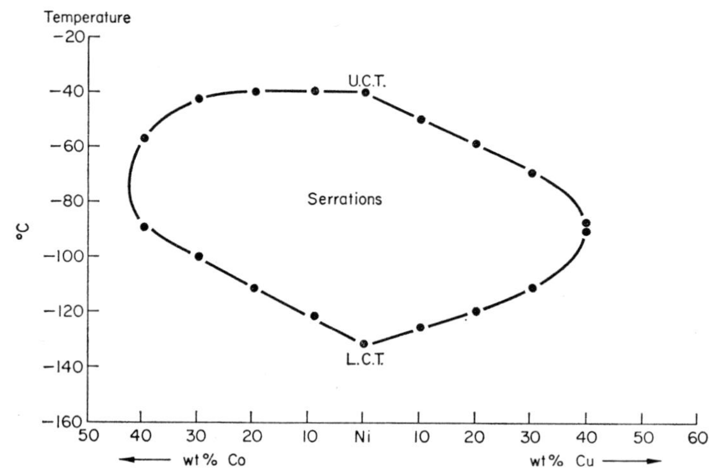

FIGURE 4.7. Region of serrated yielding in hydrogen-charged Ni-Cu and Ni–Co alloys ($\dot{\epsilon} = 1{\cdot}37 \times 10^{-4}\,s^{-1}$, $H_2 = 15\,cm^3/100\,g$) (Blakemore, 1968)

lowered (Harris *et al.*, 1966). This might lead to the possibility of Suzuki locking of the extended dislocations. However, the results (Fig. 4.7) show that the serrations die out after the addition of about 40% of copper. Since the hydrogen concentration has been kept constant, the structure and yield stress are virtually unaffected, the disappearance of the serrations can only be associated with the increasing difficulty of forming hydride, which must be the locking medium.

Similar results can be obtained in nickel–cobalt alloys, where again the stacking-fault energy falls rapidly with cobalt additions. The dislocation densities, higher in deformed nickel–hydrogen than in pure nickel, also even out as additions of alloying element are made.

The relative constancy of the upper critical temperature, as shown in Fig. 4.7, allowed the binding energy to be determined as 0·12 eV at 15 ml/100 g of hydrogen. This is slightly higher than determined by Bonizewski and Smith (ibid.) but here the Cottrell relation gives U + Q, where Q is the activation energy for diffusion, as mentioned in Section 1.5.1.

4.6 *Palladium*

Palladium is remarkable for its well-known affinity for hydrogen, for it can absorb up to 1000 times its own volume of the gas. This selective adsorption has been used in hydrogen alarm systems and hydrogen purification trains. There is, however, very little information on the effect of hydrogen on the physical properties, especially the mechanical properties of palladium. It is known that the hydride Pd_2H forms, with a similar structure and only a slightly larger lattice parameter than the metal, but only limited knowledge of the mode of nucleation of the hydride is available.

Virtually the only two papers on mechanical properties are those of Sugeno and Takagi (1964) and Takagi and Sugeno (1965). They found the yield stress (0·2% offset) was raised by hydrogen charging (cf. polycrystalline nickel) at room temperature, and that serrated yielding was noted at −196°C. However, serrated yielding was absent at −80°C. This implies palladium has a lower binding energy for hydrogen than has nickel, i.e. smaller than 0·07 eV, and bearing in mind that palladium has a lower stacking-fault energy than nickel, Takagi and Sugeno prefer an explanation based on Suzuki locking. A good deal more tensile testing and ageing kinetics remain to be done, however, before this result may be taken as final.

Some care will need to be exercised, since Hollister and Darling (1967) have shown that a pick-up of 10 p.p.m. of silicon from annealing crucibles can lead to a pronounced initial yield point.

4.7 *Titanium and zirconium*

These two metals, in contrast to palladium, have been extensively studied by metallography, for both are exothermic occluders and form stable hydrides. Hydrogen embrittlement, related to failure in or near the hydride platelets, is of considerable importance, particularly in zirconium or its alloys when used in nuclear reactors where the possibility of hydrogen ion adsorption is important (Coleman and Hardie, 1966a).

The metal–hydrogen relationships in these two metals are, however, further complicated by the presence of an allotropic phase transformation at high temperatures. The high-temperature (β) b.c.c. form can be retained by alloying, and this phase (in mixed α/β alloys) will have different properties again from the pure α. In fact, in α/β titanium the β phase is the preferred site for the hydrogen, with a partition ratio of about 5:1, not surprising perhaps since the close packed hexagonal lattice of the α is much less 'open' to receive the gas than the β.

However, the embrittlement is not of such importance here as the possibility of yield points and strain-ageing effects, and it is this aspect which has unfortunately received scant attention. As will be seen in Chapter 7, there is considerable doubt whether these two metals do show a true yield point, and many factors such as heat treatment and resultant grain size, as well as the purity level of the interstitials, play a part. Under these circumstances, it is doubtful if the role of hydrogen can be explicitly determined as yet. Westlake (1964) found yield points in certain single crystals of zirconium with high hydrogen and oxygen contents, but this was not systematically studied, although he discounts the possibility of work softening. Soo and Higgins (1968b) do not mention yield points in their single crystal studies. Weinstein (1965), on the other hand, found that very fine-grained samples of zirconium were needed to induce a yield point in the presence of hydrogen. It should be noted, however (see Chapter 7), that it is in these fine-grained samples that yield points are most commonly observed in zirconium, and Edmonds and Beevers (1968) state that the addition of hydrogen to fine-grained zirconium suppresses the yield point. In view of the inherent difficulties of obtaining comparable samples of these anisotropic materials after different heat treatments (see Chapter 7) it may be some little time before the separate roles of hydrogen and other interstitial elements can be delineated.

5

Aluminium and Its Alloys

5.1 *Introduction*

To leave for the moment the question of interstitial elements responsible for yield point effects, the next two chapters will consider the case of the possibility of locking by substitutional atoms.

The original theory of Cottrell and Bilby (1949), as was mentioned in Chapter 1, is applicable both to interstitial and substitutional solid solutions. However, in the case of carbon in iron, the dilatations resulting from one carbon atom were estimated from the lattice parameter of martensite, and corresponded (at the levels equivalent to one carbon atom/cell or per dislocation plane) to a carbon content of ~ 50 at.%. In the case of a substitutional atom, the dilatations will be much smaller varying from zero in the case of silver–aluminium to a maximum of about 20%. Thus, a higher concentration of substitutional atoms would be expected to give rise to an observable yield drop and/or strain-ageing effects in these metals.

In this chapter, aluminium and its alloys will be discussed, the major discussion centring on the aluminium–magnesium system, while in Chapter 6 other metals in the face-centred cubic system will be included. With the notable exception of the copper–tin system, many of these studies are fragmentary, but will at least serve to show that many problems of interest still remain.

But first it would be convenient to examine the small but measurable yield point which occurs on unloading and reloading **pure** face-centred cubic metals. This effect, called the unloading yield point, has been mentioned in Chapter 1, where it was compared with the yield point found in mild steel in very early stages of strain ageing; naturally, for pure metals, other theories must apply.

5.2 *The unloading yield point effect*

5.2.1 *The nature of the effect*

The unloading yield point was apparently first noted in silver by Edwards *et al.* (1943), but was not recognized as such. Subsequently, the effect was noted by Blewitt (1953) during the low-temperature deformation of copper single crystals, and in the following years a large number of papers appeared, commencing with a detailed study of aluminium and nickel by Haasen and Kelly (1957), followed more recently by Titchener and Davies (1965) on copper.

It is generally found that in face-centred cubic metals, a small yield point may be observed by applying a prior strain, unloading (or even merely interrupting the test) and then restraining. In many cases, the effect appears very rapidly, but in comparing the various results,

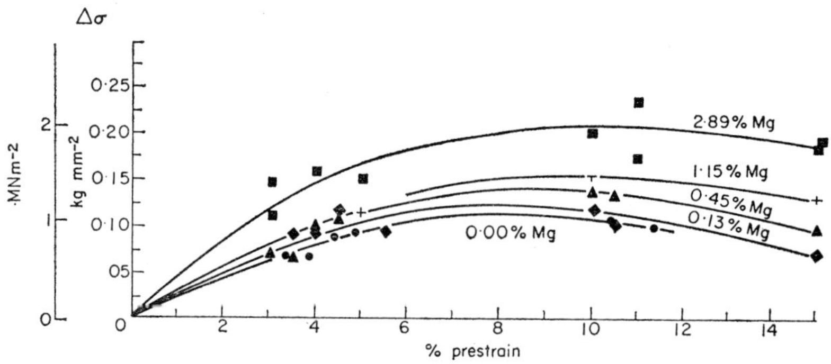

FIGURE 5.1. The magnitude of the unloading yield point (Δσ) in aluminium and its dilute alloys with magnesium at 90 K (Westwood and Broom, 1957a)

an absolute comparison is difficult, because of the wide variety of parameters used by the many investigators. These were in fact illustrated in Fig. 1.22(a) in Chapter 1, but it will be seen below that whatever the parameters employed, the magnitudes (in nominally pure metals) are all much smaller than the *f* factors which were used to describe the ageing of mild steel.

As a typical example of the effect, we will consider the early results of Westwood and Broom (1957a) for aluminium and some dilute aluminium–magnesium alloys (Fig. 5.1). There are three principal differences between this case and the cases of interstitial impurity locking described in earlier chapters:

(a) The effect exists in super-pure and even in zone-refined materials.

(b) The results increase to a maximum when plotted against strain and then die away. Similar results apply if $\Delta\sigma$ is plotted against stress.
(c) The magnitude of the effect is small. For example in Fig. 5.1 above, the maximum f value would be $\simeq 0{\cdot}02$, whereas in mild steel, values of $f > 0{\cdot}2$ are common.

General studies on the effect are made more difficult, however, by the presence of two other effects, recovery and work softening. Recovery, the slow decrease in yield stress from the thermal annihilation of dislocations, is readily detected, but its presence may complicate kinetic studies. The presence of work softening, as has already been indicated in Section 1.8.3, is obviously tied in with this effect, but will appear only at large strains and in cases where the temperature is raised during interrupted tests. It is, however, an effect to be considered in ageing studies.

The general features of the unloading yield point can now be discussed in terms of the parameters which may be used during an interrupted tensile test. We will consider only high purity metals at this stage.

(a) *Strain.* As illustrated in Fig. 5.1, the effect will increase with strain, and this is generally found in most metals. A difference does arise, however, in that single crystals do not appear to show a yield point unless extended beyond the easy glide region, as has been shown by Hauser (1961) in silver, Makin (1958) and Birnbaum (1961) in copper. The work of Thomas (1960) on aluminium single crystals gave similar results, while Haasen and Kelly (1957) did not specifically comment on this fact, although some of their crystals did have the correct orientation and the unloading yields are zero at very small strains. Birnbaum (ibid.) also noted differences resulting from orientation changes.

(b) *Stress.* Given a linear or parabolic relation between stress and strain, the effects will vary in much the same way as described in (a) above. The effects increase, in roughly a linear manner at early stages, and then flatten out. In certain temperature ranges, clear indications of a maximum are found (Titchener and Davies, 1965, Birnbaum, 1961). Plots of $\Delta\sigma$ against σ seem for some reason to be more popular than $\Delta\sigma$ against strain.

(c) *Existence of a Luders front.* There has been little research on this, but Makin (1958) has examined a repolished copper single crystal immediately after an unloading yield, and found a random distribution of fine slip bands with no indication of a single crystal Luders extension.

(d) *Effect of grain size.* Titchener and Davies (1965) examined the effect in polycrystalline copper of two grain sizes and found only marginal differences. Indeed, with single crystals the effects are comparable in magnitude with polycrystalline material, always excepting the results during the easy glide extension.

(e) *Degree of unloading.* The magnitude of the effect is enhanced as the degree of unloading is increased. Makin (1958) and Titchener and Davies (1965) both found a linear relation between $\Delta\sigma$ and the percentage of unloading, but whereas the effect was independent of temperature in Makin's copper single crystals, in polycrystalline copper Titchener and Davies found a marked temperature dependence.

(f) *Surface condition.* By etching the surface of pre-strained copper crystals, Birnbaum (1961) showed this unloading yield point was not a surface phenomenon. Feng and Kramer (1963, 1965) and Kramer and Kumar (1969) believe, however, the surface state is important. (See also Brydges, 1969.)

(g) *Combined stresses.* Birnbaum (ibid.) also examined the effects of pre-straining his single crystals in torsion (to activate another slip system) prior to straining in tension. The magnitude of the effect was markedly increased at low strains by this treatment.

(h) *Temperature.* Excluding for the moment ageing treatments at other than testing temperature, it has been found that the effect is more marked at lower temperatures (Titchener and Davies (1965) – polycrystalline copper, and Westwood and Broom (1957a) – polycrystalline aluminium). Bolling (1959) also noted an increased effect at lower temperatures. On the other hand, Makin (1958) noted little difference in his single crystals in tests carried out at 20°C and −195°C, but Thomas (1960), allowing for recovery effects, noted a more marked effect in his aluminium single crystals at liquid nitrogen temperature.

(i) *Ageing studies.* It can thus be seen that the effect shows further marked differences between the normal solute ageing effects, where the effects decrease with lowered temperature, and decrease with the degree of unloading (Chapter 2). It now remains to discuss the ageing kinetics before the various theories of the unloading yield point are treated.

All indications, as shown in the references listed above, are that the unloading yield point in high purity material appears in very short times. Yet there is still a possibility that at least part of the effect is time dependent, and much work has been done on time dependent effects. Broadly speaking, there are two groups of results, those which find short-term effects and those which do not.

In the case of polycrystalline copper aged at testing temperature (79 K), Titchener and Davies (1965) found no effect up to 100 minutes with 100% unloading, and then a slight increase. The opposite trend was observed at room temperature. On the other hand, Westwood and Broom found time dependent effects inside a few hundred seconds in straining aluminium. Makin (1958) aged samples at room temperatures, and tested at lower temperatures, and found that although the magnitude of the yield drop was increased by this ageing, the effects were independent of time. Haasen and Kelly (1957) had results which again were independent of time, and Thomas (1960) confirmed this with aluminium single crystals. But again Takamura and Miura (1962) found short-time effects in polycrystalline copper at 77 and 90 K while Birnbaum (1963a, b) and Birnbaum and Tuler (1961) also found time dependent effects in the same material. In particular, Birnbaum (1962) found indications of two peaks in his samples which were tested at 77 K and aged at 338 K.

5.2.2 *Theories of the unloading yield point*

The characteristics of this effect have now been outlined, and consist of a yield drop which appears rapidly, given correct temperature, orientation and purity conditions, and then may or may not show a time dependent effect on ageing. The presence or absence of the latter effect neatly subdivides the two main groups of theories:

(a) That the effect is due to the locking by point defects generated during the prior deformation;

(b) That the effect is due to an interaction between dislocations.

Those who find time dependent effects will not surprisingly tend to theories of the first type. It may be easily seen that the creation of vacancies (and possibly interstitials) by deformation will lead to a source of pinning points for the dislocations, for a vacancy will be attracted to a dislocation in just the same way as a solute atom, and the concentration of point defects around the dislocation will vary as $(\text{time})^{2/3}$. The elementary Cottrell–Bilby theory was thus applied to the case of aluminium by Westwood and Broom (1957a) with satisfactory results and an activation energy of 0·1 eV.

Similar results were obtained by Birnbaum and Tuler (1961) in copper; again the effect was attributed to interstitials. Overageing was put down to annihilation of point defects. The later discovery of a second ageing peak (Birnbaum, 1963a, b) was placed, by activation energy determinations, as a result of the diffusion of divacancies to the dislocations. On the other hand, it should be mentioned here that Takamura and Miura (1962), while finding a time dependent change, could not make a $t^{2/3}$ time law fit. Nevertheless, from activation energy determinations, the effect was considered as interstitial pinning.

The other group of time independent theories must be judged against this background. These should explain firstly the immediate increase in yield stress on reloading, and also the correct variation with stress, strain temperature and degree of unloading.

One of the simplest theories is to assume that on unloading, there is a rearrangement of the dislocations. These can move in slipped regions which are surrounded by barriers such as the Lomer–Cottrell type, or other dislocation tangles. During unloading, new barriers must form, or there would be a marked shape change in the specimen. These new barriers will present an obstacle to further slip on reloading, but on increasing the stress, these new barriers will dissociate and a yield point be observed. These Lomer–Cottrell barriers do not form during the easy glide region, and no yield point is found there; furthermore, since the dissociation will be assisted by thermal energy, the unloading yield point should be more evident the lower the temperature, as indeed it is.

This theory, used by Haasen and Kelly (1957), Makin (1958) and Hauser (1961) has some difficulties in predicting the variation of $\Delta\sigma$ with stress, and other fine details in the results. An attempt to rationalize the results has been carried out by Titchener and Davies (1965). They separate the effect into two: an initial yield, and a slow time-dependent part whose magnitude is dependent also on degree of unloading. They consider it possible that point defects are responsible for both. They are able to examine the maximum in the $\Delta\sigma$ curves, and suggest that the dislocations, in running back on unloading, sweep up point defects. The slower, time dependent, part is due to diffusion of the defects to the immobilized dislocations. There, for the moment, the debate rests.

5.2.3 *The effects of alloying*

Several investigators have attempted to cover the boundary between the unloading yield, and true strain ageing effects by deliberate alloying additions to the pure metal base. In particular, amongst those investigators mentioned above, Westwood and Broom (1957b) added up to 2·9% Mg to their aluminium, Thomas (1960) up to 0·4% Zn, while Bolling (1959), as part of a study on α-brass, did report on the unloading yield in several other metals and alloys.

In general, the effect of the alloy addition is to enhance any time-dependent effects (Westwood and Broom, ibid.) or to make the effect apparent over a wider temperature range (Bolling, ibid.), or again to produce a yield under stress conditions where none appeared before (Thomas, ibid.). In the case of aluminium–zinc alloy single crystals, and despite the presence of recovery, Thomas was able to show, from the shape of the stress–strain curves, that true strain-ageing yield points did appear, but only in alloy crystals at room temperature and then only at low strains. These fell to zero as strain increased, and at higher strains a more rounded yield point with all the characteristics of the unloading yield point was present.

An initial survey of the two effects in a much less dilute system has been carried out by Bolling (1959) on 70/30 α-brass. In contrast to copper, a marked time dependent increase in the yield was found at room temperature and above. Figure 5.2 shows the set of results at 25°C, showing that the magnitude is dependent also on strain, with clear signs of a maximum. These results could be analysed in terms of a $t^{2/3}$ plot; at higher temperatures serrated yielding was observed, and at lower temperatures, less than −45°C, the effect had largely disappeared. These are all the characteristics of true strain-ageing; the activation energy was determined as 0·4 eV. The rapidity of the ageing, however, suggested that diffusion was speeded up by vacancies formed during deformation (see Section 6.2.11), the vacancies increasing as

$$C_{\mathrm{v}} = B\epsilon^{m'}$$

where the exponent m' in brass appears as 1·36. The yields were also enhanced by ageing under stress.

At temperatures lower than −45°C, the unloading yield point is found to appear again; thus, in the case of this alloy, temperature, rather than strain, could differentiate between the two effects. It is possible to devise experiments where the two effects interacted; but with the tendency for the unloading yield effect to be both small and more important at low temperatures, chances of confusion are slight, and it was for that reason that unloading yields have been discussed separately in this part of the monograph.

5.3 *'Commercially pure' aluminium*

From the data given in the previous section, it can be seen that superpure aluminium (greater than 99·99% Al) is not expected to show a normal yield point or strain-ageing effects. Two cases of pseudo-yields are of interest: the first result in that by Bush and Huggins (1964) who found yield points in testing at high temperatures (between 450 and 650°C). It was found that these were the results of prior strain, either accidental or deliberate, so that the effect is merely one of work softening (see Section 1.8.3). The result is of some interest here, however, in the sense that the normal Cottrell–Lomer

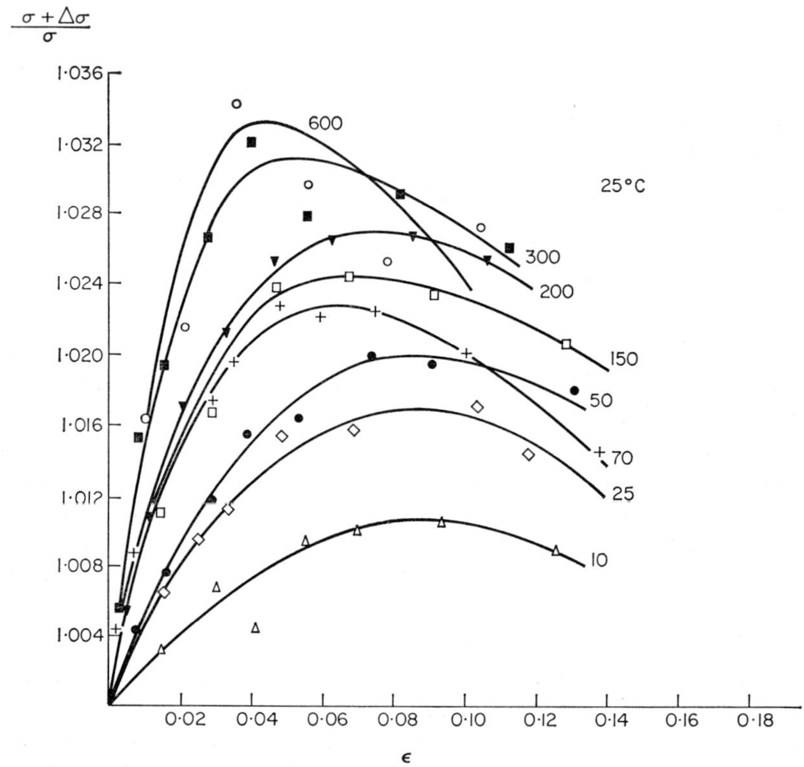

FIGURE. 5.2. Variation of $\Delta\sigma$ with strain for 70/30 brass at 25°C. Ageing times are in seconds (Bolling, 1959)

barriers were not expected to be stable at these high temperatures. Alternative explanations, in terms of modified jog distributions along dislocations, or a change in tangle network size, were put forward.

At the other end of the temperature scale, Basinski (1957) noted serrations in the stress–strain curve of super-pure aluminium at liquid helium temperatures. The results have been extended by Hosford *et al.* (1960) who, using crystals of selected orientations, found that crystals in [111] orientations exhibited irregularities when tested at 4 K, deformation appearing as closely bunched slip. This is in accord with Bazinski's (ibid.) suggestion that the material is thermally unstable at these temperatures, and local heating leads to continuing localized deformation. The effect disappears above about 12 K (see also Section 1.8.5).

Finally, the effects of electron irradiation at 80 K on the mechanical properties of super pure aluminium have been studied by Ono *et al.* (1964) and Ono and Meshii (1965). In contrast to other face-centred cubic metals (see Chapter 6), the comparatively low doses used here did produce an initial yield, but on studying the unloading yield, a marked yield point was seen, which was not observed in the unirradiated sample. In the second paper, a time-dependent raising of the whole of the stress–strain curve occurred on reloading. This was put down to pinning of dislocations by diffusion of interstitials to them.

In the less pure grades of aluminium, such as the 2S, 1100 or 1200 grades to give them their commercial designations, one might expect the impurities which can amount to a total of 1% to play a more interesting role.

The possibility of serrated yielding in air-cooled 2S aluminium was noted by McReynolds (1949) and one of his curves is shown in Fig. 5.3. It should be stressed here that McReynolds was using a dead-weight loading machine, which thus comes into the 'soft' category (Section 1.2) and his curves are therefore apparently unlike the serrated curves of mild steel, and the other curves to be displayed later in this chapter. However, the effects are undoubtedly of the same origin, and imply, in these machines, that a wave of plastic deformation, i.e. a Luders band of small ϵ_L, is moving very rapidly along the specimen at some cm s^{-1}, dependent on the rate of loading which, in this case, will determine the strain rate.

The curve also shows the characteristics of many of these serrated stress–strain curves from face-centred cubic alloys. The curve starts off smoothly, and after some pre-strain, the serrations (in this case steps) begin and grow in size as deformation proceeds. McReynolds also found that the serrations existed only over the range $-10°$ to $+50°$C. This, as stated in Chapter 1 and in Section 5.5.1 below, is probably the result of a diffusion rate enhanced by vacancies produced during the deformation.

The major impurities in the 2S aluminium observed above were Cu 0·14%, Fe 0·48%, and Si 0·11%. McReynolds also made up some pure binary aluminium–copper alloys, containing up to 0·5% Cu and found that although serrations were still present, they were sharper and appeared over the temperature range +10 to +120°C for the 0·1% Cu alloy. This might imply a different, higher binding energy in this case, and it cannot be said that the impurity responsible has yet been identified. Grain size should also be carefully controlled in making these comparisons.

McReynold's work was followed with confirmatory papers by Lubahn (1949, 1952) using 61–S aluminium sheet. Typical analyses of this material give 0·25% Cu, 0·6% Si, 1·0% Mg and 0·25% Cr, and

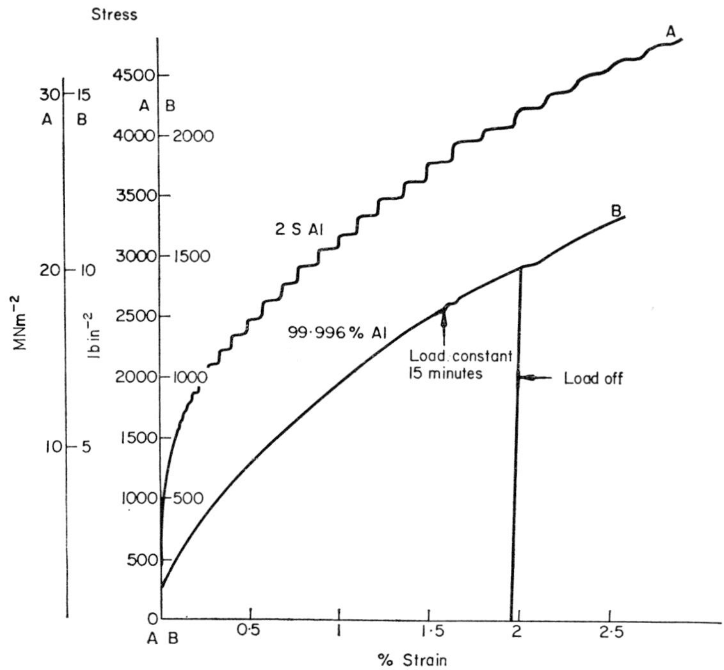

FIGURE 5.3. Stress–strain curves for high-purity and 2S aluminium. (McReynolds, 1949)

the serrations in this material are a function of prior ageing treatment (see Section 5.5). Lubahn again used a soft machine and some of the transient yield points he noted on changing strain rate could have been pseudo-yields as discussed by Bolling *et al.* (1961) – see Section 1.8.4.

It is strange that most of the work done on impure aluminium has been with 'soft' tensile machines. Later work by Bell and Stein (1962) using compression testing, Dillon (1963) using torsion, and Sharpe (1966) using tension, all come in this category, and naturally it is not a convenient method for studying ageing, either during, or separately from the tensile test. No activation energies have been determined, and no attempt, other than McReynold's, made to tie down the element responsible.

Practically the only study using a hard machine is that of Smallman *et al.* (1953), this time with single crystals of pure, impure, and dilute alloy of aluminium. The major impurities in the commercial aluminium were 0·14% Fe, and 0·11% Si, and these showed well-developed yields in tests at −196°C after straining and ageing at room temperature. So also did alloys containing 0·21% Cu, and 2·49% Zn prepared from high purity material. Kinetics were not, unfortunately, studied. Carreker and Hibbard (1957) noted some initial yield point anomalies in a 99·975% Al, tested at −196°C, provided the samples were annealed below 300°C. Again, the possibility of ageing was not explored.

Considering the commercial importance of 2S and other similar grades of aluminium, it is indeed surprising that no full scale investigation of the ageing behaviour has apparently been carried out. A detailed survey of the common impurity elements, copper, iron, and silicon, and their effect separately on the yield behaviour would be of immense value.

5.4 *Aluminium-copper alloys*

5.4.1 *Polycrystalline material*

The addition of little more than 1% impurity to aluminium will, in many cases, take us out of the range of solid solution effects, into the precipitation hardening field. One of the alloys which has received a great deal of metallographic and X-ray attention is the duralumin series, based on an alloy of 4% copper in aluminium.

It was duralumin which was one of the two alloys—the other being an aluminium–magnesium–silicon alloy—examined by Portevin and le Chatelier (1923) who found the serrated stress–strain curve typical of simultaneous straining and ageing, and their names are rightly given to this effect. They also noted the appearance of Luders bands in their specimens.

Other early work on the duralumin alloys has been carried out by Fell (1937), Elam (1938) and Stang *et al.* (1946). With the exception of Elam's work (ibid.) on RR72, where tensile tests were carried out at this date, soft machines were used, so that instead of obtaining a serrated stress–strain curve, the curves appeared stepped, similar to those shown in Fig. 5.3.

Nevertheless, certain features of the stress–strain curve are apparent from this work, which can readily be confirmed from hard machines:

(a) The initial yield is smooth and rounded, and steps (or serrations) are only seen after a clear plastic strain;
and
(b) the length of the steps, and hence the magnitude of the serrations, increases with plastic strain.

The Luders bands which are commonly observed on these duralumin specimens consist of fine parallel narrow bands which propagate rapidly through the specimen. Fell (ibid.) measured the angle these markings made on the front face of a 1″ × 0·19″ specimen, and found values of 56–59°, rather similar to mild steel. These bands were indistinct in the heat-treated condition, and more distinct in the annealed and unaged state.

The geometry of the bands in several aluminium–copper alloys has been studied more recently by Hooper (1952). He found that 'random' markings could be developed in an aluminium–copper–magnesium alloy (HS14) (4·07% Cu, 0·76% Mg, 0·56% Mn, 0·46% Si, 0·44% Si) which had been solution treated and quenched in oil. An example of these random or 'flamboyant' markings is seen in Plate 5.1. The strain range gives $\epsilon_L \sim 0{\cdot}5\%$, which is in fact much smaller than those observed in aluminium–magnesium alloys (up to 2%) and would explain why they are only observed with difficulty. Beyond the Luders strain, these random markings are replaced by parallel bands, as shown in Plate 5.1(b). The random markings can be reduced in intensity, or removed entirely, by a precipitation heat treatment, leaving only parallel bands.

We are now in a position to compare more closely the types of Luders bands in aluminium alloys with those in mild steel. The random markings may be compared with the initial Luders extension in mild steel, while the parallel bands correspond to the secondary bands in the blue brittle case of mild steel (Chapter 2). It is likely, anticipating results on aluminium–magnesium alloys, that the random markings are complex Luders bands, caused by excessive stiffness in the sample, and otherwise would be a simple primary band; moreover, if a hard machine had been used, the primary band would have been serrated. We thus expect a close analogy between the stress–strain curves in mild steel at 200°C (Fig. 1.3(c)) and the curves for duralumin. Following the nomenclature of Eborall *et al.* (1952) we will refer to the primary and secondary bands in aluminium alloys as Type A and Type B yielding respectively.

It is possible to extend the analogy further. Mild steel is a precipitation hardening alloy of iron (and carbon or nitrogen), and a 'solution treated' sample of mild steel shows no yield point effect. Type A yielding in duralumin is also suppressed by solution treating and quenching. It will also not appear if the grain size is too large, but the analogy cannot be pressed too far, for it has already been mentioned that some plastic deformation often occurs in duralumin before serrations commence. Secondly, the generally larger grain size mean smaller Luders extensions and faster Luders band velocities than in mild steel, and the lower elastic modulus will give different critical specimen sizes for simple primary bands to be present rather than complex Luders bands.

Finally, since Type B yielding can often be present without Type A, again in contrast to fine-grained mild steel, it might be expected that locking conditions are different for both Type A and Type B yields.

It now remains to define the conditions under which these two types of yielding occur, using the papers mentioned above, and two later ones by Krupnik and Ford (1952) and Phillips (1953). All these unfortunately used soft machines, but the characteristics are such that the yield types may readily be distinguished.

Krupnik and Ford (ibid.) used a commercial duralumin (the analysis was given as 4·00 Cu, 0·55 Mg, 0·60% Mn, and traces of Si and Fe) and two higher purity alloys containing 1·35 and 3·83% Cu. No Type A yield was found whatever the heat treatment, and the Type B yield tended to disappear in heavily aged or over aged material. High rates of loading (broadly equivalent to high strain rates in a hard machine) showed that the Type B steps (or serrations) were still present, but that they tended to be smaller at early stages of deformation. All these tests were at room temperature, and the results seemed independent of the copper content.

Broadly similar results were obtained by Phillips (ibid.) using a duralumin-type alloy analysing 4·78 Cu, 0·43 Mg, 0·73 Mn, 0·73 Si, and 0·35% Fe. Type A yielding was difficult to detect, despite the deliberate grain refinement, which yielded grains of 0·024 – 0·022 mm, compared with Hooper's 0·015 – 0·02 mm. However, there is a suggestion of a Type A yield in the annealed samples. Ageing for seven hours at 180°C virtually removed all Type B, and lowering the temperature again caused the steps (serrations) to disappear at a little below −60°C. Static ageing studies were also carried out, usually not under stress, which showed as expected that the serrations are tied with the capacity to age, and secondly, that serrations could be made to appear at lower temperatures than normal (say −76°C) by strain ageing at room temperature.

Phillips rightly points out that analysis of the ageing kinetics is difficult because normal age-hardening (which is not necessarily tied with dislocation locking) is proceeding together with true strain ageing. In this respect, the difficulty of determining ageing kinetics in these alloys is pronounced, because of the difficulties of separating the two. Both are diffusion controlled, and in contrast to mild steel, there may not be sufficient solute available after precipitation from age hardening (quench ageing) is complete, as higher levels of solute are required in these substitutional systems. In other words, if analogies are again sought in mild steel, one is dealing with problems of strain-ageing in a partially quench-aged sample. This, as will be appreciated from Chapter 2, is a potentially complicated study.

5.4.2 *Single crystals*

Since the presence of Type A yielding would appear to be strongly grain size dependent, it might be thought that the use of single crystals would help to elucidate some of the difficulties presented in the previous section.

Unfortunately, this does not appear to be the case. Examination of single crystals of aluminium – 3·5 wt% copper were carried out by Carlsen and Honeycombe (1954) and of a 4·5 wt% copper alloy by Greetham and Honeycombe (1960). Neither in the air-cooled, fully aged, nor overaged conditions could any signs of serrated yielding be found. Differences in rate of strain hardening were noted, and static strain ageing noted in an interrupted tensile test at room temperature. In the later paper, Greetham and Honeycombe extended the temperature range of testing at the fixed points of 77 K, 293 K, and 373 K, and while again evidence of strain ageing and some initial yield points were found in static tests, no serrated yielding was observed.

Byrne *et al.* (1961) likewise examined single crystals of a 1·7 at.% (3·9 wt.%) alloy in various heat-treated conditions, and extended the temperature range down to 4 K. Ageing studies were not carried out, and again no serrations were noticed. Analysis of the temperature variations of the yield stress were carried out in terms of thermal activation theory.

The absence of any serrated yielding is also noted from the work of Price and Kelly (1964a, b), Dew-Hughes and Robertson (1960) and Kelly and Chiou (1958). The absence continues to puzzle; it cannot be due to failure to scan the probable temperature range in an adequate manner; it may be that there is some impurity missing, which cannot in fact be identified without a proper kinetic study; or, finally, that grain boundary effects are paramount in determining the presence or absence of serrated yielding.

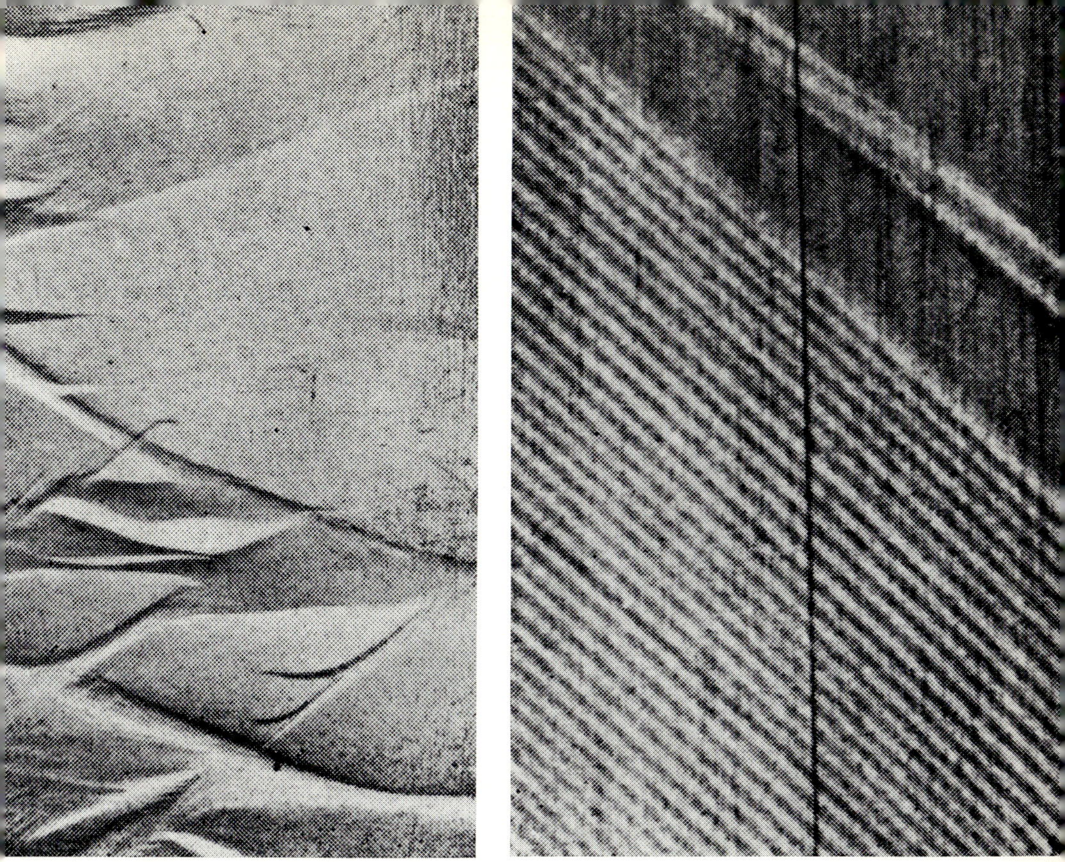

PLATE 5.1 (a) (*left*) Type A and (b) (*right*) Type B markings in Al–Cu alloy (HS14). (In (b) the vertical line represents a Talysurf trace.) (Hooper, 1952)

PLATE 5.2 (a) (*below*) Type A and (b) (*bottom*) Type B stretcher strains in annealed Al–3·5% Mg alloy strip. (Phillips *et al.*, 1952)

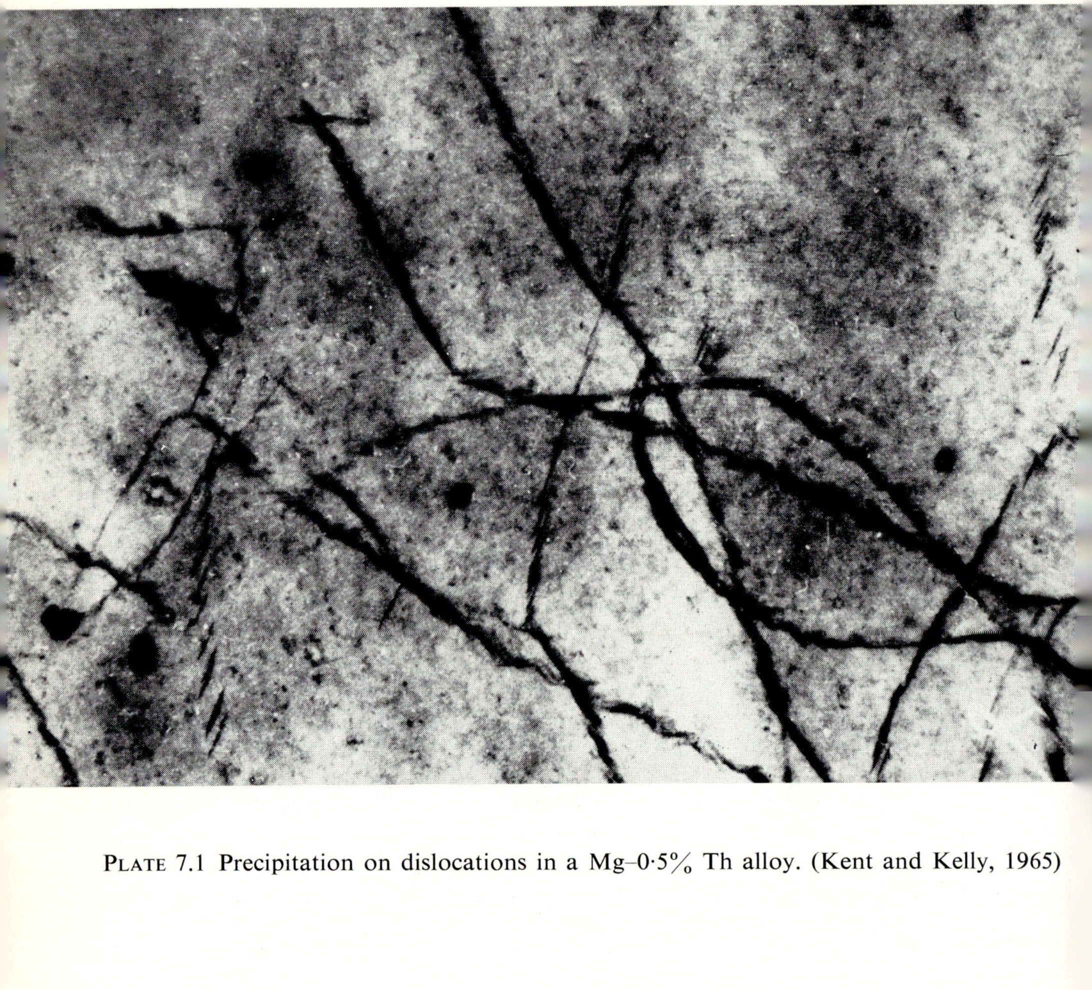

PLATE 7.1 Precipitation on dislocations in a Mg–0·5% Th alloy. (Kent and Kelly, 1965)

5.5 *Aluminium–magnesium alloys*

The second of the two alloys discovered by Portevin and le Chatelier (1923) to show serrated yielding was in the aluminium–magnesium–silicon system. Knight and Murray (1946) were able to find evidence of Luders bands in aluminium–magnesium alloys, and in recent years there has been considerable interest in the mechanical properties of these alloys, mainly because of the clear Type A and B yields, and ensuing strain markings which these alloys show.

The appearance of the Luders bands themselves was carefully examined by Chadwick and Hooper (1951). Type A yielding, with its 'flamboyant' markings, was found only in material of fine grain size, less than 0·05 mm. In the grain size range greater than 0·05 mm but less than 0·15 mm, no Type A markings were observed, while above this size an 'orange peel' effect was found, caused presumably by the grains deforming independently. Type B yielding, showing parallel bands as in duralumin, would appear at larger deformations in all but the largest grained material.

In the following year, Phillips *et al.* (1952) carried out an extensive series of mechanical tests, using a soft machine, and Krupnik and Ford (1952) performed a more limited investigation on an as-extruded 3·5% magnesium alloy. Despite the limitations of a soft machine, Phillips *et al.* investigated two high purity alloys, containing 3·06 and 3·65% Mg, and three commercial alloys.

The distinction between the two types of yield is seen in their results illustrated in Fig. 5.4. It can readily be seen that the Type A yield persists at temperatures where the smaller (Type B) steps have disappeared (cf. the case of mild steel). Photographs of the two types of yielding are shown in Plate 5.2(a) and (b); and at this point it is worth noting that Grimes (1967) claims the angle made by the Type A fronts is, *inter alia*, a function of the preferred orientation in the samples. This type of machine used by Phillips *et al.* will not show whether serrations are appearing on the primary Luders band or not (see Chapter 2). At elevated temperatures, one would expect Types A and B yields to disappear together, but this test under these grain size and heat-treatment conditions has not apparently been done. As can be seen from Fig. 5.4, the Luders strain is approximately constant, at about 1·2%, but this should be treated with caution, as the material is yielding at the upper yield or breakaway stress, and this is not a good indication of the true mechanical properties of the material.

The existence of Type A yield is also strongly dependent on grain size, as might be expected, and the yield stress results of Phillips *et al.*

(ibid.) have been shown by Armstrong *et al.* (1962) to obey the $d^{-1/2}$ law (Table 1.2). The Type B yields were much less sensitive, except when the grain size was very large.

The effect of heat treatment on the presence of Type A yielding was also carried out by Phillips *et al.* (ibid.). As in mild steel, material water quenched from 550°C did not show this type of yield. The temperature, however, must be chosen carefully to take sufficient

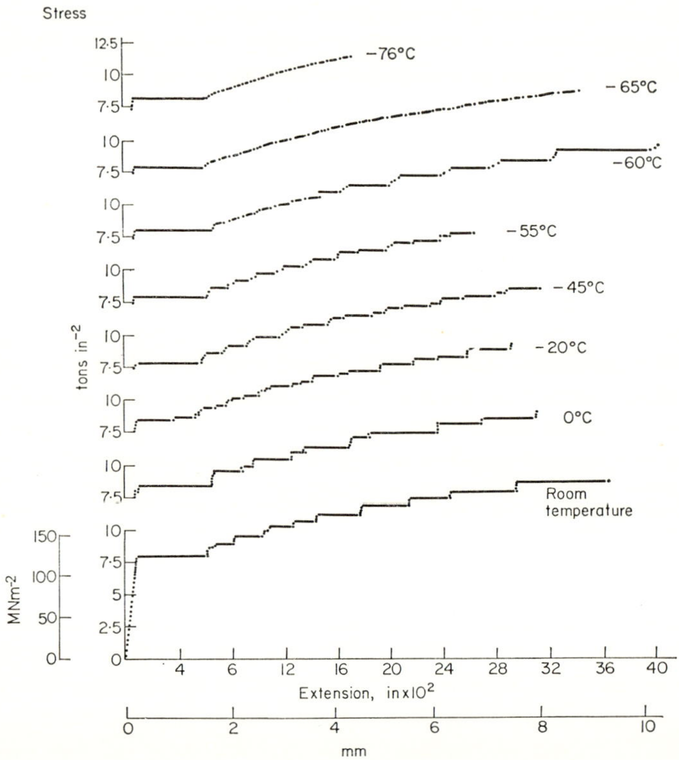

FIGURE 5.4. Stress–strain curves of a commercial Al–3·5% Mg alloy at sub-zero temperatures (Phillips *et al.*, 1952)

solute into solution, and yet not cause excessive grain growth. Type A yielding was found to reappear on ageing at 100°C, and taking the yield point extension as a parameter, not perhaps the best that could be chosen, a value for the activation energy was found as 28·5 kcal mole^{-1}, a figure which, despite its uncertainties, agrees fairly well with diffusion data for magnesium in aluminium. The reappearance of Type A yielding is thus a comparatively slow process, and the strain ageing to produce Type B yielding could be studied in this time interval.

The value of the activation energy for Type B has been determined by various investigators, with rather variable results. Sherby *et al.* (1951) found no single-valued constant, answers varying from 8–15 kcal mole^{-1} (34 to 63 kJ mole^{-1}) dependent on the arbitrary value of the yield drop chosen. Westwood and Broom (1957a, b) using water-quenched 3% Mg alloys, followed by pre-straining various amounts in liquid nitrogen, again found a marked scatter in their results, values ranging from 0·27 eV to 0·40 eV (6·2 to 9·2 kcal mole^{-1} or 26 to 39 kJ mole^{-1}). On the other hand, a 1% Mg alloy gave a more constant value of 0·38 eV (8·8 kcal mole^{-1} or 37 kJ mole^{-1}).

As will be indicated in the last section of this chapter, ageing results, to be meaningful, must be made on material where the conditions can be characterized. A variant on the interrupted stress–strain curve as a means of providing suitable parameters for activation energy determinations is to determine the critical strain ϵ_c at which the first serration appears. Again, the specimen condition at the testing temperature must be characterized carefully. Thomas (1966) has carried out a useful study on a commercial aluminium–magnesium alloy which shows this difficulty. If Cottrell's theory of solute atom locking by enhanced diffusion is correct, then one would expect (Section 1.7.4) that for serrated yielding

$$\dot{\epsilon} \simeq 10^9 D$$

and that as $\dot{\epsilon}$ increases, ϵ_c should also increase. According to Thomas, the reverse is the case. The graph of $\ln \epsilon_c$ against $1/T$ is likewise complex, and an activation energy can only be taken with confidence from the low temperature results, giving a value of 0·35 eV. The critical stress to the first serration, however, does give a $d^{-1/2}$ relation with $\sigma_0 = 140$ MN m^{-2} and $k = 0{\cdot}40$ MN m$^{-3/2}$.

Another interesting attempt to characterize the specimen condition more accurately has been made by Abbott (1964). Here the specimens were cold-drawn a small amount before ageing and, finally, tensile testing. By measuring ϵ_c as a function of ageing time and temperature, Abbott deduced an activation energy of 0·23 eV.

As in the duralumin case, single crystal studies have thrown little light on the mechanism of deformation of polycrystalline material.

Delayed yielding in these alloys has been studied by Shepherd and Dorn (1956). Because of the smooth initial yield, the material was pre-strained at 78 K and aged at 273 K, then loaded to stress levels below the known upper yield stress. A (Type B) yield front was found to propagate some minutes after the application of the load. Since this is many orders of magnitude longer than the delayed yield of mild steel under dynamic conditions, it is not surprising that the Fisher analysis (Chapter 1) does not apply. Assuming that the unlocking of dislocations was thermally activated, Shephard and Dorn showed that empirically

$$t_d = C \exp\left(\frac{E}{kT}\right) \exp-\left(\frac{\sigma}{\sigma_c}\right)$$

where σ_c and C are constants, and $E = 6{\cdot}8$ kcal mole^{-1} (29 kJ mole^{-1}).

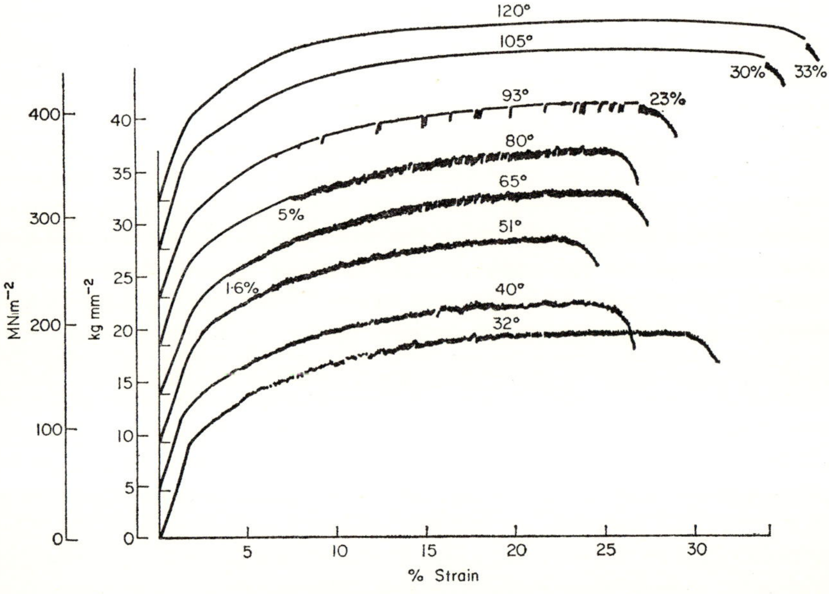

FIGURE 5.5. Stress–strain curves of an aluminium–magnesium alloy at elevated temperatures (Caisso, 1959)

The results of Abbott and of Thomas mentioned above are two of a very few studies on these alloys using hard machines. Two other extensive papers on this system exist, however, in the work of Caisso (1959), and Caisso and Guillot (1962). Although the grain size was not apparently sufficiently fine to reveal Type A yielding, they do show the disappearance of Type B serrations at temperatures around 100°C, dependent on strain rate (Fig. 5·5–cf. Fig. 5.4). The critical strain to first serration has also been measured, and unlike Thomas (ibid.), Caisso finds the value of ϵ_c to rise with increasing strain rate at room temperature. At higher temperatures, above about 50°C, the situation is reversed. This point will be discussed again in a later section. Caisso also determined the activation energy from these curves, and gives values varying from 0·25 eV at low temperatures to 0·45 eV near the upper temperature limit, using a relationship derived by Friedel (1964)

$$\epsilon_c = AT \exp [U/kT]$$

where A is a constant.

While the theories put forward to explain these effects will be covered in the last section of this chapter, it is worth noting here that the conditions for Type A yielding are to be avoided in deep drawing operations. Wright *et al.* (1962) have shown, in conformity with the tensile results discussed above, that stretcher strains can be eliminated by either solution treatment and water quenching from 500°C, or alternatively by temper rolling. These results are again analogous to the case of mild steel. The Type B yields, being more localized, do not present many problems in drawing operations.

To summarize this section, Type A yielding is found in alloys of aluminium and magnesium provided (a) that the material is a fine-grained condition, and (b) that the material has been furnace or air cooled, or alternatively quenched and lightly aged. Type B serrations increase with increasing magnesium content (Caisso (1959)), and are more readily observed up to coarser grain size, when the so-called 'orange peel' effect sets in. Activation energies determined for both Type A and Type B yielding show marked scatter, with Type A rather larger.

5.6 *Other Aluminium Alloys*

5.6.1 *Ternary alloys*

It will have been apparent from the earlier sections 5.4 and 5.5 that copper and magnesium are two potent elements for inducing yield points and strain ageing effects in aluminium, and provided the phase boundaries are not radically altered, these binary alloys can tolerate appreciable fractions of a ternary element without basically affecting the yield characteristics.

The aluminium–magnesium–silicon is a good example, for it was one of the two alloys originally examined by Portevin and le Chatelier (1923). A 0·90% Mg, 0·98% Si alloy has been studied by Phillips (1952) and a 61S alloy (see Section 5.3) by Lubahn (1949). Phillips' is the more extensive, and shows that despite fine grain size, Type A yielding is difficult to detect.

Aluminium–magnesium–manganese alloys may be an exception to this. Hall (1952b) examined cursorily an aluminium 2·25% Mg–0·4% Mn alloy and found clear evidence of both Type A and B yielding in furnace cooled samples. Thomas (1966) also claims that manganese is not essential for producing Type A yielding.

Studies by Hooper (1953) on an aluminium–copper–magnesium alloy (HS14) have been covered in an earlier section (5.4.1). The early work of Elam on RR72 (4% Cu–1·2% Mg–0·8% Mn–0·6% Si) has also been mentioned earlier.

Thomas (1966) has also examined an aluminium – 6% Zn–2·5% Mg–1·5% Cu alloy (DTD 687 type), finding only Type B yielding. Ageing appears rapid at room temperature, so that eight hours after water quenching all Type B serrations had disappeared.

The ternary additions may naturally influence the kinetics of ageing, as well as the total solute atom load. Arkharov *et al.* (1957) have clearly shown that the ageing is speeded up by the addition of silver or zinc to a binary aluminium–magnesium alloy, and this fact alone makes it often very difficult to compare the results of one project with another, even when the alloys are of nominally comparable purity.

5.6.2 *Aluminium–iron*

Chossat (1950a) shows some indicative stress–strain curves of an aluminium – 0·7% iron alloy, which seem to show steps (serrations) after annealing at approximately 350°C and above. A further paper (Chossat, 1950b) amplifies these curves and the steps (a soft machine was used) are clearly seen. However Chossat's main interest was in recrystallization, and only room temperature deformation was used, with no kinetics measured. In view of the fact that iron is a common contaminant in commercial aluminium, this system deserves further study, particularly since the iron addition can be used as a grain refiner.

5.6.3 *Aluminium–manganese alloys*

Phillips (1952) has tested an aluminium – 1·25% Mn alloy and found little evidence of strain ageing. Serrated yielding (Type B) was rarely observed.

5.6.4 *Aluminium–silicon*

Chossat (1950a, b) has also examined two aluminium alloys containing 0·5 and 0·7% Si. Slight steps indicative of serrations were found, but the effect is much less marked than with magnesium or iron additions. Again, silicon is a common impurity in commercial aluminium, and its possibilities as a cause of serrated yielding cannot be overlooked. It is not, however, a suitable addition for production of fine-grained material.

5.6.5 *Aluminium–silver*

This alloy system should be of some interest, since the size difference between these two atoms is negligibly small. Thus, one would not expect this alloy to strain-age or to show serrated yielding, if Cottrell locking is the operative mechanism. Solid solution hardening is likewise absent, but the alloys can age-harden.

Price and Kelly (1964a) did find an initial yield point in an Al–20% Ag alloy, which had been heat treated to produce either G.P. zones, or the intermediate precipitate γ'. The former showed an initial yield point, which Price and Kelly showed was too large to be accounted for by geometric softening, and they suggested that the initial movement of dislocations had to cut through the G.P. zones. Thereafter they could multiply more readily. Samples containing γ' had a normal smooth yield.

This initial yield point would not be expected to return on ageing, unless marked precipitation occurs on the dislocations. Information on this point is scanty.

5.6.6 *Aluminium–zinc*

Serrated yielding was observed as long ago as 1912 in these alloys by Rosenhain and Archbutt (quoted in Phillips, 1953), and this system is also interesting in that its phase diagram shows that aluminium can dissolve no less than 83 wt% zinc at 382°C.

The effectiveness of zinc in producing serrated yielding was demonstrated by Chossat (1950b), while Suitkin (1957) found clearly marked serrated yielding in a sample containing near maximum quantities of zinc. In neither of these cases, however, were any kinetic studies carried out.

Single crystal studies, unlike those on aluminium–copper, also show clear evidence of strain ageing and serrated yielding. Smallman *et al.* (1953) found a yield point reappearing after straining a 2·49 wt% zinc alloy at −196°C and ageing at room temperature. Thomas (1960) also found strain ageing (at small strains) in alloys containing 0·2 and 0·42 at.% zinc when testing at room temperature. At higher strains the yields were of the unloading type. Neither of these found serrated yielding, either because the testing temperature was too low, or because the atom fraction of zinc was too small. However, Price and Kelly (1964) found this in 15 wt% zinc crystals, tested at room temperature. They also found an initial yield drop on testing at −196°C, but this was shown to be due to geometrical softening.

5.7 *Theories of yield points in aluminium alloys*

5.7.1 *The Cottrell theory*

In comparing each of the aluminium systems developed in this chapter with the others, the first question to be settled is the degree of locking resulting from the differing atomic sizes of solvent and solute. These are displayed in Table 5.1.

TABLE 5.1. Goldschmidt radii of aluminium and its solutes

Metal	Al	Ag	Cu	Fe	Mg	Mn	Si	Zn
Goldschmidt radius (Å)	1·43	1·44	1·28	1·27	1·60	1·30	1·34	1·37
Δr(%)	—	+0·7	−10·5	−11·2	+11·9	−9·1	−6·3	−4·2

Thus, it may be seen that if dislocation locking is based on a Cottrell–Bilby theory, then one might expect magnesium to be most effective, atom for atom, closely followed by copper and iron, then manganese, silicon, and zinc, with silver showing virtually little difference. This, in general, seems to follow the magnitude of the serrated yielding observed in these alloys, in so far as such a comparison is valid between different investigators using different machines and samples with differing purity. The exception in the sequence is manganese, which

TABLE 5.2. Solubilities in aluminium (wt%)

Metal	Ag	Cu	Fe	Mg	Mn	Si	Zn
Solubility at 400°C	8·0	1·40	Nil	14	0·1	0·3	81
Maximum solubility	55·6	5·70	0·052	17·4	1·5	1·65	82

seems to show little effect as an additive causing strain ageing (Section 5.6.3). This, however, may in part be due to solubility effects. Table 5.2 shows the maximum solubility and the solubility at 400°C of these seven elements in aluminium.

The low solubility of manganese, even at the most favourable temperature, may make it difficult to take enough of the element into solution, or, alternatively, on quenching it is rapidly precipitated.

Table 5.2 does, however, show up another anomaly, that of aluminium–iron. The low equilibrium concentrations for this element might imply that the only reported instances of serrated yielding by Chossat (Section 5.6.2) may be due to another impurity altogether.

First attempts to explain the serrated yielding in terms of the original Cottrell–Bilby (1949) theory were by Phillips *et al.* (1953), considering that magnesium was locking dislocations in position in exactly the same way as carbon and nitrogen act in mild steel. The dependence on Type A yield on grain size was disposed of by assuming that on ageing a fine-grained solution treated specimen, the magnesium is distributed both in the dislocations and in the grain boundaries. Deformation of a Type A yield is thus analogous to mild steel, but one must explain why it does not return on strain ageing. Phillips *et al.* assumed that after straining, the dislocation density is so much higher that the degree of grain boundary segregation was less effective. Explanation of Type B yielding was more difficult, because it was appreciated that the diffusion coefficient of magnesium in aluminium was much less than carbon in mild steel at 200°C. Nevertheless, bearing in mind the short distances the solute atoms would have to move at these high concentrations, they felt the theory could apply.

This difficulty over the diffusion coefficient was resolved by Cottrell (1953b) (Section 1.7.4), who noted that Type B serrations appeared only after a clearly defined amount of plastic strain. He therefore proposed that the diffusion coefficient could be markedly increased by the creation of vacancies during deformation as Seitz (1952) had proposed that the diffusion coefficient is given by:

$$D = \nu a^2 Z C_v \exp(-E/kT) \tag{5.1}$$

where ν is the Debye frequency factor, a the lattice parameter, Z the co-ordination number, and C_v the vacancy concentration. Seitz also suggested that the vacancies could be produced according to a power law

$$C_v = B\epsilon^{m'} \tag{5.2}$$

where B and m' are constants for a given metal. As with all empirical laws there have been objections to this, summarized by Jaffrey (1966), but for the purposes of this section, equation (5.2) will be accepted as accurate, at least at small strains.

Values of the constants m' are given below in Table 5.3.

Returning now to Cottrell's (1953b) theory, it will be recalled in Section 1.7.4 that the condition for serrated yielding was there taken to be

$$\dot{\epsilon} = D\rho \tag{5.3}$$

TABLE 5.3. Values of parameter m'

Metal	m'	References
Aluminium	1	Seitz (1952)
Copper	1	Peiffer (1963) analysed by Jaffrey (1966)
α-Brass	$0{\cdot}75 \pm 0{\cdot}35$	Bolling (1959), revised by Charnock (1968)
Bronze	1·03	Russell (1963), revised by Ham and Jaffrey (1967)

Under conditions of constant temperature, the diffusion coefficient D may be replaced, and the strain to the first serration ϵ_c will be related to $\dot{\epsilon}$ by the equation

$$\dot{\epsilon} = K(\epsilon_c)^{m'}\rho_\varepsilon \quad (5.4)$$

where ρ_ε is the density of dislocations at the first serration and K is a constant. Again assuming that ρ_ε changes little over the range studied, equation (5.4) is often taken as

$$\dot{\epsilon} = K'(\epsilon_c)^{m'} \quad (5.5)$$

This implies that $\ln\dot{\epsilon}$ is linearly related to $\ln\epsilon_c$, a relation used by Russell (1963) to good effect in his study of tin bronze (Section 6.2.11).

It was, however, found by Vöhringer and Macherauch (1967b) that the value of m' (i.e. the slope of the log–log plot) is sensitive to grain size. This presumably arises from the fact that ρ_ε is a function of grain size as well as strain, and both Ham and Jaffrey (1967) and Charnock (1968) have modified the analysis given above by substituting

$$\rho = K\epsilon^\beta$$

where β is a constant, and K is a function of grain size (this equation is strictly applicable only at low strains). Thus, equation (5.5) becomes

$$\dot{\epsilon} = K''(\epsilon_c)^{m'+\beta}$$

where K'' is a function of grain size. Ham and Jaffrey (ibid.) show that if $m' + \beta = 2.2$ for Cu–Sn alloys, as Russell (1963) showed, then dislocation counts give $\beta = 1{\cdot}17$ and so $m' = 1{\cdot}03$. This corrected figure is shown in Table 5.3.

Charnock (ibid.), using α-brass, likewise revised the value of Bolling (1959), who had found $m' + \beta = 1{\cdot}36$. Values of β are a little over 1, but Charnock believes Bolling's results are in error, and is able to analyse his own experimental results to give $m' \sim 0{\cdot}75$. Further developments of this theory are to be found in the papers by Charnock (1969a, b) and Brindley and Worthington (1969a, b).

The effect of temperature enters into the discussion from the exponential term in equation (5.1) and gives reliable results in copper–tin alloys (Russell ibid.). In aluminium alloys, the situation is not as clear, as is shown below.

Objections to the Cottrell theory outlined above fall on the observed variation of ϵ_c with $\dot{\epsilon}$. Harris (1958), Caisso (1959), and Bailey *et al.* (1965) have all commented on this. A slightly different approach was therefore adopted by Caisso (1959) who suggested that if the time taken to form a serration on the stress–strain curve was greater than that required to form a Cottrell atmosphere, then the critical strain would be given by

$$\epsilon_c = AT \exp [U/kT]$$

where U is an activation energy and A a constant. Since this fails to include the vacancy concentration, dislocation density and other factors, it must be regarded only as an approximation.

Caisso (1959) found that where ϵ_c increased with $\dot{\epsilon}$ at low temperatures, as is required by theory, above about 50°C the reverse was true. Bailey *et al.* (1965) noted that at low strain rates, ϵ_c **decreased** as $\dot{\epsilon}$ increased, but that the relationship tended to the correct slope at higher strain rates, at least in 2024 aluminium. Other materials studied were an Al–1% Mg alloy at 87°C, and an Al–0·2% Cu alloy at 110°C; the agreement in these cases was, if anything, worse (Fig. 5.6).

While it may be difficult to correlate these three sets of results, because of varying purity levels of the alloys, it may yet be possible to explain this inverse relationship in general terms.

It is often forgotten that the original Cottrell (1953) theory, which yielded equation (5.3), assumes, inter alia, that the solute concentration does not vary. In these alloys it must, since they are precipitation hardening at the same time. Furthermore, as Jaffrey (1966) has mentioned, recovery may be occurring, which would effectively reduce the effective dislocation density ρ as time proceeded; this remark should apply to the Al–Cu and Al–Mg alloy shown in Fig. 5.6, and may even apply to the 2024 alloy at room temperature. Some allowance can be made for these effects arising during testing if we rewrite equation (5.3) in the following terms:

$$\begin{aligned}\dot{\epsilon} &= D\rho \\ &= K\epsilon^{m'} C_{s,t} \rho_{\varepsilon}\end{aligned}$$

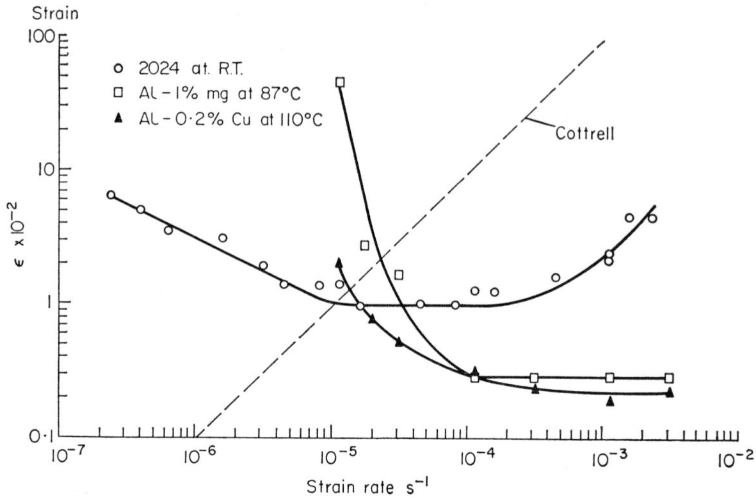

FIGURE 5.6. Plots of ϵ_c against $\dot{\epsilon}$ for three aluminium alloys (Bailey *et al.*, 1965)

where $C_{s,t}$ is the concentration of solute at time t, and ρ_ε is the dislocation density at strain ϵ. If, in addition, we assume the dislocation density falls, due to recovery, according to the law

$$\rho_t = \rho_0 \exp -(t/\tau_1)$$

and that the solute concentration falls according to the law

$$C_{s,t} = C_{s,0} \exp -(t/\tau_2)$$

where ρ_0, the density of **unlocked** dislocations, varies with strain according to the usual law

$$\rho = C'\epsilon^\beta$$

with C' a function of grain size, then one obtains the criterion for the first serration as

$$\dot{\epsilon} = K'\epsilon_c^{\beta+m'} \exp -(t/\tau_0)$$

where $1/\tau_0 = 1/\tau_1 + 1/\tau_2$. Since $t = \epsilon_c/\dot{\epsilon}$ the equation is soluble. Using computer solutions, it is possible to approximate to the curves in Fig. 5.6, but whether the parameters so determined are reasonable is difficult to say. One must also remember the possibility of vacancy–solute interaction, which will also affect the value of $C_{s,t}$ (see Westmacott *et al.*, 1961), but this factor may also perhaps be included in the parameter τ_2.

Thus, it would appear that the Cottrell theory of locking is satisfactory in its essentials, despite these anomalies which may be clarified by reasonable modifications to the theory.

There still remain difficulties in detail, however, and some of the more recent papers have attempted to re-examine other possible modes of locking. Sperry (1963a), for example, was perturbed by the appearance of recovery in these alloys near room temperature. If the dislocations are locked by solute atoms, this should not occur. He therefore proposed that in moving dislocations would act as a type of separator on the solute atoms present; small atoms and vacancies would be attracted to the upper stress field of a positive edge dislocation, large atoms and interstitials to below the slip plane. The segregation of these defects will then produce bulk lattice strains, and either relaxation by diffusion, or an additional applied stress, or both, will be needed to continue deformation, and this could lead to instability or serrations in the stress–strain curve.

This approach has been criticized by Riggs and Demer (1963) who suggested that it was energetically more likely for vacancies to combine with the solute (magnesium) atoms, and thus energy must be supplied (on Sperry's model) both to dissociate these and effectively to order these through the lattice, thus raising the configurational entropy. If segregation is possible, why does it not continue to form a Cottrell atmosphere?

In reply, Sperry (1963b) suggests that the activation energy for motion of vacancies in aluminium is higher than the binding energy of solute and vacancy. Thus, association does not occur, and complete segregation (in the Cottrell sense) is also not favoured around the moving dislocations. He also rejects precipitation on dislocations, or any form of Suzuki locking as being remotely possible, as had been suggested by Riggs and Demer (ibid.).

Finally in this section, Thomas (1966) has modified the original Cottrell approach to allow for the more recent dislocation multiplication theory of yielding, but his results also show anomalies in that a singly activated process is occurring only at sub-zero temperatures and he was unable to correlate his values of ϵ_c as Russell (1963) did in the case of Cu–Sn alloys. In view of some of the variable parameters involved in the analysis given above, this is not surprising, but it would imply that by careful choice of temperature, grain size and strain rate, the effects of recovery and solute concentration may be covered.

5.7.2 *Electron microscopy*

When one considers the vast amount of X-ray and micrographic work carried out on duralumin and similar alloys over the last twenty years, it is surprising that so little has contributed to explaining the tensile properties of aluminium alloys. With the advent of transmission electron microscopy, however, some of the factors controlling the dislocation morphology are now clearer.

To summarize from a number of papers, a selection of the salient facts which emerge are as follows:

(a) the concentration of vacancies in quenched Al–Mg alloys is remarkably high, of the order of 1% (Embury and Nicholson, 1963);
(b) the binding energy between vacancies and magnesium atoms is high, as is the jog energy. The former is 0·3–0·4 eV, the latter approximately 1·2 eV (ibid.);
(c) dislocations can be punched out by precipitate particles, which form extensive prismatic loops (Eikum and Thomas, 1964), or which climb (Embury and Nicholson, 1963) by absorbing vacancies;
(d) the dislocation density and morphology is strongly dependent on magnesium content (Waldron, 1965). The dislocation density is markedly higher in alloys with substantial (approximately 5%) magnesium; cells are absent.

While (a) is unexpected, it does receive confirmation from resistivity measurement by Pansieri *et al.* (1963), as does the binding energy of magnesium atoms and vacancies. Internal friction (Entwistle *et al.*, 1962) also confirms the latter point. The higher dislocation densities are consistent (naturally) with yielding due to dislocation multiplication, but the absence of cells is confirmation of difficulty of cross slip. This in turn is in agreement with Thomas and Nutting (1956) who showed that the number of glide lamellae per slip band decreased as magnesium was added, and with Syutkina and Yakovleva (1960) who found that slip was finer and closer spaced in aluminium–magnesium alloys. The higher dislocation densities are also consistent with the value of β required in the mathematical analysis in the previous section.

In none of the electron micrographs is there any evidence of extended dislocations, nor are there signs of substantial precipitation on dislocations, except in later stages of ageing (Thomas and Nutting (1959), on Al–4% Cu alloys) or when the solute concentration becomes high (greater than 9% Mg, Embury and Nicholson (ibid.)).

The tendency will thus be to discount any forms of precipitate or Suzuki-type locking, and to continue to try to interpret results in terms of the Cottrell theory. In this respect, it is interesting to note that Embury and Nicholson (ibid.) have found what could be direct evidence of magnesium segregation to dislocations. In some cases where their loops have suddenly moved, faint contrast effects remain, which could be the residue of a Cottrell atmosphere.

5.7.3 *Conclusions*

To this stage in the discussion, it has been shown that results on aluminium–magnesium and other aluminium-base alloys are not inconsistent with the Cottrell atmosphere theory. There do, however, remain difficulties of interpretation, and in this final section they will be discussed.

The first of these concerns the variable values of activation energy which have been found for one process or another in effects due to ageing. It must be appreciated from earlier discussion that these are complex systems, and being dependent on the composition and prior heat treatment of the alloy, the kinetics will be altered. If these kinetics are used to find an activation energy, for example, then the conditions must be carefully characterized. In addition to solute depletion by the formation of G.P. zones and intermediate precipitates, there is also the concentration of vacancies, the association of solute and vacancies, the formation of jogs of high energy on dislocations and the possibility of recovery, all of which can affect the parameter being measured, which is usually assumed in like investigations to be singly activated.

Nevertheless, in spite of these analytical difficulties, it would appear that there are copious means of locking dislocations at around room temperature in these alloys, either by pinning at frequent intervals, or by the formation of atmospheres. Careful studies of yielding and ageing, particularly over limited ranges at sub-zero temperatures, may eliminate many of the unwanted age-hardening and recovery effects. In fine-grained material, ageing could be studied at the front of a Type A band, as has been done in mild steel. Too much emphasis, perhaps, has been placed on dynamic yielding, and not enough given to static strain ageing studies under controlled conditions.

A second difficulty concerns the Type B yield. It is reasonable, from the work of McReynolds (1949) and others, to assume that these are similar to the secondary bands observed in strain–aged mild steel. In a soft tensile machine, these fronts are very fast, in a hard machine the serrations virtually form a lower yield plateau (Thomas, 1966). Once they form, however, one is dealing with a very inhomogeneous specimen, parts of which have been more heavily strained and lightly aged, others which have been heavily aged and lightly strained. Nevertheless, more studies are needed on band front velocities, and particularly on the effects of grain size. It may be that these Type B fronts will degenerate, at large grain sizes, into the diffuse fronts seen in mild steel (Plate 1.4(b)), and give rise to the orange-peel effect. This has been commented on only in early sections, since definitive information is lacking, but it may well be that the analogies between mild steel and aluminium–magnesium are closer than imagined. The transition from Type B yielding to orange-peel deformation, where each grain is virtually deforming independently, could also explain why serrated yielding seems so difficult to detect in aluminium alloy single crystals. The only case where this is known is in certain aluminium–zinc and aluminium–silver alloys. It should not be too difficult to extend these studies into the region of coarser-grained samples, and at the same time gain further information on the role of grain boundaries in the Luders deformation.

6

Other Face-Centred Cubic Metals and Alloys

6.1 *Introduction*

In this chapter, other metals and alloys belonging to the face-centred cubic system will be examined in turn. Unfortunately, in general the amount of detailed study which has been given to each of them separately does not illuminate many of the theories; nevertheless the results in themselves are interesting and deserve compilation.

Some of these metals, such as copper, exhibit an unloading yield point effect, similar to the case of aluminium discussed in the previous chapter. Its common alloying elements, such as zinc and tin, also produce examples of discontinuous yielding. Nickel–hydrogen was also covered in Chapter 4, but the introduction of other interstitial impurities, such as carbon or nitrogen, can again lead to serrated yielding, while the Nimonic alloys themselves exhibit this effect at high temperatures. The chapter concludes with a discussion of the tensile properties of some intermetallic compounds of interest.

6.2 *Copper and its dilute alloys*

6.2.1 *Single crystals of copper*

The deformation of single crystals of pure copper has been well studied, and exhibits, particularly for crystals near a [110] orientation, the phenomenon of easy glide. In this early stage of deformation, the rate of work hardening is low, and could be construed as a Luders type of deformation by slip, the parallel slip bands spreading from one end of the specimen. Similar cases have been mentioned in previous chapters.

It should also be mentioned here that the problems of yield stress at low and high temperature have been analysed, along the lines in

Chapter 3, by Koppenaal (1965) and Gallagher (1967) respectively. At low temperature, the results in irradiated material are consistent with Fleischer's (1962b) theory of lattice distortions.

Where additional hardening is achieved, either by incidental impurities, by deliberate solid solution hardening, or by the effects of irradiation, the chances of obtaining a Luders type extension are increased. For example, Cupp and Chalmers (1954), using a soft machine, found evidence of delayed yielding, which could in some

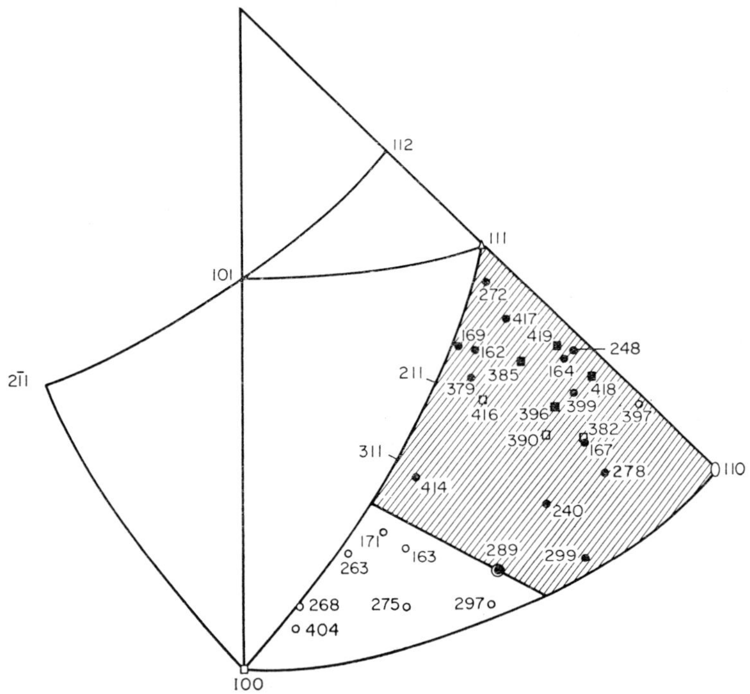

FIGURE 6.1. Stereographic projection of crystal orientations for copper single crystals, deformed at 4 K, which twin (filled symbols) or which do not twin (open symbols) (Blewitt *et al.*, 1957)

cases have been due to hydrogen. Some of the more spectacular effects have, however, been noted in pure single crystals by Blewitt and his co-workers (Blewitt, 1953, Blewitt *et al.*, 1957, 1960, Makin and Blewitt, 1962). Copper, either hardened by neutron irradiation or not, can at 4·2 K show deformation twinning or slip, either of which (dependent on orientation) can spread in a true Luders mode along the specimen.

A specific example was given in the stress–strain curve in Chapter

1 (Fig. 1.3(b)). Twinning is confined to a large area of the stereographic triangle (Fig. 6.1), but samples with orientations close to [100] showed a mixture of continuous and discontinuous slip, but no twinning (Category A). If, in addition, twinning appears, the mode of deformation is termed Category B. The effects of irradiation are to suppress the normal slip which often precedes discontinuous slip. Gold and silver were also shown to twin at 4·2 K, lead and aluminium did not.

Irradiation at higher temperatures (say 78 K) followed by testing at room temperature can also produce a yield point (Makin and Blewitt, 1962) but this is probably a work softening effect, as discussed in Section 1.8.3.

6.2.2 *Polycrystalline Copper*

(a) *Unloading yield points.* The problem of unloading yield points has already been discussed in Section 5.2 and it is not intended to enlarge on this again. However, specifically in the case of copper, the unloading effects have been studied by Lubahn (1952), Bolling (1959), Makin (1958), Birnbaum (1961), Schröder (1959), Bolling *et al.* (1961) and reviewed by Titchener and Davies (1965). The effects were also noted by Bullen and Rogers (1964).

(b) *Effects of irradiation.* A series of papers by Makin (1959), Makin and Minter (1960) and Makin and Manthorpe (1961) have shown that polycrystalline copper develops a distinct yield point and Luders deformation after irradiation (Fig. 6.2), although this is not observed in torsion (McReynolds *et al.*, 1955).

The effect has to be treated with care, for as Makin showed in his 1959 paper, testing at temperatures higher than the irradiation temperature can lead to a work softening yield point. However, when due care is taken to circumvent this, the yield phenomena shown in Fig. 6.2 is a real one and the yield stress obeys the Hall–Petch relation. The value of k_y, however, is temperature dependent (Makin and Manthorpe, ibid.) in the sense that if the point defects responsible for the effect are annealed out, the value of k_y will fall and tend to zero, whereupon the effect will disappear.

6.2.3 *Copper–aluminium*

An early study of copper – 5% aluminium single crystals by Elam (1927b) failed to locate any yield point or strain ageing effects. However Koppenaal and Fine (1961) have discovered yielding effects in copper single crystals containing 5, 10, and 14 at.% Al. The system is interesting because there is a notable atomic size difference (14%), the stacking fault energy decreases with composition, and short range order has also been detected by X-rays. Thus, possibilities exist for Cottrell, Suzuki, or Schoeck locking.

A complete study was made of the kinetics of ageing for the three alloys. In the case of the 14% Al alloy, the kinetics did not appear to obey a $t^{2/3}$ law, while the activation energy, determined from f values as described in Section 1.7.2, varied with f, and showed that

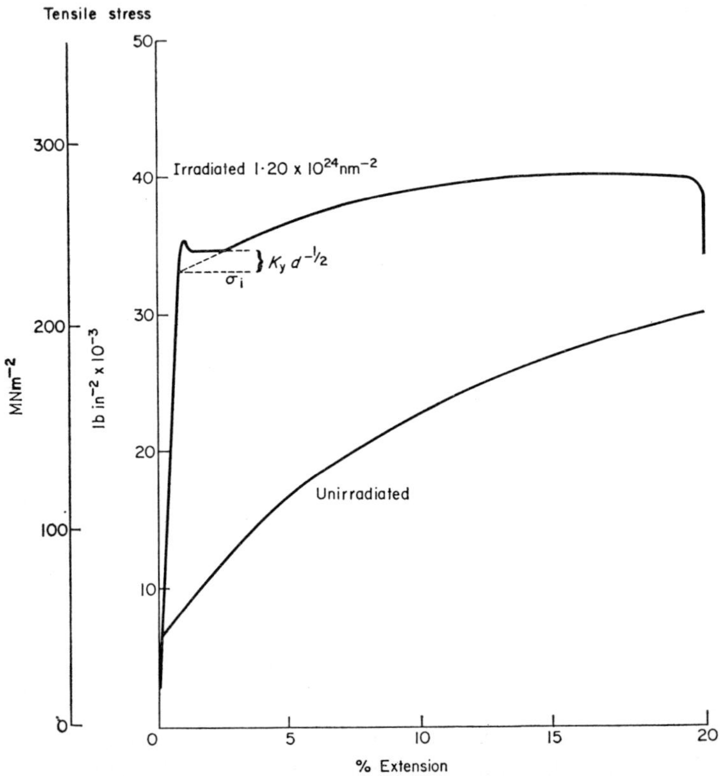

FIGURE 6.2. Stress–strain curves for unirradiated and irradiated copper tested at 20°C (Makin, 1959)

it was not a singly activated process. The evidence moves in favour of short-range order locking, since the results here are strongly dependent on the initial annealing temperature, and since the magnitude of the initial yield drop shows a minimum following annealing at about 600 K followed by air cooling. This, it is argued, is because short-range order is induced at lower temperatures, while in the higher range the higher content of point defects again contribute to the early onset of ordering. Finally, the variation in yield drop varies with composition, as shown in Fig. 6.3. According to the theories given in Section 1.5.7, Cottrell and Suzuki locking should vary with composition as $c(1-c)$, while Flinn (1956, 1958) has suggested that for ordering, the locking should vary as $[c(1-c)]^2$. While it is not possible yet to predict the actual magnitude of the induced yield, normalizing the effects for 14% Al shows that the short-range order theory does have the correct form, while the others do not.

Other investigations by Köster and Konzelmann (1963) and Köster and Speidel (1965) established that the Hall–Petch grain size equation held for these alloys. The value of k_y increased in a linear manner

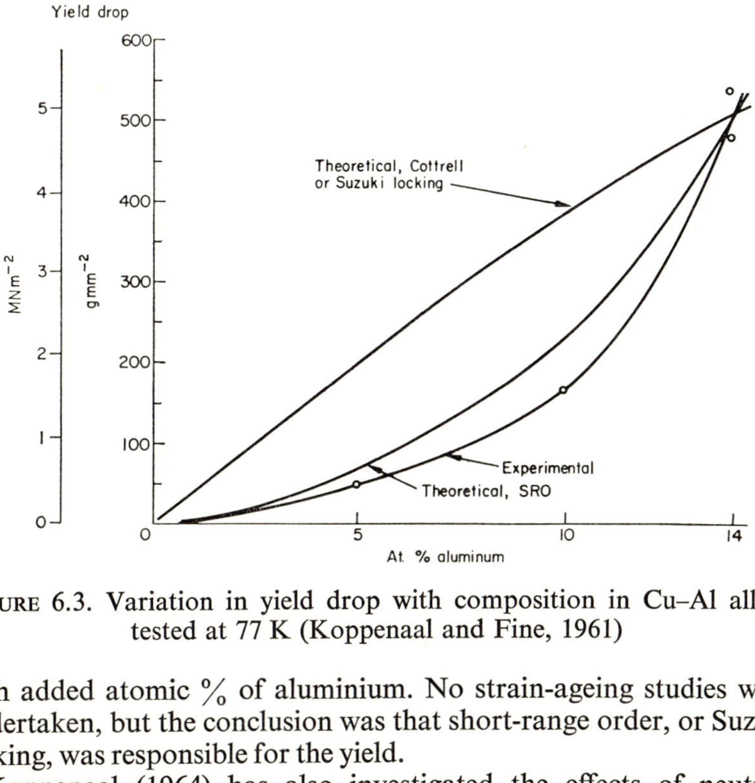

FIGURE 6.3. Variation in yield drop with composition in Cu–Al alloys tested at 77 K (Koppenaal and Fine, 1961)

with added atomic % of aluminium. No strain-ageing studies were undertaken, but the conclusion was that short-range order, or Suzuki locking, was responsible for the yield.

Koppenaal (1964) has also investigated the effects of neutron irradiation on these crystals. Qualitatively, the results are similar to those on pure Cu by Blewitt *et al.* (ibid.) described earlier in this Chapter. (Section 6.2.1.) A single crystal Luders band, comprising coarse slip on the primary slip system, propagated down the specimen; the crystal hardened, and later a secondary band appeared, forming on the conjugate system; ϵ_L for the primary band increased with neutron dose.

6.2.4 *Copper–arsenic*

Schröder (1958) examined the case of arsenical copper single crystals containing 0·36 wt% and 4·1 wt% As. In the former case, there were clear indications of strain ageing after 1 hour at 200°C, while in the higher arsenic alloy, serrated yielding was observed at 178°C, i.e. in the strain-ageing range. However, no kinetic study was undertaken, although the interpretation was put down to the formation of Cottrell atmospheres. Köster and Speidel (1965) also noted that a copper-5·9 wt% As alloy obeyed the Hall–Petch equation.

6.2.5 *Copper–beryllium*

As part of their early study on the yield point in steel, Edwards *et al.* (1943) also examined a number of non-ferrous alloys, including a copper–1·85% beryllium alloy. A yield point and yield elongation were found under widely varying heat treatment conditions; the Luders elongation varied from 0·75 to 3%. Strain ageing was also evident.

Jones and Phillips (1961) have investigated the case of a copper–1·77% beryllium alloy. This is an age-hardening system, and their initial deformation curve on fine grained material water quenched from 800°C showed no yield but after straining and ageing for up to 1 hour at 100°C a typical Luders extension was observed. Ageing at higher temperature before straining showed no yield point effect. The results were thus discussed in terms of Cottrell atmospheres.

The case of single crystals was more extensively covered by Price and Kelly (1963). Electron microscopy was used to supplement the tensile tests, when it was found that crystals with orientations near [110] and containing G.P. zones showed a single crystal Luders band extension; those containing γ precipitate did not, but the rate of work hardening was much increased. By a consideration of the rates of work hardening in the former case it was concluded that the Luders extension was the result of 'geometrical softening'; but this does not contradict the fact that some form of weak locking is present in this system.

6.2.6 *Copper–chromium*

Yield points were also noticed in a copper–1% chromium alloy by Jones and Phillips (1961). The effects are small, and as stated by these authors they could be simply the result of an unloading yield point effect.

6.2.7 *Copper–gallium and copper–germanium*

Haasen and King (1960) have examined the deformation of copper–gallium and copper–germanium single crystals. In these systems, the stacking fault energy is rapidly reduced to very low values, in which case deformation may proceed by the generation of stacking fault bundles. The alloys are also prone to deformation twinning at room temperature and below, giving rise to a serrated yielding effect.

The existence of dislocation pinning in these alloys has been considered theoretically by Fiore and Bauer (1964). The relative binding energies of Cottrell, Suzuki and electronic locking (Section 1.5) are estimated as 0·25, 0·01, and 0·05eV respectively, giving a composite total of 0·31eV. Internal friction studies on these alloys gave an experimental value of 0·28eV.

Köster and Speidel (1965) also noted a yield point in a Cu–15 wt% Ga alloy, which followed the grain size relationship, and which, in electron microscopy, showed numerous dislocation pile-ups near grain boundaries but Li (1969) proved that their presence is not related to the decrease in stacking fault energy.

6.2.8 *Copper–indium*

Brindley *et al.* (1962) have made an interesting study of Luders band propagation in copper–10% indium single crystals. Initial yield point drops were noted, and the Luders strain was quite marked, up to 50% in some cases. Being single crystals, the Luders bands were of the form of parallel slip with, in this case, a clearly defined front. Serrated yielding was observed at about 200°C.

It is difficult, in single crystal studies, to avoid the problems of geometrical softening (Section 1.8.2). However, as shown in Fig. 6.4, Brindley *et al.* were able to show that, although orientations favourable for yield points were around the [110] orientation, the extent was greater than that expected by geometric softening alone. Thus, some form of locking must be present. Ageing kinetics were not studied.

6.2.9 *Copper–nickel alloys*

See under Nickel alloys (Section 6.5).

6.2.10 *Copper–silicon*

The existence of solute atom–dislocation interaction in copper–silicon single crystals has been investigated by Oren *et al.* (1966). As in the case of the work of Fiore and Bauer (1964) on copper–germanium (Section 6.2.7), the theoretical binding energy was estimated from the sum of the misfit, electronic and chemical locking, and the result compared with an experimental determination using internal friction. The experimental result (0·16 eV) is less than in copper-germanium (0·22 eV) or copper–tin (0·44 eV), but should still be sufficient to generate yield point effects. However, as far as is known, no tensile studies on yield points in this system have yet been carried out.

6.2.11 *Copper–tin*

The occurrence of serrated yielding in copper–tin alloys shows many similarities to the deformation of aluminium–magnesium alloys discussed in the last chapter, and the results for this alloy are particularly interesting for the study of vacancy-enhanced diffusion.

The existence of initial yield points at room temperature were first noted many years ago by Bach and Baumann (1921), and strain

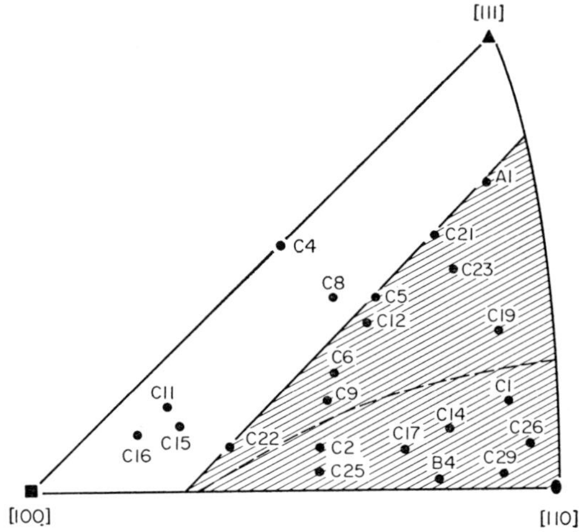

FIGURE 6.4. Stereographic projection of crystal orientations showing yield point effects in Cu–10 at.% In alloys (shaded area). (The dotted line shows the limit of the yield due to geometric softening.) (Brindley *et al.*, 1962)

ageing and serrated yielding at elevated temperatures were confirmed qualitatively in a paper by Polakowski (1952). The material in this latter case was a commercial phosphor bronze (5·9% Sn – 0·15% P), and a similar material was used by Jones and Phillips (1961). Both these papers report the suppression of the yield point in samples quenched from 600°C and above, but the ageing kinetics were not studied and no activation energies were quoted.

The binding energy of a solute (tin) atom to dislocations in copper has been estimated by Oren *et al.* (1966) (see Section 6.2.10) as 0·44 eV and is thus comparable with carbon in mild steel. One therefore expects a marked yield point and ageing effect, and an exhaustive experimental survey of the system has been carried out by Russell (1963, 1965a), by Russell and Vela (1963b), and Russell and Jaffrey (1965). Their results on radiation damage will be discussed later.

High purity samples were used, containing between 1 and 7 at.% tin. Typical stress–strain curves for a water quenched 3·2% alloy are shown in Fig. 6.5. The similarity with the case of aluminium–magnesium is immediately evident; Type A yielding (as defined in Chapter 5) occurs at lower temperatures; Type B serrations at higher temperatures. There is some indication of an initial yield, and a

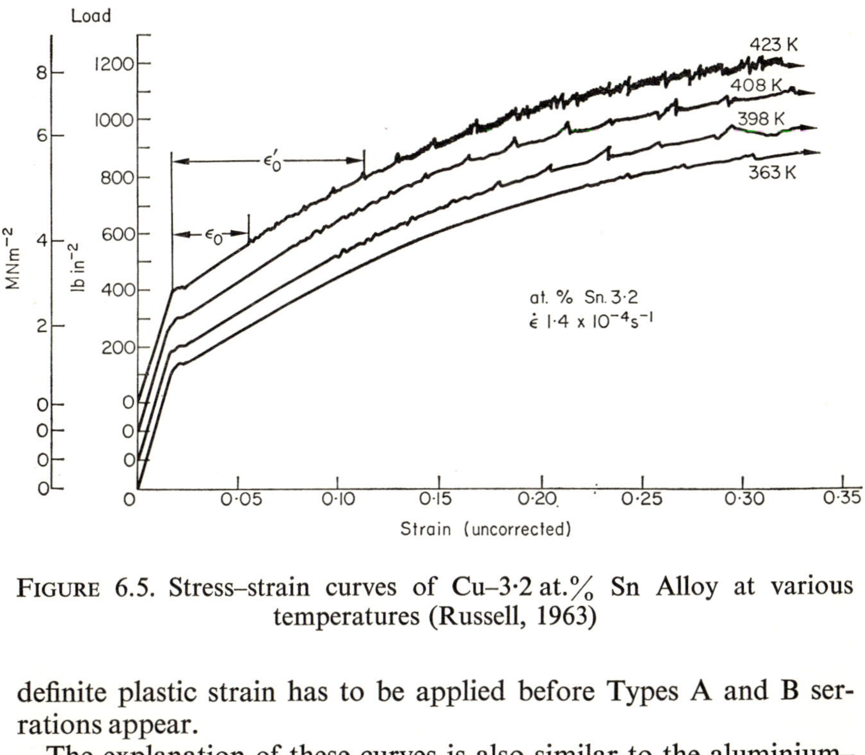

FIGURE 6.5. Stress–strain curves of Cu–3·2 at.% Sn Alloy at various temperatures (Russell, 1963)

definite plastic strain has to be applied before Types A and B serrations appear.

The explanation of these curves is also similar to the aluminium–magnesium case. It is suggested that the diffusion coefficient is increased during deformation as the concentration of vacancies increases, its relation being

$$C_v = B\epsilon^m$$

where C_v is the vacancy concentration, ϵ the strain, and B and m' constants. It was also found that the minimum diffusion coefficient for Type A yielding was $5{\cdot}7 \times 10^{-10}\dot{\epsilon}$, and for Type B, $5{\cdot}3 \times 10^{-9}\dot{\epsilon}$, the latter agreeing remarkably well with the Cottrell condition

$$\dot{\epsilon} = 10^9 D$$

for discontinuous yielding.

In a following paper, Russell and Vela (1963a) examined static strain ageing, and found that although the results are complicated by an unloading yield point effect, the results of this and the previous

TABLE 6.1. Determination of activation energy for ageing of Cu-Sn Alloys

Method	Property measured	Assumptions	E (eV)
Repeated yielding	Strain to the first repeated yield Type A	$C_v = B\epsilon^{2{\cdot}2}$	$0{\cdot}79 \pm 0{\cdot}05$
Repeated yielding	Strain to the first repeated yield Type B	$C_v = B\epsilon^{2{\cdot}2}$	$0{\cdot}79 \pm 0{\cdot}05$
Ageing under stress	Yield drop, h	The process is rate controlled	$0{\cdot}75 \pm 0{\cdot}05$
Ageing under stress	Rise in flow stress, $\Delta\sigma$	—	$0{\cdot}75 \pm 0{\cdot}05$
Ageing under stress	Rise in flow stress, $\Delta\sigma$	Dt = const., for constant f and $C_v = B\epsilon^{2{\cdot}2}$	$0{\cdot}71 \pm 0{\cdot}05$
Ageing under stress	Rate at which excess vacancies annihilate ϕ	Excess vacancies annihilate according to a natural law.	$0{\cdot}75 \pm 0{\cdot}05$
		Average	$0{\cdot}75 \pm 0{\cdot}05$

paper could be made self-consistent using the Cottrell–Bilby approach. The activation energy (E) for any one of six different estimation methods was virtually identical at $0{\cdot}75 \pm 0{\cdot}05$ eV, being considered as the activation energy for the interchange of a tin atom and vacancy. The methods used and the assumptions are listed in Table 6.1.

According to Russell, Type A yielding occurs when dislocations left behind the Luders band have time to strain age before the passage of the next band, so that their magnitude depends on the time allowed between successive yields (i.e. is related to strain rate), as well as on testing temperature and the diffusion coefficient which is itself a function of plastic strain. The Type B yielding occurs when the dislocation drift velocity matches the solute atom drift velocity. It is not, however, possible to observe these markings individually, as the strains are probably small and hence the front velocities very high.

Doubt has been cast on this self-consistent analysis by the work of Vöhringer and Macherauch (1967a, b) who showed that the strain to the first serration was a function of grain size, and consequently values of m' and E increased as the grains coarsened. Ham and Jaffrey (1967) have allowed for the increase in dislocation density during

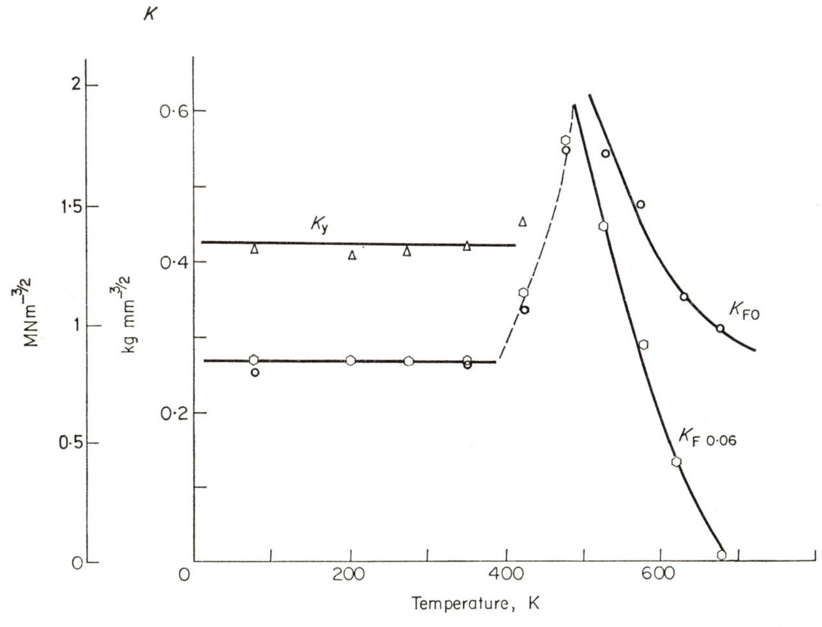

FIGURE 6.6. Variation of the grain-size parameter k with temperature in Cu–Sn alloys (Russell, 1965a)

straining, a fact omitted by Russell (ibid.), and so obtain a value of $m' = 1{\cdot}03$. In addition, Charnock (1968) has considered the dislocation density as a function of grain size, and so endeavoured to make the picture again self-consistent (see also Section 5.7).

Jaffrey (1966) also criticises Russell's detailed explanation of the Type B yielding. If the deformation fronts are moving rapidly, then it should not prove possible to lock these rapidly moving dislocations by solute diffusion. However, this explanation was also criticized on similar grounds in the case of aluminium–magnesium alloys (Section 5.7).

In a later paper, Russell (1965a) examined the effects of grain size. The value of the parameter k_y was determined both at yield, and at higher values of plastic strain (called k_f by Russell). It was found that the parameters were independent of temperature up to the point where serrated yielding occurred ($\sim$ 400 K), when there was a rapid increase followed by a fall (Fig. 6.6). The result was interpreted in terms of unpinning, but the results now need to be considered in the light of multiplication theory.

This change in the value of k_y was not confirmed in independent work by Köster and Speidel (1965) although their temperatures may have straddled the critical temperature region. They did, however, show that k_y increased with added tin content, while their electron micrographs showed numerous pile-ups at grain boundaries, as well as grain boundary nucleation of the dislocations.

Finally in this section, the effects of radiation damage have also been examined by Russell (1965b, c). As in the case of mild steel, the initial yield is suppressed by irradiation. However, in contrast to the case of mild steel, the increased point defect concentration leads to the appearance of serrated yielding at lower strains. This implied that the initial rate of formation of vacancies (from the heavily jogged sources) was very high, but after a 5% strain, the rate was virtually as in unirradiated material. The activation energy for repeated yielding was, however, unchanged at $0{\cdot}78 \pm 0{\cdot}1$ eV, showing that the kinetics, but not the nature of the process, are altered by radiation damage.

Russell's interpretation of the more detailed part of the results has been questioned by Koppenaal (1966), who would prefer to analyse the results (as he did for irradiated Cu–Al alloys) in terms of an activation volume v^*. As Russell (1966) points out in reply, the interpretations are based on different criteria, and which is correct is open to question.

6.3 *Brass*

6.3.1 α-brass

First references to the possibility of yield point effects in brass is by Elam (1927a) and by Köster (1927). Ergang and Welz (1953) also noted irregularities and strain-ageing effects in a 63/37 brass, but the annealing conditions are not fully stated. Hundy (1954) found serrations in the stress–strain curve of 70/30 brass, and similarities with the stress–strain curves of aluminium–magnesium and tin bronzes just discussed above.

The small serrations noted by Hundy were somewhat increased by unloading and ageing, so that like other metals in this group, one might expect the effects to be either masked or otherwise complicated by this unloading yield point effect. This unloading yield point may also be responsible for the effect observed by Adams (1958) in single crystals of copper carrying a layer of zinc on the outer surface. The results suggest that the first dislocation sources to operate are those in the outer layer, and these become locked. Bolling (1959) has examined this, as indicated in the previous chapter. In the case of brass, the investigation was extensive, and it was shown that the stress increment initially increased according to $t^{2/3}$. The rapidity with which the effect appears suggested that enhanced diffusion is taking place, and Bolling found the exponent for vacancy creation (m') as 1·36 (Section 5.7). The activation energies determined in a similar manner to the copper–tin results, gave 0·4 eV, which is however not consistent with the diffusion of zinc or copper, or with vacancy creation.

It is now known that these results are a function of grain-size, and Charnock (1968) has examined the strain to the first serration, as well as determining the dislocation density as a function of strain and grain size. Although the smaller serrations in α-brass make high accuracy impossible, the results could be made self-consistent and gave $m' = 0{\cdot}75 \pm 0{\cdot}35$, which is comparable to exponents for other f.c.c metals (Table 5.3).

The yield points were followed as a function of grain size, composition, and temperature by Köster and Konzelmann (1963) and Köster and Speidel (1965). These papers showed that k_y was virtually independent of temperature, and increased slowly with added zinc content. Heubner and Schumacher (1968) also plotted Luders extension against grain size, and showed it is a maximum with $d \sim 6\mu$, the Luders strain then being just over 1%. The yield is seen only when $80\mu > d > 2\mu$, and stretcher strains are also visible in this range.

Feltham and Copley (1960) examined the effects of composition and temperature, and it was found that the yield stress varied with temperature according to the equation

$$\sigma_y = A/(T+B)$$

where A and B are constants for a given composition. This result had been proposed by Suzuki (1957) but in this case the experimental values were strongly dependent on zinc content, which the theory did not predict if Cottrell locking were present. One of the features in this system is the rapid fall in stacking fault energy with added zinc. One might thus expect Suzuki (chemical) locking to play an important part. However, Feltham and Copley also reject this theory on grounds of strong temperature dependence, and produce a new theory based on the width of the dislocations. No kinetic studies were undertaken.

Ardley and Cottrell (1953) noted yield points in single crystals of α-brass containing nitrogen. This work has been extended by Piercy *et al.* (1955) and by Brindley *et al.* (1962). The former was mainly concerned with the geometry of deformation, although the formation of single crystal Luders bands was duly noted. The latter paper examined this more extensively, using compositions up to 20 at.% solute. 1% Zn samples at room temperature showed no Luders band; at high zinc contents the Luders extension occupied roughly half of the easy glide region, which increased with increasing zinc content. At high temperatures, the Luders strain stayed roughly constant, while at 200°–400°C, serrated yielding was noted.

The propagation of a Luders front in these crystals was not necessarily restricted to the crystal orientations exhibiting geometric softening (see Section 6.2.8 on copper–indium) and it must be concluded that some form of solute locking is present, which, in turn, must be dependent on orientation.

The nature of the dislocation locking is not much clarified by the few papers on strain ageing. Evers (1959) undertook a study of ageing on brass single crystals in the range 50°–140°C, and concluded that locking was the result of local ordering. The existence of serrated yielding already noted above has been further investigated by Koch and Troiano (1964). The curves of ductility against temperature show two minima, the lower temperature corresponding to the region where serrated yielding is observed. Unfortunately, it is not possible from the strain rate effects to determine a reliable activation energy, and thus distinguish between locking and short-range ordering.

The weight of evidence, however, is in favour of ordering as the locking mode, particularly since Heubner and Schumacher (1968) show that strain ageing is less pronounced at higher ageing temperatures. Heat treatment also affects the yield point characteristics, and both these effects are difficult to explain on any other theory.

6.3.2 *β–brass*

See Section 6.7.3.

6.4 *Silver and its alloys*

6.4.1 *The unloading yield point in silver*

Information on silver and silver-base alloys is even more sketchy than might be expected. The existence of an unloading yield point was first noted by Edwards *et al.* (1943), although it was not recognized as such. The effect was later codified by Bolling (1959) in his lengthy paper on the subject. The unloading yield point in high purity silver was only noted at temperatures where recovery was not taking place, i.e. at sub-zero temperatures. The effect was thus small. Hutchison and Honeycombe (1967) also examined polycrystalline silver containing in addition up to 6 at.% of various solutes. The unloading yield was most obvious in silver and in Ag–Au alloys. Solutes of higher valencies caused the effect to be lost, except in Ag–Sb and Ag–As alloys, where a true yield effect existed (see Section 6.4.6).

The unloading yield point in single crystals of silver has been examined by Hauser (1961), who showed that no effect is observed until the end of the easy glide region, i.e. when a secondary slip process is operating. The magnitude of the unloading yield is thereafter virtually proportional to yield stress, similar to the result by Haasen and Kelly (1957) on aluminium crystals. This result, however, does not help to distinguish between the two theories of dislocation–dislocation interaction or dislocation–point defect interaction, as both possibilities may occur in the linear hardening region, where both Lomer–Cottrell barriers and point defects form. Hauser attempted this distinction by applying compressive stresses in various crystal directions prior to testing in tension, and came down on the side of dislocation–dislocation interaction. He was unable, however, to activate Cottrell–Lomer locks by the geometry of the system, so that the explanation may well be considered not proven.

6.4.2 *Silver–aluminium alloys*

As mentioned in Chapter 5, the silver–aluminium system is an interesting one for yield point study, for the atomic sizes of the atoms are closely similar, so the possibility of Cottrell–Bilby locking is small. On the other hand, the stacking fault energy is probably low and conducive to Suzuki type locking.

Hendrickson and Fine (1961) examined single crystals of silver containing up to 6 at.% aluminium. The analysis is difficult, since the limits of Suzuki locking have not been fully worked out; experimentally the time exponent varies from $\sim 2/3$ at early stages of ageing to $\sim 0{\cdot}4$ at longer times, the form of this curve is remarkably similar to mild steel as a completely different system, while the activation energy for the process increases from 0·41 eV at early stages to 0·55 eV when half complete. The locking stresses, however, are not inconsistent with the theoretical increments calculated by Suzuki (1957) and Flinn (1958).

The stress increments are all small, and of almost the same order as the unloading yield point effect. However, Hendrickson and Fine (ibid.) found no unloading yield point in their single crystals, in contrast to Hauser's results observed above.

6.4.3 *Silver–gold alloys*

Single crystals of these alloys, tested at low temperatures, were found by Suzuki and Barrett (1958) to form deformation twins. The general characteristics of the deformation were very similar to the results for copper crystals described in Section 6.2.1; unfortunately the tensile machine used was a pendulum type, and while it could detect and record twins, it could not detect serrated yielding. It is highly likely that this does occur, however, but the paper does at least show some very clear examples of twins spreading in Luders band fashion. At temperatures higher than 100 K, none of these anomalies is seen (Ramaswami *et al.*, 1965) although a small unloading yield point may be present (Kloske and Fine, 1969).

6.4.4 *Silver–tin*

Peters *et al.* (1965) have tested single crystals of silver–4 at.% tin alloy, and in common with the silver–aluminium system, found that ageing effects appeared at the end of the easy glide system and beyond. The activation energy was determined as 0·6 eV; the effect in this case is larger than Ag–Al or Ag–Zn alloys. A comparison of these three systems suggested to Peters *et al.* that solute valence was clearly playing a part; this could then allow the effects to be interpreted in terms of chemical (Suzuki) locking.

6.4.5 *Silver–zinc*

Similar experiments to those just described have been carried out by Tardiff and Hendrickson (1962) on silver–zinc single crystals. The activation energy is about 0·44 eV, again very similar, suggesting that vacancy complexes rather than isolated vacancies were responsible. The effects are in fact somewhat smaller than in silver–aluminium alloys, although the size factor is greater – Fig. 6.7 correlates these results. Since the stacking-fault energy of silver is lowered more readily by zinc than aluminium additions, it is concluded that Suzuki locking is the operative mechanism.

A further study on single crystals of this system, but using higher zinc contents (30% instead of a maximum of 9·5%), has been carried out by Phillips (1963). Although strain ageing was noted, the kinetics were not studied. One interesting feature, however, is the presence of an initial yield, in conflict to the results cited above. This, too, is not examined or discussed.

6.4.6 *Other silver–base alloys*

Hutchison and Honeycombe (1967), in the course of a study on unloading yield points in silver base alloys, noticed that Ag–6 at.% Sb and Ag–6 at.% As alloys showed a true yield point extension at about 100°C. The effect is, however, unusual in that it dies out at lower temperatures (approximately −30°C). The effect was put down to Suzuki locking, overshadowed, at low temperatures, by a rise in the friction stress. However, other examples are known (see Chapter 7) of yield point disappearing at low temperatures, and are put down to changes in slip morphology.

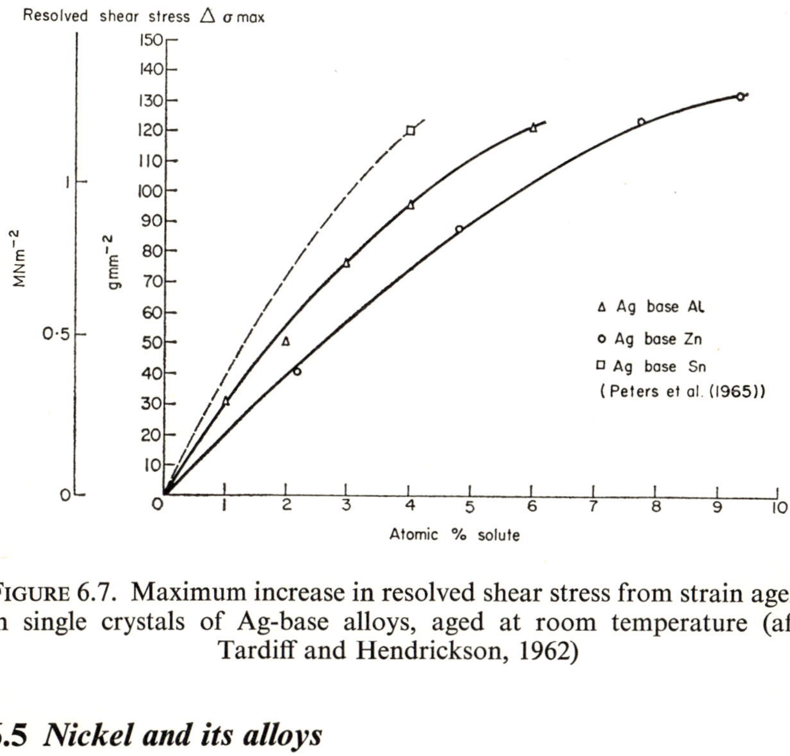

FIGURE 6.7. Maximum increase in resolved shear stress from strain ageing in single crystals of Ag-base alloys, aged at room temperature (after Tardiff and Hendrickson, 1962)

6.5 *Nickel and its alloys*

6.5.1 *Nickel*

The unloading yield point in nickel was first noted by Haasen and Kelly (1957) and assigned to dislocation–dislocation interactions, while Bolling (1959) confirmed the existence of the effect. Apart from this effect, which has been dealt with in Chapter 5, and the special case of nickel–hydrogen discussed in Chapter 4, there is now good evidence that other interstitial elements can cause serrated yielding and strain effects in nickel. Sukhovarov *et al.* (1962) noted serrated yielding in a temperature range around 200°C, the effect disappearing after a wet hydrogen anneal. It was thus ascribed to carbon and/or nitrogen. In other papers, Sukhovarov (1962) and Popov and Sukhovarov (1964) studied the kinetics of the ageing process, and deduced an activation energy of $30{\cdot}7 \pm 2{\cdot}4$ kcal (or 129 ± 10 kJ) roughly equal to the activation energy for diffusion of carbon in nickel (35·7 kcal or 150 kJ). Blakemore (1968) has confirmed that nitrogen-charged nickel wires do not show strain ageing effects, while nickel–carbon do, although in the latter case the temperatures needed were somewhat higher than Sukhovarov's.

At very low temperatures (4·2 K), Wessel (1957a) found serrated yielding, but this again could be due to mechanical twinning, or to thermal softening (Section 1.8.5). Makin and Minter (1960) noted that neutron irradiated nickel, like copper, developed a yield point, and then obeyed the grain size relation.

6.5.2 *Nickel–base and 'nickel silver' alloys*

The special case of nickel–copper–hydrogen alloys has been covered in Chapter 4. Blakemore (1968) has also investigated the nickel–copper–carbon and nickel–cobalt–carbon alloys, charging the binary alloy with carbon by annealing in a static vacuum with a sample of plain carbon steel. The results are in many ways similar to the effect of hydrogen (see Section 4.5.3) but serrated yielding occurs at much higher temperatures, because of the lower diffusion coefficient of carbon. The serrations die out as either copper or cobalt is added to the nickel (Fig. 6.8 – cf. Fig. 4.7). In the case of nickel–cobalt alloys, the results are analogous to hydrogen-charged material; the upper critical temperature is roughly constant, and gives a binding energy of approximately 0·26 eV. The disappearance of serrated yielding may be attributed to the increasing difficulty of forming carbide, which is known in nickel but not in cobalt or copper.

The case of nickel–copper alloys is anomalous, in that the upper critical temperature rises with increasing copper content before serrated yielding disappears. On the simple Cottrell theory (Section 1.5.1) this could imply an increase in the binding energy, or could be associated with a structural change in the carbide.

The effects of radiation damage on nickel and nickel–copper alloys containing carbon has also been studied by Blakemore and Hall (1968b). Neutron irradiation has been shown to eliminate serrated yielding and strain ageing in these alloys. The explanation here is presumably similar to the case of mild steel (Section 2.6.2) – the point defects associate with the carbon atoms and so immobilize them, stopping the drift to the dislocations.

The commercial 'nickel silver' alloys are also known to show extensive Luders bands and yield points. Typical compositions are around 60% Cu, 20% Ni, 20% Zn (all wt%). The anomalous tensile properties were first noted by Ergang (1952) and Ergang and Welz (1952, 1953). Since nickel–silver is often used as the basis of flat-ware, and if the rough shapes are formed by pressing, then Luders bands are commonly noticed. Ergang (ibid.) found that these Luders extensions could be suppressed, as in the case of iron, by either

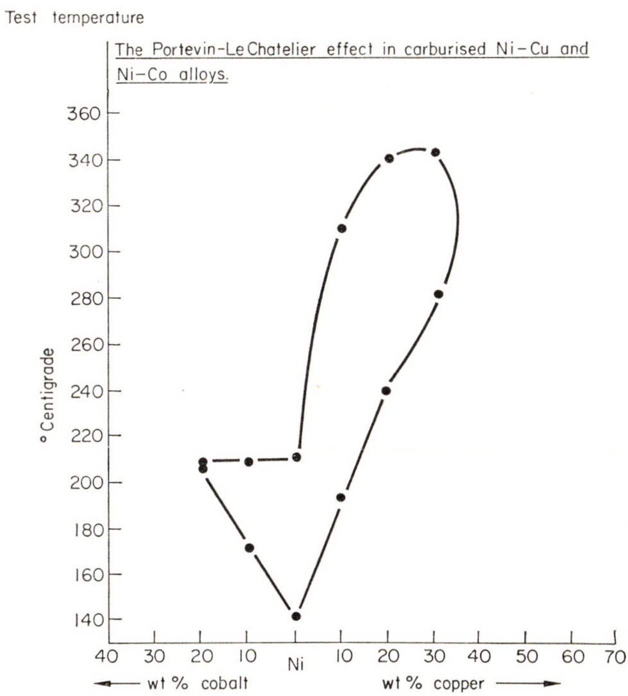

FIGURE 6.8. Region of serrated yielding in Ni–Cu–C and Ni–Co–C alloys ($C \sim 0{\cdot}2$ wt%, $\epsilon = 1{\cdot}37 \times 10^{-4}s^{-1}$) (Blakemore, 1968)

quenching from heat-treatment temperature or by roller levelling. It was also shown that grain refinement increased the yield stress, and later Köster and Konzelmann (1963) and Köster and Speidel (1965) showed that the Hall–Petch relation applied to these alloys. k_y in this case increased slightly at low temperatures.

It might be inferred from these results that ageing is arising from some interstitial impurity, but this connection was not tested. In another paper, Phillips and Jones (1961) attributed the main yield point to Cottrell locking by nickel and zinc; however, there were complications in that alloys higher in nickel showed clear indications of ordering, particularly long-range order hardening due to formation of Cu_2NiZn (a Heusler-type $L2_1$ structure). This leads to a grain size dependent and a grain size independent part to the initial yield point, particularly at higher Ni contents, but still does not explain the origin of the former. Further work is needed in this interesting system, particularly since Blakemore (1968) has shown that pure nickel–zinc alloys can show small strain ageing and serrated yielding effects.

A commercial alloy known as 'Duranickel' (Ni–4·2% Al) has been shown to give all the necessary evidence of blue-brittleness on testing at around 600°C by Jenkins and Willard (1962). This is put down to precipitation of Ni_3Al on dislocations.

6.5.3 *Nickel-manganese*

Indications of an initial yield point in a nickel–2·5% manganese alloy were found by Edwards *et al.* (1943). The system was examined more closely by Jones and Phillips (1961) with a 3·3% manganese alloy. The initial yield point could be suppressed by quenching, but it reappeared after ageing for one hour at 100°C and above. The effects of grain size and strain ageing were also cursorily examined.

6.5.4 *Nichrome and similar alloys*

A brief report on serrated yielding in nichrome type alloys (6–7% Fe, 15–16% Cr, balance Ni) at around 200°C has been given by Duffaut and Lacombe (1965). The four alloys used also carried between 0·014% and 0·093%C, but the activation energy obtained from strain rate studies ($\sim$ 24 kcal g^{-1} or 101 kJ g^{-1}) is well below that for carbon in nickel. Strain ageing, as might be expected, gave a similar value for the activation energy. Electron microscopy revealed precipitation of Cr_7C_3 on glide planes in the matrix, but decorated dislocations are not mentioned.

A series of similar oxidation resistant alloys have also been studied by Franz *et al.* (1967). An 80 Ni–20 Cr alloy showed serrated yielding in the range 400–800°C, together with a minimum in ductility in this range. Similar ductility minima were found in other Ni–Cr–Fe alloys, but no activation energies were determined. However, electron microscopy revealed the presence of $Cr_{23}C_6$ on unextended dislocations and sub-boundaries, so that carbide locking, as in austenitic steels is again possible.

In the more complex high-temperature alloys, such as Hastelloy X (nominally Ni–20 Cr–17 Fe–8 Mo–0·05 C) and Inconel 625 (nominally Ni–20 Cr–5 Fe–8 Mo–3·5 Nb–0·05 C), Kotwal (1968) has found that in the latter case the stacking fault energy of nickel had been sufficiently lowered to allow slow precipitation of NbC carbide on the faults, as in austenitic steels. In the former case, the dislocations are not extended, but particles of $M_{23}C_6$ carbide may be found on the dislocations after ageing.

6.5.5 *Nimonic alloys*

No data appears in the literature on the occurrence of serrated yielding in the Nimonic alloys at high temperature. Information kindly

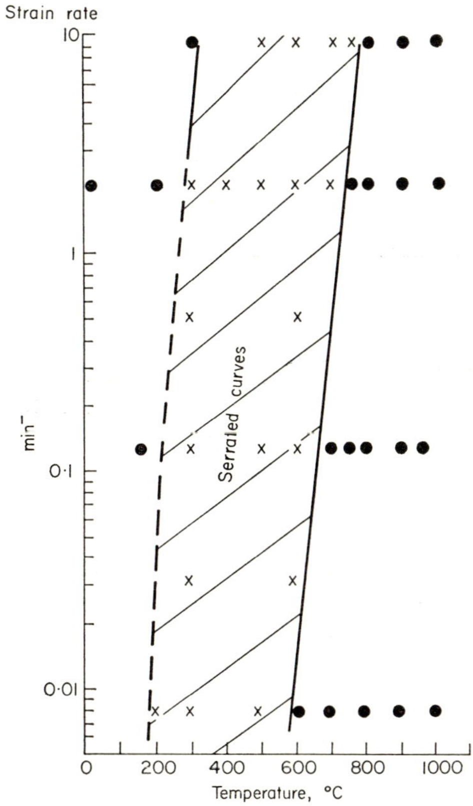

FIGURE 6.9. Region of serrated yielding in Nimonic 90 (results by courtesy of Henry Wiggin & Co. Ltd.)

supplied by Henry Wiggin & Co (for which I am indebted to Dr. C. H. White) shows that serrated yielding is known in at least two of the alloys of this series.

Nimonic 90 (20% Cr–2·5% Ti–1·4% Al–17% Co, balance Ni) shows serrated yielding in the temperature range 200°–800°C. The curves show the serrations increase with strain (cf. aluminium–magnesium and copper–tin) and the temperature range of serration varies, as would be expected, with strain rate. Figure 6.9 shows the temperature range of serrated yielding plotted against strain rate for this alloy; from this curve it is possible to calculate the activation energy as 18·3 kcal g^{-1} (74·8 kJ g^{-1}). This is, in fact, much less than the activation energies for carbon, cobalt, or aluminium in nickel, while diffusion data on the other metals is lacking. Alternatively, one may show that all three elements have diffusion coefficients $\sim 10^{-12}$ cm^2 s^{-1} at 500°–700°C, and may thus be responsible, if the results in Fig. 6.9 are inaccurate.

Nimonic PK31 also shows serrations and a very limited ductility if tested in the region of 600°C. It should be noted that both these are behaving in common with some austenitic steels, although this does not prove in the present case that carbide precipitation on dislocations or stacking faults is the operative step. An alternative explanation might result from the precipitation of γ' (Ni_3 (Ti, Al)) on dislocations, or on sites which could otherwise lock the dislocations.

6.6 *Thorium*

The metal thorium has been shown to possess a yield point (Milko *et al.*, 1958; also Westlake in discussion on Peterson, 1961, and Peterson and Skaggs, 1968). These writers strongly suggest it is due to carbon pick-up, especially as the former note that strain ageing and the yield point effect may be eliminated by alloying with 5% titanium.

The low temperature tensile properties of the metal have been analyzed by Zerwekh and Scott (1967). The results were compared with the several theories of the yield strength (Section 3.9) and it was considered that the random solute barrier mechanism proposed by Friedel (1963) was most appropriate, although material of only one purity was studied.

6.7 *Ordered Alloys*

6.7.1 *Introduction*

In Chapter 1, a brief mention was made of the short-range order theory of locking of dislocation. It is now convenient to extend this survey and to cover ordered alloys. Where long-range ordering occurs, other strengthening mechanisms become possible, and deformation of these crystals may lead to some destruction of order and a possible drop in yield stress. To examine this, attention will be given first to general theories of dislocations in ordered lattices, and then to examine in turn the few superlattices which show yield point effects. Finally, brief mention will be made of some I-type intermetallics, whose study did not fall naturally in Chapter 2, while the hexagonal intermetallics are treated separately in Chapter 7.

In recent years, the general properties of intermetallic compounds have been admirably reviewed in the two books edited by Westbrook (1960, 1966), while Stoloff and Davies (1966) have specifically covered the case of mechanical properties of the compounds. Only a brief résumé of the factors influencing mechanical strength will be therefore

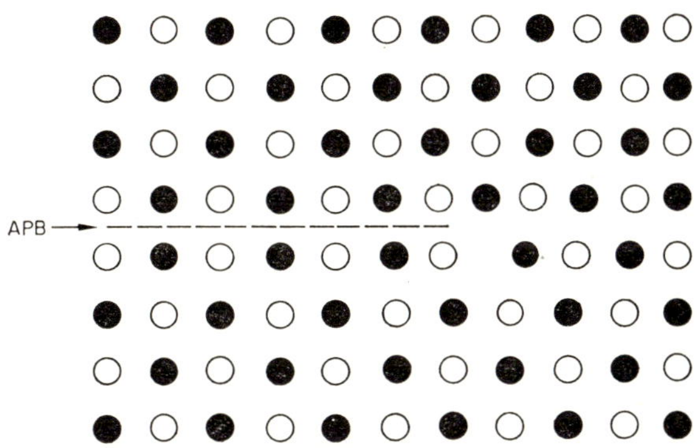

FIGURE 6.10. Antiphase boundary created by the passage of a unit dislocation in an ordered crystal (Beeler, 1966)

included. There are four possible ways in which ordering may influence the mechanical properties of pure binary compounds; vacancy (or defect) hardening; solid solution hardening; and short and long range ordering effects, the latter two of particular importance in possible yield point studies. If, in addition, minor impurities are present, then the possibilities of strain ageing or serrated yielding will also arise.

In an ordered alloy, the passage of a simple dislocation is complicated by the fact that a single dislocation will destroy the local order across the slip plane (Fig. 6.10). This was first noted by Koehler and Seitz (1947), and to overcome this disordering effect, dislocations will tend to move in pairs, separated by a region of anti-phase boundary (A.P.B.). Such a configuration is known as a superlattice dislocation; the spacing of the dislocations (r) is given by an equation of the type

$$r = \frac{a_0^2 \sqrt{\mu}\, b^2}{2\pi\, kT_c\, S^2} f(\theta)$$

where a_0 is the lattice parameter, μ the shear modulus, b the Burgers vector, T_c the critical (disordering) temperature, S the long-range order parameter, and $f(\theta)$ a function of the angular dependence of the stress field of the dislocation and Poisson's ratio. The separations predicted are of the order of several tens of Ångströms, and have now been observed in numerous examples using transmission electron microscopy. In certain examples, the superdislocations are complicated by the formation of stacking faults at the leading and trailing edges.

The change of dislocation pattern from simple dislocations in the disordered case to these superlattice dislocations also follows an observable change in the optically observed slip patterns on the surface. Disordered alloys in general show coarse bunched slip, with cross slip marked, while the ordered alloys exhibit fine, dispersed slip.

The (partially) ordered structure itself will exhibit a characteristic antiphase domain (A.P.D.) of size L, and this has led Cottrell (1954) to suggest that the yield strength in these alloys is given by the equation

$$\sigma = \frac{\gamma}{L}\left(1 - \frac{\alpha\, a}{L}\right)$$

where γ is the APB energy, α is a constant ($\sim$ 6) dependent on domain shape, and a is the boundary width. Assuming α and a remain constant,

$$\frac{d\sigma}{dL} = -\frac{\gamma}{L^2} + \frac{2\alpha a}{L^3}$$

which will be zero when

$$L = 2\alpha \mathrm{a} \quad \text{and then} \quad \sigma_m = \frac{\gamma}{4a\alpha}$$

Thus, domain hardening should show a maximum, typically for values of $L \sim$ 20–40 Å. Thereafter, a decrease in L, such as might occur through the passage of single dislocations, will lower L and an instability in yield stress (i.e. a yield point) may result. The formula has been extended by Ardley (1955), Logie (1957), and Marcinkowski and Fisher (1963) and a hardness maximum has indeed been observed in several cases.

This formula has been criticized on the grounds of two assumptions it makes, firstly that the domains are considered to remain perfectly ordered, irrespective of L, and secondly that the disordered material is softer than the ordered. Both these assumptions are incorrect, and in discussing the yield strengths of these alloys it is necessary to have a clear knowledge of both the short-range and the long-range order parameters. The variation in yield strength with temperature leads to a further interesting phenomenon, namely that a maximum is observed in the yield stress at temperatures close to (but below) the critical temperature T_c (see also Brown, 1959). Rudman (1962) has suggested the yield stress will vary as $(\rho' - S^2)$, where ρ' is the S.R.O. parameter, but again this is of little value in discussing the possibility of yield points unless there is a clear indication of how the parameters vary with strain.

Stoloff and Davies (1966), in reviewing these results, suggest that at low ageing temperatures, increases in strength arise from unit dislocations cutting ordered regions, and the maximum in stress corresponds to a change from deformation by unit dislocations to superdislocations. The variation in yield stress with temperature, however, is still a subject for speculation.

A limited number of studies have also been made on the effects of grain size, even where the material does not show a yield point effect, and confirmation of the equation

$$\sigma_\varepsilon = \sigma_{0,\varepsilon} + k_\varepsilon d^{-1/2}$$

has been found. The subscripts apply to a given strain ϵ. It would appear that k is increased by ordering, and at least in Ni_3Mn, is roughly independent of strain (Johnston *et al.*, 1965). In the $L2_0$ type lattice FeCo, however, k increases to a maximum, and then falls again (Marcinkowski and Fisher, 1965). Bearing in mind the interpretation of k_y given in Chapter 1, Marcinkowski and Fisher suggest that the new sources generate unit dislocations, severally, which in this alloy need a marked additional stress to move them. As deformation proceeds, the disordering lowers the local stress and k falls.

6.7.2 *Yield points in $L1_2$ and $D0_3$ type lattices*

The previous brief introduction does at least suggest that opinions vary on the salient features controlling the yield stress of these particular alloys. When we now consider the possibility of yield point effects, the picture becomes even more diffuse.

One of the classic examples of an order–disorder transformation is found in Cu_3Au (Type $L1_2$) and the mechanical properties have been examined by Ardley (1955). Biggs and Broom (1954) had earlier noted jerky flow in this compound, but Ardley carried this further with single crystal studies. Jerky flow was widely observed, and persisted to temperatures well above the critical temperature T_c. Figure 6.11 shows his results – there is a maximum in the shear strength near T_c, as mentioned in the previous section, but after a fall the strength rises yet again, and jerky flow persists to a further maximum T'. Thereafter, jerky flow disappears and the stress falls.

The difference between T' and T_c, nearly 200°C, would suggest that ordering has little to do with the case. Indeed, this second peak is strain rate sensitive, and from Ardley's results it is possible to

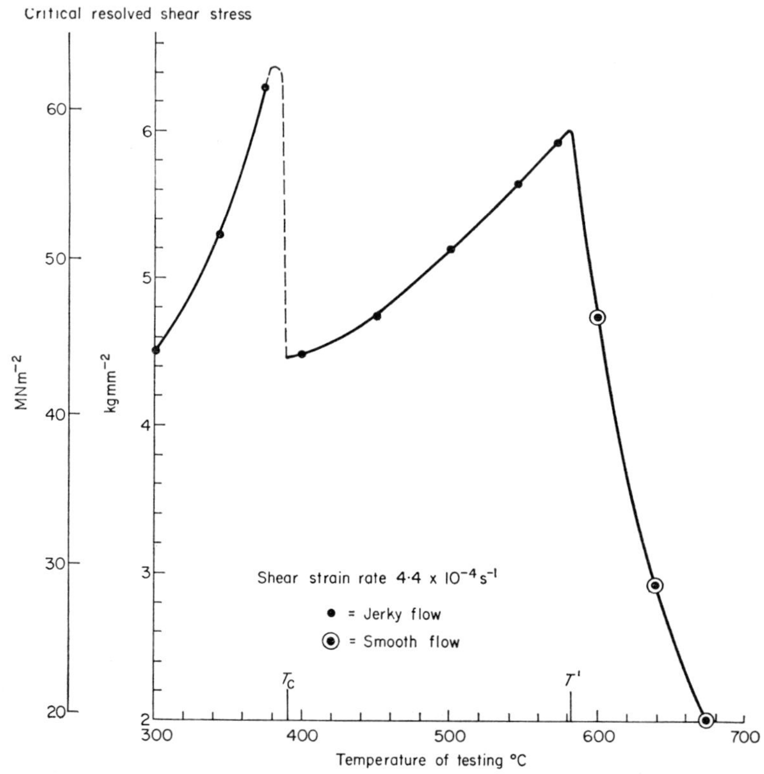

FIGURE 6.11. Critical resolved shear stress plotted against temperature for single crystal Cu_3Au, previously ordered at 300°C (Ardley, 1955)

estimate the activation energy for the process as about 76·2 kcals mole^{-1} (320 kJ mole^{-1}). Ardley was not, however, able to identify the element responsible, nor indeed to suggest unequivocally the mode of locking, and the activation energy is much higher than expected from interstitials. His materials, however, were supposedly of 99·999% purity. The position is further confused by the paper of Hordon (1963) looking specifically for an unloading yield point effect in these alloys. The unloading yield point was very marked in disordered material, but not in the ordered alloys. The effect is also sensitive to plastic strain. Again, unfortunately, he did not correlate his results with Ardley's, but explained his effects in terms of unit and superlattice dislocations. Neither does a study on microyielding in Cu_3Au by Wagner *et al.* (1962) illuminate this effect.

This question of purity is important, as the same high level of purity can hardly be expected in other alloys of this type such as Ni_3Mn, Ni_3Fe, Ni_3Al, or Pt_3Fe. In certain of these cases, strain ageing is known to occur, and Vidoz and Brown (1962) and Vidoz

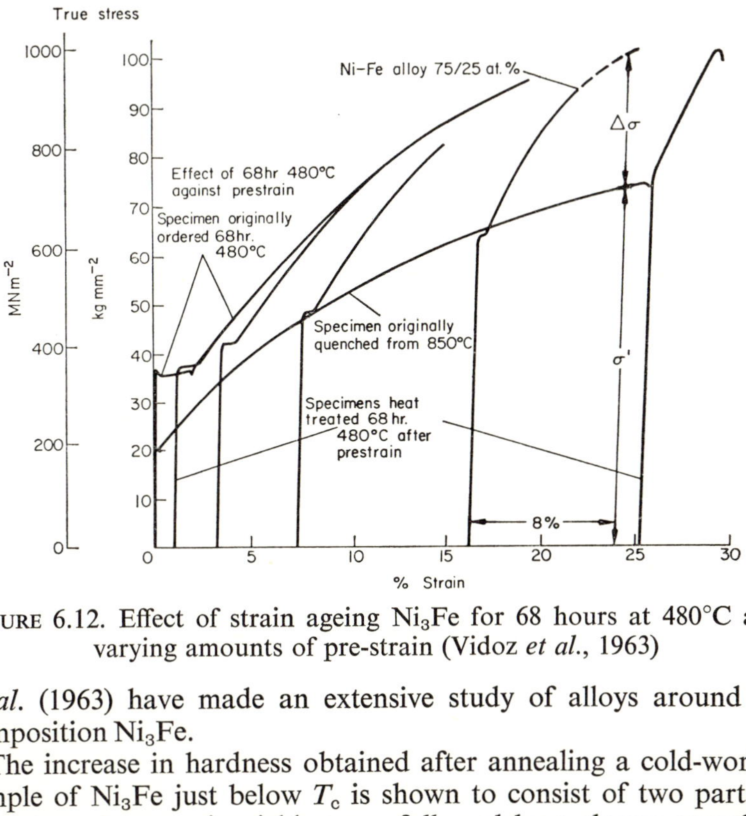

FIGURE 6.12. Effect of strain ageing Ni_3Fe for 68 hours at 480°C after varying amounts of pre-strain (Vidoz *et al.*, 1963)

et al. (1963) have made an extensive study of alloys around the composition Ni_3Fe.

The increase in hardness obtained after annealing a cold-worked sample of Ni_3Fe just below T_c is shown to consist of two parts: a short term increase in yield stress, followed by a slower growth in the rate of work hardening. Figure 6.12 shows results for long-term annealing and clearly indicates that a yield plateau has developed, which is, in fact, almost as marked after only two minutes at 480°C. Unfortunately, this effect has been put down to an unloading yield point, although the effect appears larger than that normally associated with this. No upper yield points were apparently observed, and it is worth stressing that the initial curve is free of any yield point effect.

The interpretation (in the general field of increased work hardening) was based on the formation of jogs on superdislocations, caused by intersections. These then trail, creating additional APB material. Stoloff and Davies (1966) on the other hand, suggest that the APB formed on ageing lie along {100} planes in the $L1_2$ structure planes, which act as barriers to newly created glissile {111} dislocations. There are thus many barriers which mean new dislocations have to be activated at a high stress level with a marked increase in work-hardening rate. It is, however, to be noted in passing that this latter theory would also provide a simple explanation for the yield point effects observed.

Ni_3Al has been examined by Guard and Westbrook (1959) and by the use of hot hardness measurements only, has shown that strain ageing is likely for some of the anomalous peaks observed in the range 0–800°C. Despite the use of doped samples, however, the element responsible could not be identified.

The compound Fe_3Al ($D0_3$ type) with varying purity has been examined by Lawley *et al.* (1961a), Kayser (quoted in Stoloff and Davies, 1966), and Morgand *et al.* (1968). These note the usual maximum in strength found in quenching these alloys from near T_c, and in the latter paper studies show that this is most marked in alloys around 25 at.% aluminium. A more detailed study of tensile properties is, however, contained in the paper by Schmatz and Bush (1968). This shows that a well-developed, rounded yield is obtained from samples tested near T_c equal to 550°C in these alloys. No serrated yielding was seen. Electron microscopy revealed a low initial density of dislocations in the slow cooled alloys, and the yield was put down to dislocation multiplication. No strain-ageing studies were apparently made.

Studies on the isomorphous phases Fe_3Si, Fe_3Be, and Cu_3Al are without note for yield point investigations, although Fe_3Be is anomalous in that it deforms by twinning (Bolling and Richman, 1965).

In conclusion, it would be surprising if the phenomenon of strain ageing were characteristic only of the $L1_2$ type structures, and not of the $D0_3$ or similar types. There is some indication that the effects are impurity controlled, but no clear answer exists in the case of Cu_3Au single crystals formed in a highly pure state.

6.7.3 *The $L2_0$ type lattices*

As mentioned earlier, superlattices based on a body-centred cubic structure are out of context in this chapter, but to avoid repetitive introduction, will be discussed here.

The alloys to be discussed in this section all belong to the CsCl $L2_0$

type lattice in the ordered state, and of the many examples of this structure, which is often a 3–2 electron compound, we will consider only those where yield or other anomalies are of value to us. These include β-brass, FeCo, NiAl, NiTi and AgMg. In common with many others from the extensive list of known examples, these compounds can also show very rapid changes in physical properties with departures from stoichiometry, caused in many cases by a high concentration of point defects (vacancies). In general, there is a minimum in strength at the stoichiometric composition, with increases on either side due, on the one hand, to vacancies and on the other to substitutional alloying effects.

First investigations on single crystals of the compound β–brass (CuZn) were apparently carried out by Elam (1936). She noted the existence of an initial yield drop, and clear metallographic evidence of single crystal Luders type extension. In view of later results by Ardley and Cottrell (see Chapter 7), this yield could well be due to the presence of dissolved gases. Other indirect evidence of strain ageing from dissolved nitrogen is provided by the internal friction evidence of Clareborough (1957), and by the work of Kramer and Maddin (1952), with delayed yield studies. Other investigations using gas-free crystals have found little evidence of any initial yield or strain ageing. Green and Brown (1953), for example, showed that ordered β–brass is softer than disordered, in line with other alloys.

There remains the anomalous results of Wessel (1957a) who noted serrated yielding in β–brass tested in liquid helium temperatures. As explained in Section 1.8.5, this could well be the result of thermal instability, since at these temperatures, the yield stress is very sensitive to temperature, and, moreover, the specific heat is low.

Experiments on the compound AgMg are contradictory. Wood and Westbrook (1962) had noted the mechanical properties, like NiAl, are strongly temperature dependent, and in the temperature range 150–300°C, found clear indications of serrated yielding. This was ascribed to dislocation solute interaction, although the activation energy varied each side of the stoichiometric composition. The variation of yield stress with temperature also showed anomalies in this temperature range, as occurs in mild steel. On the other hand, Terry and Smallman (1962) and Smallman and Terry (1963), while agreeing with the general variation of mechanical properties with temperature, failed to find any evidence of serrated yielding, initial load drop, or strain ageing in static tests. Wood and Westbrook (1963) after further work, suggest that oxygen contamination could be responsible, but all this investigation shows only that the compound is capable of strain ageing.

The mechanical properties of AgMg are themselves very sensitive to composition. For example, a 50·3 at.% Mg alloy is brittle at room temperature, a 49·8 at.% Mg alloy quite ductile. Microhardness studies suggest that excess magnesium leads to segregation of some gaseous impurity at the grain boundaries, and hence to a rise in transition temperature. The results are quoted by Burke (1963).

Gavert and Mack (1967) report yield points and serrated yielding in several $L2_0$ lattices, including AgMg, (the compounds were in fact AuZn, CuAu, AgZn, AgMg, AuCd and NiAl) provided the temperature is greater than $0{\cdot}45T_m$. The Portevin–le Chatelier effect thus obviously exists in these alloys, but controlling factors are not yet known.

The compound NiAl has also been extensively studied since its mechanical properties are, like AgMg, sensitive to vacancy hardening on the high aluminium side of the stoichiometric composition. Yet again there are conflicting results on the existence of a yield point effect.

Ball and Smallman (1966) carried out an extensive investigation on the properties of both polycrystal and single crystal samples. A marked increase in ductility was found above $0{\cdot}45T_m$ in polycrystalline material, but below this temperature samples rapidly failed by intercrystalline cracking. Single crystals were more ductile at room temperature, and in compression showed kinking (if compressed along [100]), but no true yield point as defined in this monograph. On the other hand, Rozner and Wasilewski (1966) found clear indications of discontinuous yielding in the range 25–500°C, with Luders strains of the order of 1% (Fig. 6.13; see also Gavert and Mack, ibid). Above this temperature, the yield point effect disappears, the yield stress rapidly falls, and elongation increases. It was also noted that increasing the strain rate increased the temperature at which the yield point effect disappeared (cf. the case of mild steel, Chapter 2). Rozner and Wasilewski did not apparently test for strain ageing, which would have tied this effect to impurities, but instead suggested that the yield is due to localized disorder, leading to a drop in stress. It is interesting to note that the yield stresses recorded are about half those shown by Ball and Smallman (ibid.); this might imply that the latter did not have exactly stoichiometric alloys, and the yield effect may have been masked by the higher stresses employed.

Other extensive papers on the mechanical properties of NiAl are by Wasilewski *et al.* (1967) and Pascoe and Newey (1968). Both show that in compression tests kinking can lead to pseudo yield points, but the latter also show a true yield point in near-stoichiometric polycrystalline samples in the range 100–500 K. Since these anomalies were found to reappear after straining and ageing at 350 K for one hour, they were put down to Cottrell locking. However, no activation energies were determined, nor was ordering apparently thought responsible.

Rozner and Wasilewski (ibid.) also examined the compound NiTi, which has been considered ismorphous with NiAl. However, its plastic properties are very dissimilar; a small yield drop and extensive

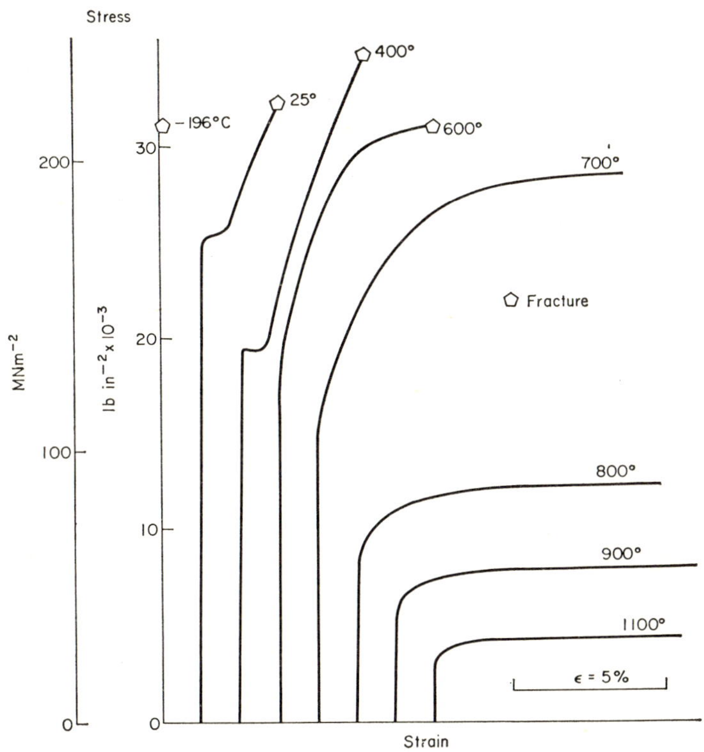

FIGURE 6.13. Stress–strain curves for NiAl at various temperatures ($\dot{\epsilon} = 1.1 \times 10^{-3}$ s^{-1}) (Rozner and Wasilewski, 1966)

Luders elongation (4–7%) were noted in the range −196° to +70°C, with a minimum in yield stress at about 50°C. Above 70°C, the curve becomes rounded, the yield stress rises to a maximum at about 150°C, and then starts to fall again. The two structures cannot now be compared directly, as it is known that NiTi undergoes a martensitic transformation below about 70°C, and the yield extension may simply be an artefact induced by the internal stresses set up in the transformation (Sastri and Marcinkowski, 1968, Ball *et al.*, 1968).

The only other compound of interest at this stage is FeCo, which was found by Marcinkowski and Chessin (1964) to have a small initial yield point (in compression) in the ordered state, but was absent in the disordered material which was, as generally mentioned above, considerably stronger than the former. The significance of these yields, generally ~ 0·5% in strain, is discounted.

A later paper by Marcinkowski and Fisher (1965) examines the validity of the Hall–Petch relation to this compound, both in the ordered and disordered states. As mentioned above, the parameter k_y increases to a maximum with strain, and is generally larger in the ordered than the disordered alloy. On the other hand, σ_0 was initially some six times higher for the disordered alloy. The interpretation in the case of σ_0 is due to the difficulty of cross-slip in the ordered alloy. The changes in k_y are discussed in terms of the theories outlined in Chapter 1 – it is concluded the evidence is against Li's (1963) concept of slip being nucleated at grain boundary ledges, as the stresses required to do this in the ordered alloy would be much too high. Rather similar results were obtained by Jordan and Stoloff (1969), using FeCo containing 2% vanadium.

To summarize this rather diffuse section, it may be said that we are at last seeing some understanding of this difficult field of the mechanical properties of ordered alloys. Where the problem is additionally complicated by yielding or strain ageing effects these compounds have hardly been examined in a systematic manner at all, and much valuable work remains to be done.

7

Miscellaneous Materials

7.1 *Introduction*

In this chapter will be discussed cases of yield points arising in metals and alloys which do not fall conveniently into the previous chapters. For example, the hexagonal metals do not, by and large, appear readily to form systems showing kinetics which lead to dislocation locking, and while no doubt more will be uncovered in the future, only a few have been noted to date and these examined only briefly.

The majority of cases studied in this chapter arise from crystals with low ingrown densities of dislocations. These, as discussed in Chapter 1, will show an initial yield point effect, but they will not show strain-ageing phenomena; the Portevin–le Chatelier effect is likewise absent. However, in many cases the initial yield is quite spectacular, and many of these cases occur in whiskers.

7.2 *Whiskers*

7.2.1 *Structure and growth of whiskers*

The occurrence of metals in the thin filamentary form known as whiskers is leading to important new emphasis on the properties of 'strong solids' (Kelly, 1966). The basis of the remarkable strength of composite materials is in the fibre reinforcement itself, for these fibres, some 1–10 μm in diameter, and several hundreds of micrometres long, have strengths approaching the theoretical limit.

Both metallic and non-metallic fibres are known; the metallic ones are often grown by a process of extrusion, or growth from the base, and both metallic and non-metallic ones may be formed by condensation or by reduction, from the tip. The actual mechanism of crystal growth is still subject to speculation, but it is certain in many cases to be assisted by the presence of a single screw or edge dislocation running down the length of the whisker. The total dislocation density is generally very low.

7.2.2 *Strength of whiskers*

The mere fact that whiskers have such a small diameter would be expected to enhance their strength. Since the yield stress of a segment of dislocation line of length l is given by

$$\sigma = \alpha \mu b / l$$

where $\alpha \sim 1$, μ is the shear modulus, and b the Burgers vector, l is typically $\sim 10^{-3}$ mm in most metals. This is the diameter of some

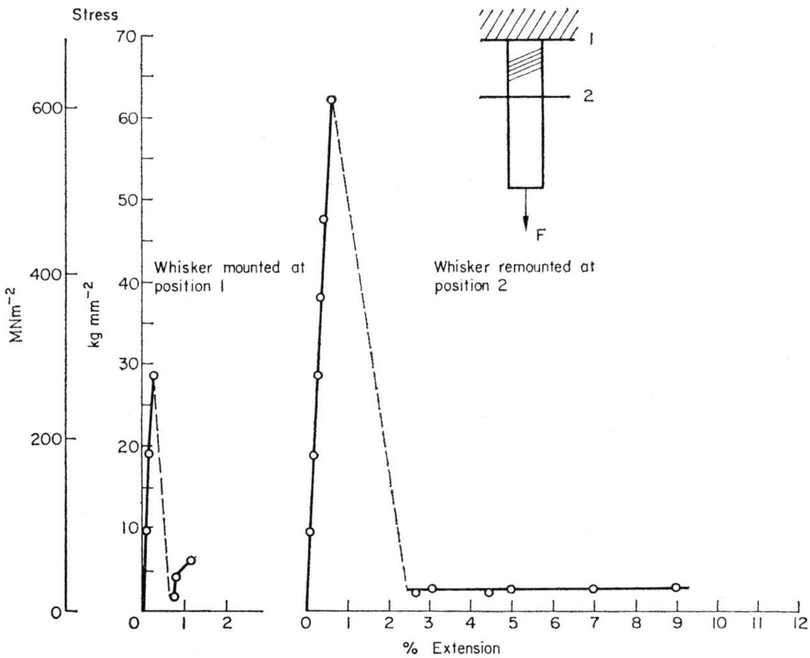

FIGURE 7.1. Stress–strain curve for a copper whisker (Brenner, 1958)

of the strongest whiskers, so that it becomes increasingly impossible to nucleate active dislocation sources. In these crystals, their strengths approach those of the theoretical strength ($\sim \mu/30$) and may be classed as the strongest materials known today.

Whiskers can yield or fail in a variety of ways, by brittle fracture, or kinking (Jackson, 1965), but where the opportunity occurs for multiple slip systems, as in the face-centred cubic and body-centred cubic metals, a notable yield point is obtained. Figure 7.1 shows a stress–strain curve (Brenner, 1958) of a copper whisker. Note the elastic elongation of about 1%, and the very large ratio of σ_U/σ_L, much higher than in mild steel. Zinc whiskers are also known to have yield point effects (Cabrera and Price, 1958, Price, 1961) as do nickel whiskers (Bailon and McQueen, 1967). In the latter case, not all samples showed a yield drop, a fact which the authors put down to surface defects in the whiskers which were formed from dissociation of halides.

7.3 *Ionic crystals*

7.3.1 *Lithium fluoride*

The study of the deformation of transparent crystals has been actively pursued for many years, for they possess the dual advantages that the slip systems are in general simple, and, using polarized light, the stress systems within the crystal may be examined. In recent years, the dislocation networks within these crystals have been made visible by doping, as in the work of Hedges and Mitchell (1953), but the relevance of this work to metals has been rendered a signal service in the work of Johnston and Gilman (1959) and Johnston (1962a, b) on lithium fluoride.

In this work, the dislocations are made visible on the surface of the crystal by selective etching, so that for the first time dislocation velocities were measured as a function of stress and temperature, giving the relation

$$v = \sigma^{m^*} \exp\left[-E/kT\right]$$

and as indicated in Chapter 1, this formula has found wide application. In addition, the strain is shown to be a linear function of dislocation density (ρ), so that

$$\epsilon = \alpha\rho$$

and the stress–strain curve may be synthesized from an empirical knowledge of the constants involved.

The effects of impurities on the mechanical properties are likewise very similar to the effects of, say, carbon and nitrogen in mild steel; delayed yielding and strain-ageing effects are both noticed. In fact, one of the most satisfactory ways of observing the yield point phenomena in this material is to strain and age at around 115°C in situ; the yield points are very sensitive to axiality of loading, as in the case of metal single crystals, and the pre-loading helps to equalize the loading eccentricities.

Figure 1.3(a) showed a typical yield point observed in these LiF crystals; the magnitude of the yield can be very effectively accounted for from a knowledge of the strain rate $\dot{\epsilon}$, the initial dislocation density, and the dislocation velocity parameter m^* (Fig. 7.2). The value of m^* ($\sim$ 20 for these crystals) thus gives a controlling magnitude to the yield drop, and Fig. 7.3 shows how this varies with the initial dislocation density.

The only point of difficulty remaining with this theory is the mechanism of strain ageing. Undoubtedly in these crystals the impurities are playing an important role, and Johnston (1962a) suggests that as the principal impurity was Mg^{2+} there would be a strong tendency to form vacancy-impurity atom pairs, but even if there is this tendency, the interaction of these with moving dislocations is more difficult to fathom. Nevertheless, small amounts of impurity ($\sim$ 0·01%) do harden these crystals markedly and the hardness is also very sensitive to the heat-treatment history.

The kinetics as studied do not, however, enable us to distinguish between this and other modes of locking. There are basically two theories which have been put forward, one an extension of the Cottrell–Bilby theory by Bassani and Thomson (1956); the other depends on the fact that an edge dislocation in an ionic crystal carries a core charge, and so attracts vacancies (see Eshelby *et al.*, 1958; Brown, 1961; Koehler *et al.*, 1962). The possibility of electronic locking (Section 1.5.4) cannot also be ruled out, but it is difficult to distinguish between these theories at present.

7.3.2 *Rock Salt*

The rock-salt structure has given its name to the whole group of cubic ionic bonded solids of which LiF, MgO, and NaCl form an important subsection. These crystals all have similar slip systems $\{110\}$ $[1\bar{1}0]$ and rock salt itself has been studied for many years. Indeed, first evidence of jerky flow predates studies on many metallic materials (see Klassen–Nekludova, 1929).

The two theories of strain ageing in ionic crystals have been outlined in the previous section, and an attempt was made to distinguish between these two by Brown and Pratt (1963) using rock-salt doped with $\sim 5 \times 10^{-5}$ mole fraction of $CdCl_2$. Strain ageing and jerky flow, as well as a type of Luders band, were observed; the yield at higher strain values is more of the unloading yield type (Chapter 5) and at very high strains the crystal loses the capacity to strain age altogether. Ageing out the point defects by a high temperature recovery anneal still allows jerky flow but no strain ageing is observed. This means that unlocking occurs in some other way, perhaps by a Johnson–Gilman model.

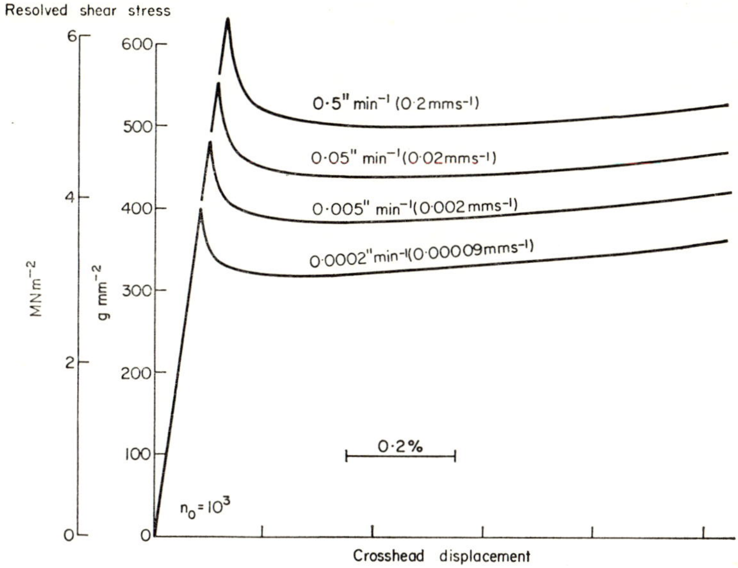

FIGURE 7.2. Effect of strain rate on the calculated stress–strain curve for LiF crystals (Johnston, 1962b)

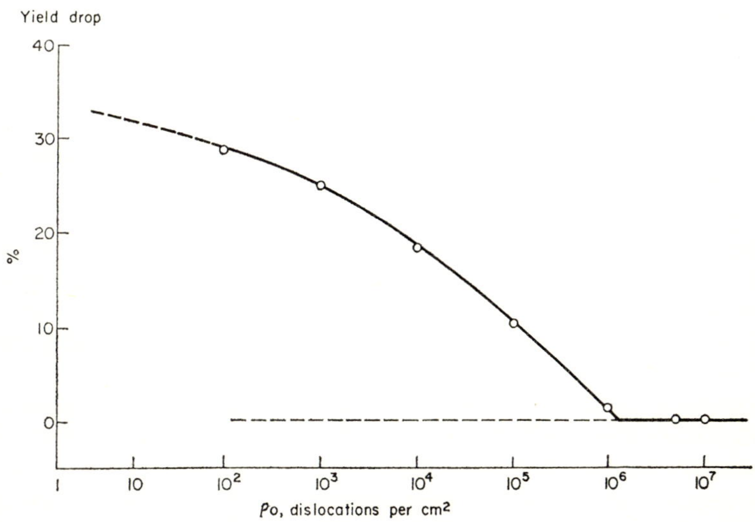

FIGURE 7.3. Effect of free dislocation density on the magnitude of the yield drop in LiF (Johnston, 1962b)

The kinetics of ageing were studied by measuring $\Delta\tau$, the increase in shear stress, against ageing times at various temperatures. The results, taken up to about 50% of total ageing, fit well with a $t^{2/3}$ time law, and although the accuracy is not high, the activation energy for the process is 0·65 eV, if vacancies are held responsible, or 0·92 eV if vacancy–impurity pairs are held responsible. Thus clear evidence seems to be obtained by impurity flow (by means of the vacancy–interstitial complexes) to the core of the dislocation and gives support to the theory of Bassani and Thomson.

It is difficult, however, to see why agreement is so good, and Brown and Pratt give thought to this. They point out that any interaction potential that varies inversely as the distance from the dislocation will give a $t^{2/3}$ ageing law; the difficulty in this case, because of the charges involved, is that the potential is only varying as $1/r$ at time $t = 0$; later the effective potential is reduced, and the exponent in the time law should be less than 2/3. Nevertheless failing any better theory, they maintain the migration of complexes will be the better theory to consider.

7.3.3 *Magnesium oxide*

Magnesium oxide is again, like LiF and NaCl, of the rock salt structure, and again deforms by slip on the {110} [1$\bar{1}$0] system. However, its mechanical properties have not been as fully examined as in the case of the latter two, although there is sufficient known to observe clear yield point effects from dislocation multiplication.

For example, Stokes (1962) has examined freshly cleaved and chemically polished single crystal blocks of MgO and finds that the former are much weaker than the latter. By studying the etch pitting he concludes that fresh dislocations created by cleavage, or by sprinkling silicon carbide powder on the surface, were of the normal glissile type, while the grown in dislocations were formed perhaps from collapsed discs of vacancies and were sessile. After high temperature annealing at $\sim$ 2000°C strengths become extremely high and approach within one-tenth of the theoretical shear strength. Miles (1964) has also examined yield point drops in bending.

The explanation of these yield points can then be followed directly from the Johnson–Gilman theory. It is unfortunate, however, that no kinetic studies appear to have been carried out to see if mobile dislocations can be locked after doping of the crystals. The increase in strength following high temperature annealing is thought to result from the solution of incoherent precipitates which act otherwise as local stress raisers (Henderson, 1964; Lewis, 1966).

7.3.4 *Silver chloride*

Another isomorphous crystal, silver chloride, has now been investigated, both in terms of the deformation substructure (Nye, 1949), and in terms of the strain-ageing kinetics (Kabler *et al.*, 1963). Strain ageing has in fact been known since 1935, but the latter paper is the only one of substance.

The strain-ageing curves illustrated in that paper do, however, show some anomalies. Over-ageing is apparent after less than one hour at 40°C, and the maximum hardness also falls as the ageing temperature increases – in this respect the curves match more nearly the case of quench ageing in iron. The ageing is considered due to pinning by divalent impurities, with an activation energy of 0·46 eV but the problem needs further detailed investigation.

One of the additional problems associated with work in single crystals of these structures is their marked sensitivity to the surface condition of the crystals. This problem, although not strictly relevant to yield point phenomena, has been well reviewed by Westwood (1963). Single crystals of as-cleaved LiF, for example, show a low yield stress and high rate of work hardening; chemically polished crystals, where the active surface dislocations have been removed, show a yield point and a lower rate of work hardening. On polished crystal surfaces, new dislocations may be engendered by sprinkling with silicon carbide powder. Some extreme cases have been found in MgO by Stokes and Li (1963), where polished and annealed crystals had very high strengths (greater than 1 $GN\,m^{-2}$) and small elongation, while similar crystals sprinkled with SiC had yield points a tenth of this and considerable extension.

7.3.5 *Sapphire*

Sapphire crystals (Al_2O_3) at elevated temperatures can show slip on the normal hexagonal system (0001) $[11\bar{2}0]$. For appreciable flow to occur, the temperature must be in excess of 900°C. From studies on the tensile creep behaviour of single crystals, Wachtman and Maxwell (1954) showed that there is an appreciable delay time and similar results have been noted in bend tests, which are widely used on these ceramic materials. Delay times for creep are often indicative of delayed yielding and hence of yield point effects, and Kronberg (1962) has demonstrated this in flame polished sapphire single crystals.

Figure 7.4 illustrates his results, showing in (a), the variation of upper and lower yield stresses with temperature, and in (b) the variation of these two parameters with strain rate. These sets of curves yielded Arrhenius plots, with activation energies of 95 and 85 kcal (400 and 350 kJ) for the upper and lower yield points respectively.

Conrad *et al.* (1965) have obtained similar tensile results which they analyzed in terms of thermally activated slip (see Chapter 3).

Unfortunately, the crystals used by Kronberg were not checked for the existence of strain ageing, so it is not possible to distinguish between yielding resulting from impurity locking or from a plain deficiency of mobile dislocations. It is also highly likely that many other inorganic crystals, so often brittle at room temperature, are capable of showing yielding and ageing if conditions of purity, testing temperature, and machine stiffness are suitable.

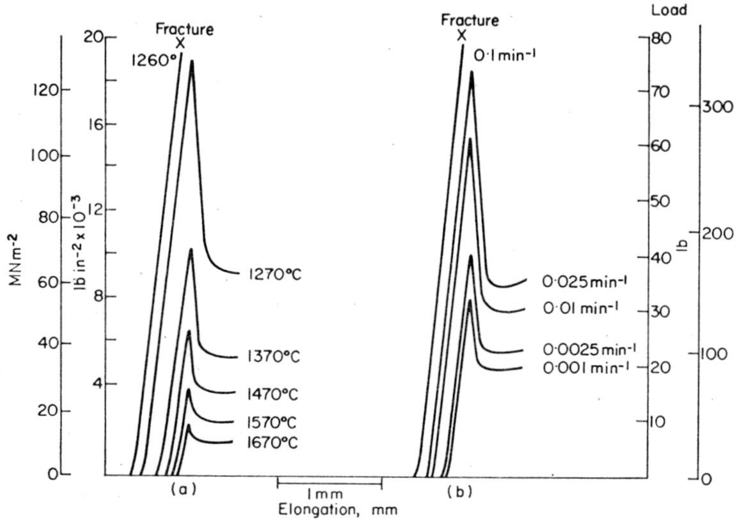

FIGURE 7.4. Stress–strain curves for sapphire crystals: (a) at varying temperatures, (b) at 1420°C at varying strain rates (Kronberg, 1962)

As implied above and as was discussed in the case of crystals with rock–salt structure, surface treatment may affect the results markedly. For example, Weil *et al.* (1963) showed that surface grinding could be beneficial in controlling the fracture strength of alumina at room temperature; samples tested after annealing the ground samples were invariably weaker.

7.4 *Semiconducting materials*

7.4.1 *Germanium and silicon*

The semiconductor metals, germanium and silicon, both possess the diamond cubic structure, and can deform at elevated temperatures by slip on a {111} $[1\bar{1}0]$ type system. These metals are produced to a very high degree of purity by zone-refining techniques, and as a consequence of the repeated passage of the molten zone, often possess very low densities of grown-in dislocations.

The situation is thus favourable for the generation of yield points by dislocation multiplication – the exponent m^* of the dislocation velocity equation is known from Chaudhuri *et al.* (1962) as 1·3 to 1·9 for a variety of semiconductors. (See Table 1.1.)

The possibility of a yield point effect had been noted as early as 1952 in the work of Gallagher (1952), who observed that to obtain deformation of these two metals, the specimen temperature had to be > 500°C for germanium, and 900°C for silicon, but at the lower temperature limits of germanium there was an induction period before deformation commenced. As indicated in Section 7.3, this may be a clear indication of delayed yield and hence of a yield point. However, despite doing tensile tests, Gallagher did not notice or mention a yield drop despite the presence of clustered slip. A similar occurrence was noted by Penning and de Wind (1959).

In silicon, the existence of a true yield point was noted during bending tests of whiskers by Pearson *et al.* (1957) and confirmed by the work of Patel and Chaudhuri (1962). However, it is also found that the effect is critically dependent on prior heat treatment, and after soaking at 1000°C, the upper yield stress is markedly reduced (Fig. 7.5). This result is put down to oxygen affinity; the silicon–oxygen clusters so formed are shown by selective etching to be prolific sources of dislocations.

This oxygen effect is not the only possibility in silicon however, for Pearson *et al.* noted a distinct strain-ageing phenomena in their silicon whiskers and in their rod samples. It was confirmed that 0·1% B was present in these samples, which, at the dislocation densities involved, is more than sufficient for atmosphere locking. Similarly, Siethoff (1969) recorded yield points and Luders bands in single crystals of silicon doped with phosphorus and tested at around 1000°C.

Germanium likewise shows a well-defined yield drop—Fig. 7.6 shows a typical stress–strain curve from the paper by Chaudhuri *et al.* (1962). Little systematic work has been done on the possibilities of strain ageing; since impurity levels are more important vis à vis electrical properties, it is understandable that this is a neglected corner of the semiconductor market, although Kurtz and Kulin (1954) have showed that Cottrell-type binding may exist between impurity atoms and dislocations in germanium, increasing the carrier lifetime.

7.4.2 *Indium antimonide*

This compound is one of a large group of potentially important semiconducting materials, made from alloying elements from Groups III and V of the periodic table, and hence often known as III–V compounds. They all possess the same diamond cubic structure of silicon and germanium, although possibilities exist for studying the effects of non-stoichiometry.

The only compound in this group whose mechanical properties appear to have been studied in detail is InSb; for in 1957, Allen noted

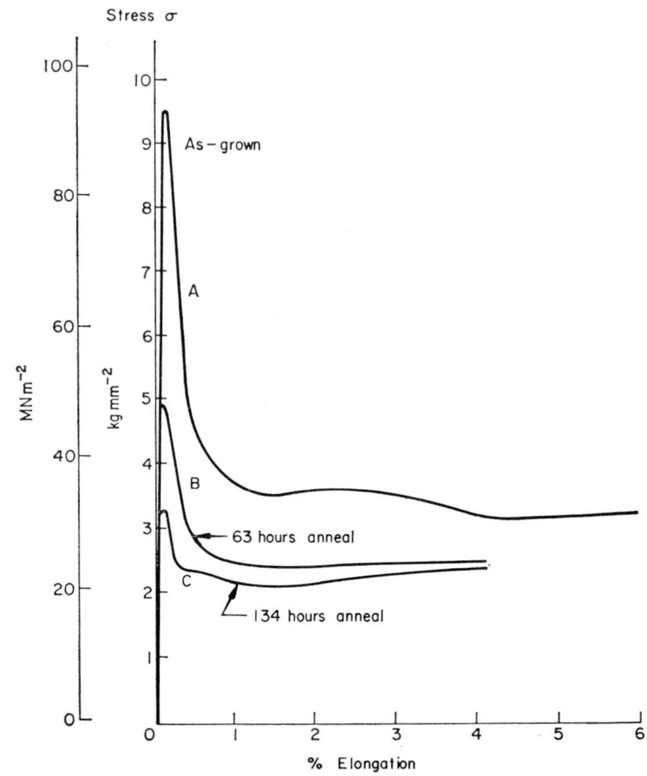

FIGURE 7.5. Effect of heat treatment at 1000°C on the yield point in silicon tested at 800°C (Patel and Chaudhuri, 1962)

a delay time in the yielding of single crystals under three-point bending, and followed this with a later paper using compression (Allen, 1958) where the delay time was found to follow the law:

$$t_{\mathrm{d}} = C\sigma^{-n}T \exp\,[E/kT]$$

Discussion of this type of formula, which was in fact developed before the dislocation multiplication theories of yielding, is found in Chapter 1.

The mechanical properties were still more extensively examined by Abrahams and Liebmann (1962), using single crystal tensile tests. The grown-in dislocation density was only 50 cm^{-2}, and as might be expected from this result, a marked, well-rounded yield was seen at the commencement of deformation. The yield stress fell linearly with temperature, extrapolating to the melting point at 525°C,

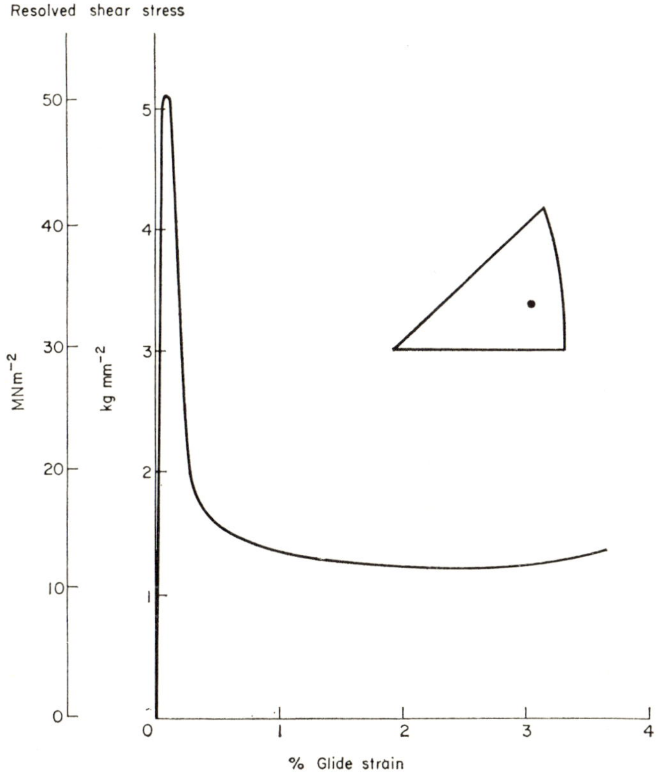

FIGURE 7.6. Yield point in a germanium single crystal at 500°C (Chaudhuri *et al.*, 1962)

although there was a noticeable deviation at 450°C and above, due probably to oxygen pick-up. Strain ageing may also occur as a result of this, for as Abrahams (1962) has shown, specimens deformed and then annealed in moist oxygen do show this effect. On the other hand, if annealing is carried out in helium, this is not noticed. At these low dislocation densities, surface condition is very important, and oxide nuclei on the surface may be a prolific source of new dislocations.

7.5 *Hexagonal metals and alloys*

7.5.1 *Introduction*

In contrast to the cases of the face-centred and body-centred cubic metals, studies of yield point phenomena arising in the hexagonal metals have received relatively sparse attention. These metals are of course close packed structures, their stacking fault energies are often high, so the possibility of serrated yielding may be difficult to locate and in this sense cases of substitutional hexagonal alloy structures showing yielding are rare. Some attention has been paid to interstitial locking, as seen below, but in the substitutional cases examples are limited to isolated cases of intermetallic compounds with the hexagonal structure.

7.5.2 *Cadmium and zinc*

The existence of yield points in zinc single crystals was noted many years ago by Orowan (1934, 1940) where the phenomenon was first called thermal hardening. Subsequently, Smith (1947) showed the phenomena could be obtained in cadmium crystals.

The explanation of the effect follows from Cottrell and Gibbons (1948) who noted that in cadmium crystals, yield points were not observed if crystals were grown from a melt under an argon atmosphere, but were present when a nitrogen atmosphere was used.

Similar results were found for zinc by Wain and Cottrell (1950), and the kinetics of the reaction were further studied by Wain (1952). The difficulties of handling and mounting these very soft crystals are great – often the yield point is observed only after pre-straining, to remove off-axis loading, and in studying the kinetics, the problem of recovery of the zinc is a further complication. Figure 7.7 shows a typical set of ageing curves for zinc – the yield stress at short ageing times (curves 4–9) fall, resulting from normal recovery, and thereafter the yield point tends to reappear. This is in marked contrast to the effect in mild steel; Wain concludes it is the result of limited amount of interstitial nitrogen in solution, which gas analysis showed was only 0·0022 wt%, and hence the number of free dislocations had to decrease during the recovery process before there was a sufficient amount of nitrogen for each dislocation to be locked. These crystals are also very sensitive to surface oxide – but both upper and lower yield are raised together. The yield points seem independent of the angle of the slip plane. The latest research on this system, however, is by Evers (1959) who, in some detail, has studied the kinetics of ageing. Using the usual ageing factor, $f = (\sigma_u - \sigma_L)/\sigma_L$ he has shown that this factor varies with time according to $t^{3/2}$. This, instead of the Cottrell–

Bilby factor $t^{2/3}$, he claims is evidence of precipitation of Zn_3N_2 on dislocations, so locking them in position. No electron microscopy was used in confirmation, although it might be doubted if the dislocations could be observably decorated.

Chiao and Gordon (1965), using ultrasonic attenuation studies, claim that the yield is due to multiplication of dislocations, but that during recovery nitrogen is responsible for repinning, the activation energy being very low.

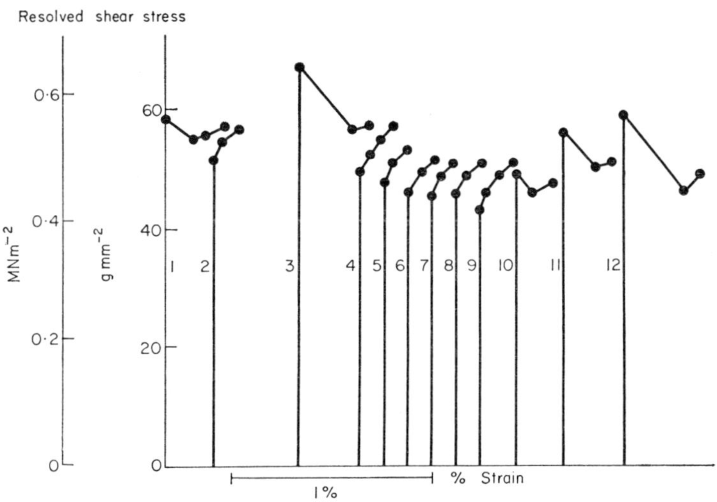

FIGURE 7.7. Recovery followed by development of yield point in zinc single crystals charged with nitrogen (Wain, 1952)

Curve No.	1	2	3	4	5	6	7	8	9	10	11	12
Ageing time at 50°C (hr)	1	0	20	0·5	1	1·5	2	2·5	3	6	17·25	39

7.5.3 *Magnesium and its dilute alloys*

Similar studies to those described above have been carried out on magnesium by Geiselman and Guy (1959). Magnesium does readily form a nitride, and has a maximum solid solubility of 0·0048 wt% N_2; however, these two workers have contended that their results are in accord with atmosphere formation rather than precipitation. Although the ageing kinetics were not studied, their view is based on the argument that the stable nitride Mg_3N_2 is formed during slow cooling from the melt, and these crystals show no yield. Solution treatment followed by water quenching likewise gives no yield point, but straining followed by ageing now produces a yield effect. With repeated straining and ageing, the yield point tends to disappear. The absence of a yield in slow cooled crystals is a good argument, but it is likewise difficult to ensure perfect axiality and a high surface finish for the specimen on the first straining, as noted above.

Other experiments have been carried out on Mg–12·5 at.% Li alloy single crystals by Quimby *et al.* (1962). The main experiment here was to examine the effect of the lithium additions, which push the ductile–brittle transition temperature to very low levels. The alloying markedly increased the critical resolved shear stress to ~10 times that for pure magnesium, and at the same time a small yield point appeared, which persisted over the temperature range 77–232 K, and was absent above this. The writers suggest Suzuki locking or short-range ordering effects, as the yield stress is virtually independent of temperature over the range 215–600 K.

Finally, in magnesium–cadmium alloy single crystals, Boček *et al.* (1965) found a small initial yield, which increased in magnitude up to 1·85 at.% Cd (the maximum used), and became more noticeable at lower temperatures.

None of the yield point effects noted so far in this section exist in polycrystalline specimens, since the possibilities of twinning followed by basal slip in the twin, or of non-basal slip, are always present. However, there are cases of magnesium alloys known where polycrystalline samples show notable yield point effects.

Couling (1959) reported that in tensile tests on a Mg–0·5 wt% Th alloy, those in the range 100–300°C showed observable yield point effects. This point was taken up by Kent and Kelly (1965) who made an extensive investigation of the effect, varying grain-size, temperature, and strain rate. The results show many similarities to mild steel: yield points were more marked in the finer grain sizes, and in fact disappeared if this became larger than ~ 35μm; serrated yielding was observed in the upper temperature range of 240–285°C. The activation energy for the return of the yield point was 28·9 kcal (121 kJ); the kinetics of ageing correspond to a $t^{2/3}$ time at short ageing times. After prolonged ageing, the dislocations are decorated with a fine dispersion of an unknown precipitate; later, overageing is evident (Plate 7.1).

The important difference between this alloy and mild steel is that here the yield point disappears below ~ 100°C. This was explained on the basis of pyramidal slip at these temperatures, and the rate of work hardening is high; the stress required to propagate yield is then much greater than the stress to initiate yield, and a yield drop is not observed.

Much the same argument would be applied to the apparent absence of a yield point in polycrystalline samples of zinc, cadmium, and magnesium. It must also be remembered that the hexagonal metals, unlike the cubic, are anisotropic and have considerable internal stresses remaining merely from cooling (Boas and Honeycombe, 1947). These in themselves may act to mask any yield point effect.

Kent (1962) also investigated the occurrence of yield points in several other dilute alloys of magnesium, while Couling (1959) mentioned, but did not describe, the effect in magnesium rare earth alloys. Kent has portrayed results from several commercial alloys, e.g. Magnesium Elektron's ZRE, (2·7% rare earths–2·2% Zn–0·6% Zr), ZTY (0·5% Zn–0·6% Zr–0·7% Th), ZA (0·5% Zr) and AM5035 (1·5% Mn) as well as some similar alloys from the U.S.A. In most cases, yield point drops again occurred, generally around 200°C, although the effect was smaller than in the Mg–0·5% Th. The AM5035 showed only strain age hardening and no yield drops as did a Mg–15% Pb alloy. The fine dispersion of the precipitate, noted in Platc 7·1, gives these alloys some particular creep properties which are of importance commercially.

7.5.4 *Hafnium*

Johnson (quoted by Goldman, 1960) reports some cases of apparent yield points in hafnium strip, where the effect was most apparent at around 200°C. This was ascribed to geometric softening (Section 1.8.2), but in view of the results of the previous section this may be too naïve a view.

7.5.5 *Titanium and its alloys*

In the two remaining hexagonal metals to be considered, the situation is further complicated by an allotropic transformation which these metals undergo in the solid state. Both undergo a body-centred cubic → hexagonal martensitic transformation; titanium at 882°C, zirconium at 862°C. The acicular product introduces further internal stresses, over and above those generated by the crystalline anisotropy, and both metals are also very sensitive to the dissolved gas content.

For example, in titanium single crystals, which are largely free of these complications, Churchman (1955) found a yield point present when the dissolved (O_2+N_2) content was 0·14 wt% but none when the gas content was 0·01%. The situation in titanium crystals is further complicated by the multiplicity of slip systems, prismatic as well as basal (Churchman, 1954) leading to a high possibility of geometric softening (Section 1.8.2). If crystals were not favourably oriented for duplex slip, then an extensive system of kink bands was formed, which migrated rather like the edge of a Luders band. However, the yield point reappears after strain-ageing, showing that some form of locking due to dissolved gases is present, but the difficulties of interpretation are increased by the ancillary metallographic factors, and no kinetic study was attempted.

Churchman also examined a few polycrystalline samples of his less pure material, and again found yield discontinuities, but only if his testing was carried out above 90°C. In this respect the yield is

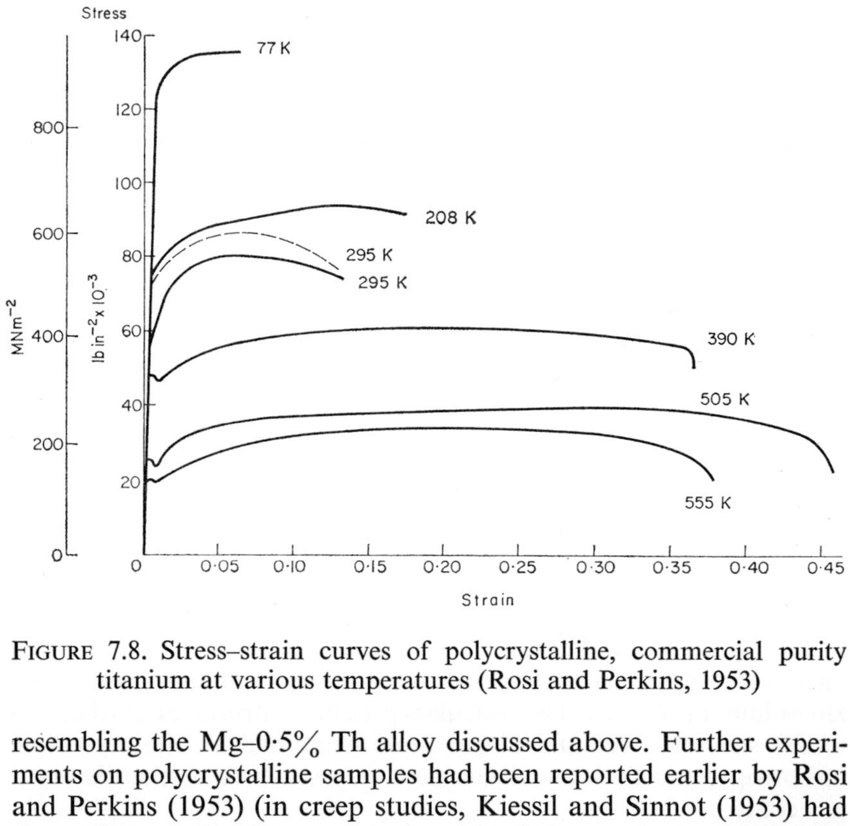

FIGURE 7.8. Stress–strain curves of polycrystalline, commercial purity titanium at various temperatures (Rosi and Perkins, 1953)

resembling the Mg–0·5% Th alloy discussed above. Further experiments on polycrystalline samples had been reported earlier by Rosi and Perkins (1953) (in creep studies, Kiessil and Sinnot (1953) had shown impure titanium can strain age); their sample was of 99·72% purity, and contained 0·08% N_2 and 0·05% C as interstitial impurities. The oxygen content was not stated. A yield point was observed with specimens tested at 117°C and above, in conformity with Churchman, reached a maximum at about 230°C, and disappeared over 350°C (Fig. 7.8). Although some rather gross irregularities appear in the curves at still higher temperatures, these are probably not true serrated yielding. Strain ageing is also apparent around 200°C, although compared with mild steel, the yield drops are small. Luders bands were noticed as bunched slip on $\{10\bar{1}0\}$ planes in some coarser grained samples used for this study. No explanation was offered for the disappearance of the yield below 117°C, but the explanation offered in the Mg–0·5% Th case could well apply again. Turner and Roberts (1968) have likewise noted weak serrations in testing impure titanium at elevated temperatures.

Rather conflicting results were obtained by Wasilewski (1963) on a series of grades of titanium, and carrying in the purer cases about 500 p.p.m. of impurities, mainly O_2. In these samples, yield points were developed on the initial test over a temperature range of −196°C to +350°C, but only provided the specimen had been annealed in the temperature range 500–550°C and quenched. Higher temperature anneals or longer time anneals produced grain coarsening and the yield point disappeared again. The less pure the sample the more difficult to produce an effect, and a complex double heat-treatment had to be used. Strain ageing is likewise difficult to detect, and an explanation of the effect (which has a very low activation energy of 0·0033 eV – from strain rate effects) was sought in terms of dislocation superstructures. Another possible effect however, might be in a massive contribution by twinning to the deformation early in the piece, and this would mean, reconciling this and earlier researches, that yielding in titanium could arise from a variety of sources. It would likewise have been interesting to compare specimens both water quenched and furnace cooled after 1 hour at 500°C, the latter avoiding many of the difficulties due to locked up stresses in these very anisotropic materials.

This work on the low temperature properties of titanium shows a marked increase in yield stress at low temperatures (see Fig. 7.8), rather similar to the b.c.c. metals which were discussed in Chapter 3, and this fact has prompted a further analysis along the same lines as was treated there. For example, Levine (1966) has studied slip and yield stresses for both basal and prismatic slip in electron beam melted and zone refined single crystals, and has endeavoured to show that thermally activated slip controlled by Peierls–Nabarro forces, are controlling up to 220 K. Beyond this and up to 300°C, a second thermally activated process takes over, while above 300°C the deformation process is athermal. His interpretation has been strongly criticized by Westlake (1967), claiming that the results cannot be considered singly activated in either temperature range.

The case of polycrystals has been fully explored by Conrad (1966) and his co-workers (Orava *et al.*, 1966, Jones and Conrad, 1967, 1968). Conrad (1966) himself pointed out that unlike the b.c.c.

metals, low temperature yield stresses are strongly dependent on both temperature and impurity level. Determination of τ^* the thermal component of flow stress, is linear when plotted against $T^{1/2}$, which implies the activation energy is given by

$$H = A\,(\tau_0^* - \tau^*)^2$$

where τ_0^* is the value of τ^* at $T = 0$, and A is a constant. It can then be found that τ_0^* is proportional to $C^{1/2}$, where C is the impurity level, and is thus in accord with Fleischer's (1962b) model of solid solution hardening (see Chapter 3).

Makin and Minter (1956) have examined high purity titanium irradiated to 5×10^{19} nvt (total) at 100°C. Mechanical tests at 200°C showed the irradiated sample had developed a yield point which was absent in unirradiated material tested at that temperature, and absent in tests at 20°C. This result was put down to the usual mobility of point defects at that temperature – but it is uncommonly high, and could be the result of enhanced diffusion of one of the impurities.

The applications of titanium alloys are not likely to be in the pure metal, despite its favourable strength-weight ratio, but in dilute alloys which can offer improved creep resistance at high temperatures. Tests by Bishop *et al.* (1953), for example, used Ti–150–A (1·3% Fe–2·7% Cr) and RC–140–B (0·24% C–3·8% Al–3·8% Mn). Both these now showed a yield point at room temperature, but a soft testing machine was being used, and the kinetics were not studied.

7.5.6 *Zirconium and its alloys*

Both titanium and zirconium have a remarkable affinity for oxygen, which makes them generally useful for gettering vacuum systems. Their similarities in structure have also been remarked in the previous section, but it is the favourable nuclear properties possessed by zirconium which have made this metal and its alloys of such value in the canning of nuclear fuel elements.

The mechanical properties of zirconium and zirconium–oxygen alloys were reported in 1953 by Treco (1953), who failed to notice any apparent yield drops in his material. In the discussion of this paper Keeler claimed to have detected a yield point, and his own results were published later (1955). Using fine-grained consumable arc-melted material, Keeler obtained yield points over the range 25–300°C, but not at −195°C (Fig. 7.9). With less pure material, the yield was only noticed at −200°C. However, the active impurity responsible was not sought, or identified, nor was strain ageing noticed.

Other papers on zirconium metal are by Wessel (1957a), who reported serrated yielding at 4·2 K but not at −78 K and higher temperatures. This is almost certainly the result of repeated twinning or thermal softening (see Section 1.8.5). Makin and Minter (1956) examined both unirradiated and neutron irradiated zirconium – in neither case was a yield point observed. Coleman and Hardie (1966b) mentioned that their specimens showed yield points, but not at −195°C, and that the lower yield points obeyed the Hall–Petch grain-size relation.

This necessary condition of fine grained material for yield points to be present has been stressed in a review by Weinstein (1965); and

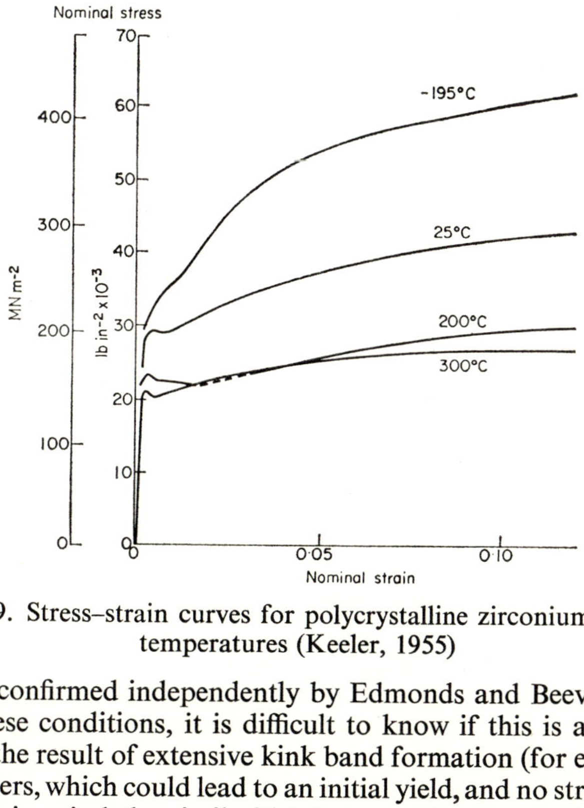

FIGURE 7.9. Stress–strain curves for polycrystalline zirconium at various temperatures (Keeler, 1955)

has been confirmed independently by Edmonds and Beevers (1968). Under these conditions, it is difficult to know if this is a true yield point, or the result of extensive kink band formation (for example) in grain corners, which could lead to an initial yield, and no strain ageing.

Bearing in mind the similarities between titanium and zirconium it would be tempting to believe that their low temperature yield stresses are controlled by similar mechanisms, particularly since Westlake (1965) showed that the yield strength increased as the square root of the oxygen concentration. However, Soo and Higgins (1968a, b) show that Conrad's analysis, applicable to titanium, cannot be used for single crystals of zirconium. Certainly a

Peierls–Nabarro approach cannot apply, because of the sensitivity to impurity level, and Soo and Higgins (ibid.) suggest that at low temperatures a dislocation–oxygen interaction is responsible, with a different process being rate controlling above 300 K.

As with titanium metal, zirconium itself has poor high-temperature creep properties, and for the containment of nuclear fuel elements, subject to considerable dimensional changes, this physical property is a *sine qua non*. The most extensively used are the Zircalloy series, the main additives for creep strength being Sn ~ 1·36%, Fe ~ 0·13%, Cr ~ 0·10% and Ni ~ 0·05%. C, N_2 and H_2 are all < 100 p.p.m., O_2 < 1200 p.p.m.

Typical stress–strain curves for this material are given in Howe and Thomas (1960). Unirradiated samples of Zircalloy, whether in the annealed, cold-worked, or tempered and cold-worked states, did not show yield points at room temperature. However, annealed material, irradiated and tested at 280°C did show a yield point. This isolated result is difficult to explain.

In recent years, information on improved zirconium base canning material has filtered out – one of the most promising apparently being 2·5 wt% Nb alloy (Cupp, 1962). In this paper, no yield point is shown under a variety of testing and irradiation conditions. However, when 0·5 wt% Cu is added to the alloy, to improve corrosion resistance, Ells and Sawatsky (1965) found a yield point present in alloys which had been slow cooled from the $\alpha+\beta$ range. Alloys slow cooled from the β field, or quenched from either did not show the effect, bringing to mind the comment on zirconium made above. If the yield was present before irradiation, it persisted. However, the effect was not examined in great detail, more interest being paid to irradiation effects, but it would be of interest to examine zirconium –0·5% copper as a possible alloy for yield point studies.

7.5.7 *Hexagonal intermetallic compounds*

As mentioned in the previous chapter, intermetallic compounds are often thought of as hard, brittle materials, usually being ordered, and often are quite intractable to tensile testing. However, in recent years, some examples of ductile compounds have been studied. A general discussion of intermetallics, and some cases of cubic materials have been quoted in Chapter 6; it remains here to consider the few hexagonal examples which are known.

The interest in hexagonal compounds lies in the fact that with alloying, it should be possible to make structures with the ideal c/a for close-packing, $\sqrt{\frac{8}{3}} = 1{\cdot}633$. One example of this case is a Ag–33 at.% Al alloy, with $c/a = 1{\cdot}61$, examined by Mote *et al.* (1961).

Like the pure hexagonal metal, single crystals of this phase undergo (0001) [11$\bar{2}$0] basal slip, (10$\bar{1}$2) twinning, and like magnesium, also undergo prismatic slip. Unlike the pure metals, however, the single crystals have a marked yield point, and an enormous Luders strain (Fig. 7.10). The value in the diagram is 135%, a Luders strain exceeded only by polymers and ζ-Cu–Ge, mentioned below.

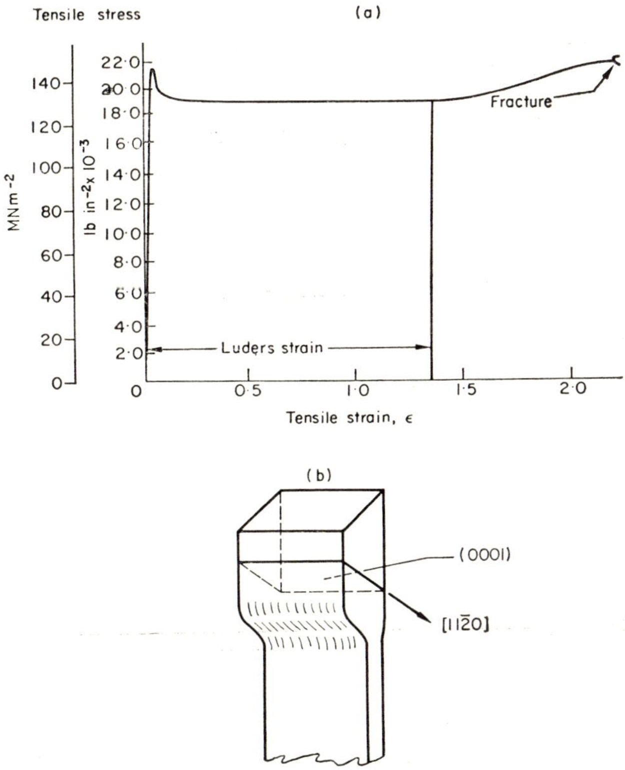

FIGURE 7.10. (a) Stress–strain curve for Ag–33 at.% Al alloy single crystal, and (b) diagrammatic sketch of the Luders front (Mote *et al.*, 1961)

Critical stresses for basal and prismatic slip were examined over a wide range of temperatures, and the modes of deformation and fracture studied. Their view of the existence of the yield point is clear; Cottrell locking is impossible, as the atomic radii of the elements are virtually identical; Suzuki locking is unlikely, for although partials are probably present only a small locking can be expected, and they plump for Fisher ordering. Certainly σ_U and σ_L are virtually independent of temperature, but the actual agreement with theory is not good.

This problem was looked at again by Okamoto and Thomas (1967), who, while confirming the mechanical properties described by Mote *et al.* (ibid.), suggested on the basis of electron microscopy that a high density of dislocation loops was left behind in the slip bands of these alloys. This means that subsequent deformation in these regions is difficult, and slip must then be nucleated in other areas of the crystal, leading to serrations on the lower yield stress.

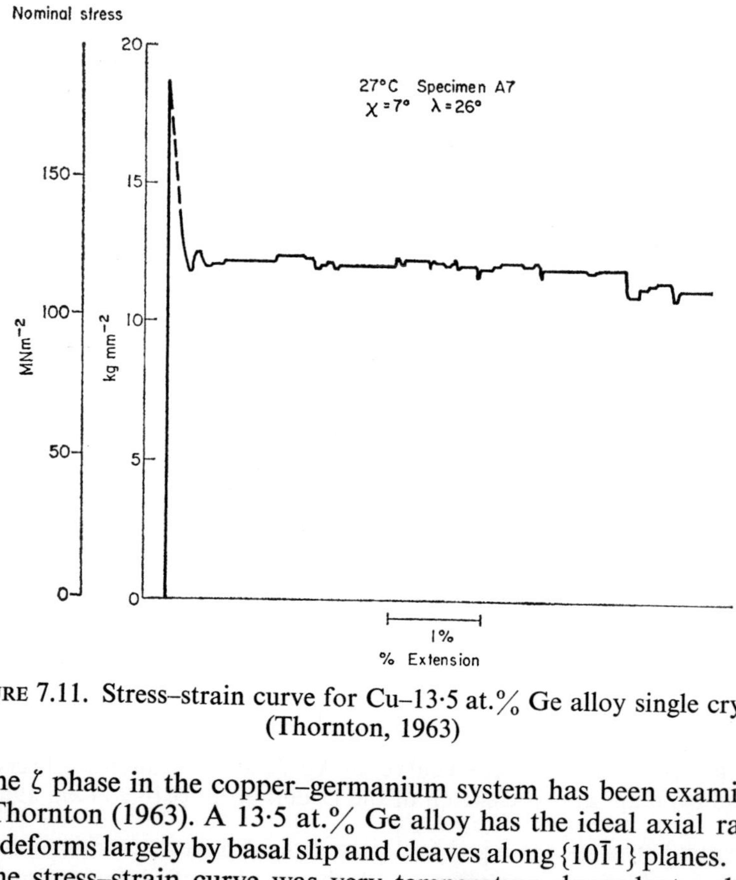

FIGURE 7.11. Stress–strain curve for Cu–13·5 at.% Ge alloy single crystal (Thornton, 1963)

The ζ phase in the copper–germanium system has been examined by Thornton (1963). A 13·5 at.% Ge alloy has the ideal axial ratio, and deforms largely by basal slip and cleaves along $\{10\bar{1}1\}$ planes.

The stress–strain curve was very temperature dependent – above 450° the curve was smooth, serrated yielding was very pronounced between 150–450°C, below 150°C a single yield drop occurred (Fig. 7.11) with the formation of a narrow band of deformation in which the shear was estimated at 500%. A true single crystal Luders band was not formed. The crystals also strain-aged and although no long-range order was detected, the yield effect was put down to short-range order or Suzuki interactions.

In a later paper, Thornton (1965) compared these results with a second alloy containing 16·2 at.% Ge. Whereas the former alloy did not deform by non-basal slip, in the latter alloy crystals suitably oriented were able to deform by extensive prismatic slip, and these showed no yield point. The interpretation thus presents difficulties, but it can certainly be said that these compounds show such remarkable deformation properties and peculiarities that much more research will be carried out on them.

The compounds $TlBi_2$, and In_2Bi also possess a complex hexagonal (B8) structure, and their mechanical properties have been examined by Robinson and Bever (1966) and Grierson and Parkins (1965) respectively.

Both compounds are interesting because of the marked strain rate dependence of the stress–strain curves. At high strain rates, the elongation to fracture is approximately 2%, but at strain rate less than 0·010 per minute in $TlBi_2$, a well rounded yield point hump is observed, and the elongations reach 70%. The yield stresses in $TlBi_2$ obey the Hall–Petch relation, which suggests that yielding occurs first within the grains, but the low work-hardening rate and other metallographic evidence suggests that soon after the yield hump, the formation of voids due to grain boundary sliding is taking place. This will complicate any detailed study of yielding in these materials.

Other extensive studies of the hexagonal compound Mg_3Cd have been carried out by Davies and Stoloff (1964), Stoloff and Davies (1964) and Noble *et al.* (1968). While these confirm that the yield stress of the ordered material is lower than the disordered, in line with results quoted in Chapter 6, this material does not show a yield drop as might have been expected. It must therefore be stressed that the examination of intermetallic compounds as an exercise in yield point studies will need to take into account not only the general impurity levels, but also stoichiometry, heat treatment and initial dislocation density. These factors will greatly complicate any fundamental study of yielding effects.

8

Discussion

In reviewing the theoretical and experimental results of the previous chapters, one might be forgiven for feeling from time to time that the background to date is too scattered and diffuse to warrant the development of any general conclusions. This is, however, too pessimistic an attitude. It is true that serious deficiencies remain, but the progress over the past decade can only be considered as sound, and most encouraging.

In the theoretical aspects, the difficulties remaining are severe, and it would be presumptuous to assume that all factors which contribute to dislocation locking have yet been discovered or discussed. Even on current models many difficulties exist, tied in part to the problems of dealing mathematically with reactions close to the dislocation core, where elastic theory must necessarily break down. For example, the Cottrell–Bilby theory (Section 1.5.1) needs obvious and detailed extension to the case of precipitate nucleation on dislocations, and it is just here that the problems of the nucleation conditions and subsequent diffusion transport become intractable. This is part of the wider problem of heterogeneous nucleation which is still not fully understood, particularly where the morphology and coherency of the precipitate are also closely involved.

In similar vein, the Suzuki or chemical locking theory has to some limited extent found verification in the occurrence of serrated yielding in austenitic steels at high temperatures. Yet the mode of formation of the precipitate does not seem to be that which the theory would predict. No simple theoretical approach appears to be available to estimate diffusion rates and nucleation rates on or near extended faults (Section 2.8.3), and even if it transpires that it is the partial dislocations which are responsible, then this is an extension of the Cottrell–Bilby theory that has been little investigated.

The theory of electronic locking seems to have its major importance in the deformation of ionic solids, and yet the theory as reviewed by Fiore and Bauer (1967) appears to have been mainly applied to the case of metals, where its effect, in general, is small.

Where ageing effects occur over short time intervals, the short-range order theories, also known as Snoek ordering (Section 1.5.3) are generally applied. This theory is also not well developed, although the recent work of Frank (1967a, b) has put it on a more quantitative basis.

Theories of the lower yield again are deficient, particularly in single crystals where the mode of propagation of the single crystal Luders band must be a sensitive function of many variables. Even in polycrystals, the geometry and propagation of the Luders front is not well understood, and this in turn means that the interpretation of the parameters in the Hall–Petch equation is still open to doubt (Section 1.6.4). Our knowledge of grain interfaces remains sketchy, and whether k_y depends critically on the grain boundary morphology or the deformation geometry is again a moot point. Indirect studies, such as the measurement of σ_0 and k_y for dilute alloy systems, are of limited value in so far as one must attempt to compare equitable systems, and cognizance should be taken of any change in mode of, say, slip step distribution.

Any complete interpretation of these parameters will have to be in accord with acceptable theories of solid solution hardening, and the variation of the yield strength of these alloys. These are fields which are largely outside the range of this book, but enough has been written on the latter point (Section 3.9.1) to show that a lively controversy exists and is directly related to yield point theories.

Finally, mention must be made of the current state of strain-ageing theories. The great success of the Cottrell–Bilby theory, and its prediction of the conditions for serrated yielding in many alloy systems, both substitutional and interstitial, still is unable to explain all the details of strain ageing, even in mild steel. In one sense, this is due to our lack of understanding of the transport and nucleation conditions on dislocation cores, as mentioned above, but the main problem as always is lack of a suitable model. This, strangely, we do not have in strain-aged mild steel. It may be that careful electron microscopy or field ion microscopy of partially quench aged samples will provide this, and while we may understand the multiplication of dislocations at the yield we certainly do not understand the nature of the relocking – other than the fact that it is a diffusion controlled phenomenon.

On the subject of experimental investigation, one is entitled to be more dogmatic on the lines of approach which should be adopted.

Too often, even in recent publications, results are reported where the existence of a yield point is merely noted and little or no attempt made to examine the kinetics of the problem and to identify the elements responsible. For example, there are only about two cases of yield point phenomena due to substitutional atoms in the face-centred cubic series which have been systematically investigated. It should be possible, without appearing presumptuous, to refer to a systematic line of approach which in turn should ensure that all the relevant variables are covered. Excluding for the moment single crystal deformation studies, the experimental investigation of yield points falls into two broad groups, those where there is an initial yield point present and others where the yield is initially smooth, but where yield points can later develop after straining or ageing.

The general line of investigation in these two cases could well develop along the set of flow lines as shown in the chart opposite. This should ensure, by kinetic and other structural studies, that the factors responsible for the deficiency in free dislocations may be found, and will, in general, allow a reasonably complete picture of the performance of the alloy system to be built up.

In single crystals, the same general lines of approach could be followed, but here there should be adequate cover of the range of orientations through the unit triangle, and care taken to ensure that geometric softening is not taken as a true yield point effect.

Many face-centred cubic systems might be cited where results to date have been haphazard, incomplete, or both. This is of course no fault of the original investigators, who were often studying other facets, but it should be apparent that many of the alloy systems mentioned in Chapters 5 and 6 are now ripe for review. In particular, there is urgent need to correlate the electron microscopy studies of many of the precipitation hardening systems with the mechanical properties. Too often electron microscopists are content to use hardness determinations as a measure of mechanical properties changes. The existence of, or changes in, yield point effects are equally noteworthy as changes in structure, and correlation may even elucidate some of the theories of strain ageing.

One field of mechanical properties studies looms large as a possible field for much exciting new work. This is the field of alloy studies with yield points engendered by minor interstitial levels – the study of nickel-base alloys with hydrogen or carbon is a good example. There must be other examples where deliberate alloying can be used to study the disappearance or enhancement of the yield. A natural extension of this is into those intermetallic compounds, whose mechanical properties are often affected in many ways across their phase field by

Initial yield

- Does not reappear on strain ageing
  - Investigate grain size, strain rate and temperature effects
  - Determine σ_0, k_y and m^*
  - Luders band studies: determination of v_L, ϵ_L and m^*
  - Determine dislocation densities
  - Determine if yield due to (a) deficiency of free dislocations or (b) unlocking of dislocations
  - If (b) determine locking factor by s.e.d. or purification
  - Correlation with theory
- Develops on strain ageing

No initial yield

- Develops on strain ageing
- Does not develop on strain ageing
  - No interest here

Develops on strain ageing

- Develops after small strain
  - Misalignment possible: proceed first as in l.h. column
- Develops after significant strain
  - Vacancy enhanced diffusion? Determine ϵ_c, m', β

Both then lead to:

- Investigate strain-ageing kinetics: Static studies: determine E and time laws. Kinetic studies: determine E, U.C.T., L.C.T., and binding energy
- T.e.m. studies on aged material
- Determination of element responsible and effects of alloying additions
- Correlation with theory

excess point defects or other reasons. Interstitial impurities can again react with these defects to produce many interesting effects (Section 6.7).

Thus it is possible to lay down some general guide lines, both on the theoretical and experimental fronts, which may help to sustain interest in this important area of crystal deformation.

Appendix

Variation of Yield stress with temperature

From equation (3.1) written in terms of shear components

$$\dot{\epsilon} = \rho b s \nu^* \exp\left(\frac{-H(\tau, T)}{kT}\right) \tag{A1}$$

and if $\tau = \tau^* + \tau_\mu$

then
$$H(\tau, T) = kT \ln(\nu/\dot{\epsilon}) \tag{A2}$$

where
$$\nu = \rho b s \nu^*$$

Since H, according to Conrad (1963b), is primarily a function of τ^*, then differentiating (A2) gives

$$-\frac{\mathrm{d}H}{\mathrm{d}\tau^*} = \mathrm{k}T\left(\frac{\partial \ln(\dot{\epsilon}/\nu)}{\partial \tau}\right)_T \tag{A3}$$

Alternatively,

$$-\frac{\mathrm{d}H}{\mathrm{d}\tau^*} = \left(\frac{\partial H}{\partial T}\right)_\nu \frac{\mathrm{d}T}{\mathrm{d}\tau^*}$$

which, from (A2) yields, again with the same assumption

$$-\frac{dH}{\mathrm{d}\tau^*} = k\frac{\ln(\dot{\epsilon}/\nu)}{(\partial \tau^*/\partial T)_{\dot{\epsilon}}} \tag{A4}$$

Equating (A3) and (A4)

$$T\left(\frac{\partial \tau^*}{\partial T}\right)_{\dot{\epsilon}}\left(\frac{\partial \ln(\dot{\epsilon}/\nu)}{\partial \tau}\right)_T = \ln(\dot{\epsilon}/\nu) \tag{A5}$$

Substituting in (A2), it is now found that

$$H = -kT^2\left(\frac{\partial \tau^*}{\partial T}\right)_{\dot{\epsilon}}\left(\frac{\partial \ln(\dot{\epsilon}/\nu)}{\partial \tau}\right)_T \tag{A6}$$

(cf equation (3.3)).

Again, since from (A2),

$$\ln(\dot{\epsilon}/\nu) = -H/kT$$

$$\left(\frac{\partial(\ln(\dot{\epsilon}/\nu))}{\partial(1/T)}\right)_{\tau^*} = -\frac{H}{k}$$

so that

$$H = -k\left(\frac{\partial \ln(\dot{\epsilon}/\nu)}{\partial(1/T)}\right)_{\tau^*} \tag{A7}$$

(cf. equation (3.2)).

By assuming the activation energy is a function of the thermal component of stress τ^* per unit volume then the derivative $-\partial H/\partial\tau^*$ may be defined as the activation volume v^*. Hence, using (A3)

$$v^* = kT\left(\frac{\partial \ln(\dot{\epsilon}/\nu)}{\partial\tau^*}\right)_T \tag{A8}$$

(cf. equation (3.4)).

Finally, H itself may be derived by substitution in (A6), which gives

$$H = -v^*T\left(\frac{\partial\tau^*}{\partial T}\right)_{\dot{\epsilon}} \tag{A9}$$

(cf equation (3.5)).

Tensile data may be used to obtain the values of these parameters, by assuming $\tau = \rho/2$, while the shear strain (which has been used in this analysis) will be 0·7 times the tensile strain.

A series of stress–strain curves at varying strain rates over the necessary temperature range will thus allow these parameters H and v^* to be determined.

Bibliography

ABBOT, B. (1964) B.Sc. Thesis, University of Newcastle

ABRAHAMS, M. S. (1962) *Acta Metall.*, **10**, 989

ABRAHAMS, M. S. AND LIEBMANN, W. K. (1962) *Acta Metall.*, **10**, 941

ADAIR, A. M. AND HOOK, R. E. (1962) *Acta Metall.*, **10**, 741

ADAIR, A. M. AND HOOK, R. E. (1963) *Acta Metall.*, **11**, 81

ADAMS, M. A. (1958) *Acta Metall.*, **6**, 327

ADAMS, M. A. (1959) *J. Sci. Instrum.*, **36**, 444

ADAMS, M. A., ROBERTS, A. C. AND SMALLMAN, R. E. (1960) *Acta Metall.*, **8**, 328

ADCOCK, F. AND BRISTOW, C. A., (1935) *Proc. R. Soc.*, **A153**, 172

ALERS, G. A., ARMSTRONG, R. W. AND BECHTOLD, J. H. (1958) *Trans. A.I.M.E.*, **212**, 523

ALLEN, B. C., MAYKUTH, D. J. AND JAFFEE, R. I. (1963) *Trans A.I.M.E.*, **227**, 724

ALLEN, J. W. (1957) *Phil. Mag.*, **2**, 1475

ALLEN, J. W. (1958) *Phil. Mag.*, **3**, 1297

ALLEN, N. P., HOPKINS, B. E. AND MCLENNAN, J. E. (1956) *Proc. R. Soc.*, **A234**, 221

ALLEN, N. P. (1963) *Iron and its Dilute Solid Solutions* (Interscience, N.Y.), p. 271

ALMOND, E. A. AND HULL, D. (1966) *Phil. Mag.*, **14**, 515

ALTSHULER, T. L. AND CHRISTIAN, J. W. (1966) *Acta Metall.*, **14**, 903

ANDERSON, R. W. and BRONISZ, S. E. (1959) *Acta Metall.*, **7**, 645

ANDERSON, E., KING, L. AND SPREADBOROUGH, J. (1968) *Trans. A.I.M.E.*, **242**, 115.

ANDERSON, E., BODE, R., DORN, J. E. AND SPREADBOROUGH, J. (1969) *Metal Science J.*, **3**, 201

APPLETON, A. S. AND WADDINGTON, J. S. (1965) *Phil. Mag.*, **12**, 273

ARDLEY, G. W. AND COTTRELL, A. H. (1953) *Proc. R. Soc.*, **A219**, 328

ARDLEY, G. W. (1955) *Acta Metall.*, **3**, 525

ARGON, A. S. AND MALOOF, S. R. (1966) *Acta Metall.*, **14**, 1449.

ARKHAROV, V. I., BERENOVA, I. P. AND MAGAT, L. M. (1957) *Physics of Metals and Metallog.*, **5** (3), 123.

ARMSTRONG, R., CODD, I., DOUTHWAITE, R. M. AND PETCH, N. J. (1962) *Phil. Mag.*, **7**, 45.

ARROWSMITH, R. (1924) *J. Iron Steel Inst.*, **110**, 317

ARROWSMITH, J. M. (1959) *J. Iron Steel Inst.*, **192**, 286

ARROWSMITH, J. M. (1963) *J. Iron Steel Inst.*, **201**, 699

ARSENAULT, R. J. (1963) *Acta Metall.*, **11**, 1111

ARSENAULT, R. J. AND WEERTMAN, J. (1963) *Acta Metall.*, **11**, 1119

ARSENAULT, R. J. (1964a) *Acta Metall.*, **12**, 758

ARSENAULT, R. J. (1964b) *Trans A.I.M.E.*, **230**, 1570

ARSENAULT, R. J. AND KOPPENAAL, T. J. (1965) *Appl. Phys. Letters*, **6**, 159

ARSENAULT, R. J. (1966) *Acta Metall.*, **14**, 831

ARSENAULT, R. J. (1967a) *Acta Metall.*, **15**, 501

ARSENAULT, R. J. (1967b) *Acta Metall.*, **15**, 1853

ARSENAULT, R. J. AND LAWLEY, A. (1967) *Phil. Mag.*, **15**, 549

ARSENAULT, R. J. (1968) *Scripta Metall.*, **2**, 99

ASKEW, B. A. AND WELLS, T. C. (1967) *J. Iron and Steel Inst.*, **205**, 869

BACH, C. AND BAUMANN, R. (1921) *Festigkeitseigenschaften und Gefügebilder der Konstruktionmaterialen* (Springer, Berlin), p. 137

BACON, D. J. (1968) *Acta Metall.*, **16**, 403

BACON, D. J. (1969) *Scripta Metall.*, **3**, 735

BAILEY, D. J., FLANAGAN, W. F. AND MILLER, G. E. (1965) *Acta Metall.*, **13**, 436

BAILEY, D. J. (1967) *Phil. Mag.*, **16**, 861

BAILEY, D. J. AND FLANAGAN, W. F. (1967) *Phil. Mag.*, **15**, 43

BAILEY, D. J., AND FLANAGAN, W. F. (1969) *Phil. Mag.*, **19**, 1093

BAILON, J. P. AND MCQUEEN, H. J. (1967) *Acta Metall.*, **15**, 580

BAIRD, J. D. (1963) *Iron and Steel*, **36**, pp. 186, 326, 368 and 400

BAIRD, J. D. AND JAMIESON, A. (1963) N.P.L. Symposium, No. 15, 'Relation between structure and mechanical properties of metals', p. 361 (HMSO)

BAIRD, J. D. AND MACKENZIE, C. R. (1964) *J. Iron Steel Inst.*, **202**, 427

BAIRD, J. D. (1966a) *J. Iron Steel Inst.*, **204**, 44

BAIRD, J. D. (1966b) *J. Iron Steel Inst.*, **204**, 1122

BAIRD, J. D. (1967) *J. Iron Steel Inst.*, **205**, 718

BAIRD, W. R. AND HARTLEY, C. S. (1963) *J. Inst. Metals*, **92**, 181

BAKER, R. G. AND NUTTING, J. (1959) Iron and Steel Inst. Special Report No. 64, p. 1

BAKER, T. N. (1967) *J. Iron Steel Inst.*, **205**, 315

BALASUBRAMANIAM, N. (1969) *Scripta Metall.*, **3**, 21

BALDWIN, W. M. Jnr. (1958) *Acta Metall.*, **6**, 139

BALL, A., AND SMALLMAN, R. E. (1966) *Acta Metall.*, **14**, 1349

BALL, A., BERGERSEN, S. G. AND HUTCHISON, M. M. (1968) *Trans. Jap. Inst. Metals Supplement*, **9**, 291

BALL, C. J. (1959) *J. Iron Steel Inst.*, **191**, 232

BARNBY, J. T. (1965) *J. Iron Steel Inst.*, **203**, 392

BARNBY, J. T. (1966) *J. Iron Steel Inst.*, **204**, 23

BARNES, R. S. AND HANCOCK, N. H. (1958) *Phil. Mag.*, **3**, 527

BARRAND, P. AND LEAK, G. M. (1963) *Acta Metall.*, **11**, 158

BARRETT, C. S. AND BAKISH, R. (1958) *Trans A.I.M.E.*, **212**, 122

BARTON, P. J., HARRIES, D. R. AND MOGFORD, I. L. (1965) *J. Iron Steel Inst.*, **203**, 507

Bassani, F. and Thomson, R. (1956) *Phys. Rev.*, **102**, 1264

Bazinski, Z. S. (1957) *Proc. R. Soc.*, **A240**, 229

Bazinski, Z. S. and Sleeswijk, A. (1957) *Acta Metall.*, **5**, 176

Bazinski, Z. S. (1960) *Aust. J. Phys.*, **13**, 354

Bazinski, Z. S. and Christian, J. W. (1960) *Aust. J. Phys.*, **13**, 299

Beardmore, P. and Hull, D. (1965) *J. Less-Common Metals*, **9**, 168

Bechtold, J. H. and Scott, H. (1951) *J. Electrochem. Soc.*, **98**, 495

Bechtold, J. H. (1953) *Trans A.I.M.E.*, **197**, 1469

Bechtold, J. H. (1954) *Trans Am. Soc. Metals*, **46**, 1449

Bechtold, J. H. (1955) *Acta Metall.*, **3**, 249

Bechtold, J. H. (1956) *Trans A.I.M.E.*, **206**, 142

Bechtold, J. H., Wessel, E. T. and France, L. J. (1961) *Refractory Metals and Alloys* (Interscience), p. 25

Beeler, J. R., Jnr. (1966) *Intermetallic compounds* (Wiley), Ch. 14

Berggren, R. G., and Wilson, J. C. (1957) Oak Ridge National Lab., Report CF-56-11-1

Berlec, I. (1963) *Acta Metall.*, **11**, 69

Bernstein, I. M. and Gensamer, M. (1968) *Acta Metall.*, **16**, 987

Bernstein, I. M. (1969) *Acta Metall.*, **17**, 249

Besag, F. M. C. and Bullen, F. P. (1965) *Phil. Mag.*, **12**, 41

Besag, F. M. C. and Bullen, F. P (1966) *Phil. Mag.*, **14**, 1259

Beshers, D. N. (1958) *Acta Metall.*, **6**, 521

Biggs, W. D. and Broom, T. (1954) *Phil. Mag.*, **45**, 246

Biggs, W. D. and Pratt, P. L. (1958) *Acta Metall.*, **6**, 694

Birkbeck, G. and Douthwaite, R. M. (1968) *Trans. A.I.M.E.*, **242**, 1595

Birnbaum, H. K. (1961) *Acta Metall.*, **9**, 320

Birnbaum, H. K. and Tuler, F. R. (1961) *J. Appl. Phys.*, **32**, 1403

Birnbaum, H. K. (1963a) *J. Appl. Phys.*, **33**, 750

Birnbaum, H. K. (1963b) *J. Appl. Phys.*, **34**, 2175

Bishop, S. M., Spretnak, J. W. and Fontana, M. G. (1953) *Trans. Am. Soc. Metals*, **45**, 993

Blakemore, J. S. (1968) Thesis, University of Newcastle

Blakemore, J. S. and Hall, E. O. (1966) *J. Iron Steel Inst.*, **204**, 817

Blakemore, J. S. and Hall, E. O. (1968a) *Trans. A.I.M.E.*, **242**, 333

Blakemore, J. S. and Hall, E. O. (1968b) *J. Nucl. Mater.* **29**, 238

Blanchard, P. and Troiano, A. R. (1960) *Mem. scient. Revue Metall.*, **57**, 409

Blewitt, T. H. (1953) *Phys. Rev.*, **91**, 1115

Blewitt, T. H., Coltman, R. R. and Redman, J. K. (1957) *J. Appl. Phys.*, **28**, 651

Blewitt, T. H., Coltman, R. R., Jamison, R. E. and Redman, J. K. (1960) *J. Nucl. Mater.*, **2**, 277

Boas, W. and Honeycombe, R. W. K. (1947) *Proc. R. Soc.*, **A188**, 427

Boas, W., and Schmid, E. (1929) *Z. Physik*, **54**, 16

Boček, M., Scharf, H., Mannel, P. and Zouhar, M. (1965) *Z. Metallkunde*, **56**, 353

Bolling, G. F. (1959) *Phil. Mag.*, **4**, 537

BOLLING, G. F., HAYS, L. E. AND WIEDERSICH, H. W. (1961) *Acta Metall.*, **9**, 622
BOLLING, G. F. AND RICHMAN, R. H. (1965) *Acta Metall.*, **13**, 709
BONISZEWSKI, T. AND SMITH, G. C. (1961) *J. phys. Chem. Solids*, **21**, 115
BONISZEWSKI, T. AND SMITH, G. C. (1963) *Acta Metall.*, **11**, 165
BOULANGER, C. (1950) *Revue Metall.*, **47**, 547
BOWEN, D. K., CAPUS, J. M. AND SILVERSTONE, C. E. (1967) *Phil. Mag.*, **15**, 1041
BOXALL, T. D. AND HUNDY, B. B. (1955) *Metallurgia*, **51**, 52
BRADFORD, S. A. AND CARLSON, O. N. (1962) *Trans. A.I.M.E.*, **224**, 738
BRANDON, D. G. AND NUTTING, J. (1960) *J. Iron Steel Inst.*, **196**, 160
BRENNER, S. S. (1958) *Growth and Perfection of Crystals* (Wiley), pp. 157 ff
BRENTNALL, W. D. AND ROSTOKER, W. (1965) *Acta Metall.*, **13**, 187
BREYER, N. N. AND POLAKOWSKI, N. H. (1962) *Trans. Am. Soc. Metals*, **55**, 667
BREYER, N. N. (1966) *Trans. A.I.M.E.*, **236**, 1198
BRINDLEY, B. J., CORDEROY, D. J. H. AND HONEYCOMBE, R. W. K. (1962) *Acta Metall.*, **10**, 1043
BRINDLEY, B. J. AND BARNBY, J. T. (1966) *Acta Metall.*, **14**, 1765
BRINDLEY, B. J. AND BARNBY, J. T. (1968) *Acta Metall.*, **16**, 41
BRINDLEY, B. J. AND WORTHINGTON, P. J. (1969a) *Acta Metall.*, **17**, 1357
BRINDLEY, B. J. AND WORTHINGTON, P. J. (1969b) *Phil. Mag.*, **19**, 1175
BRITTAIN, J. O. AND BRONISZ, S. E. (1960) *Trans. A.I.M.E.*, **218**, 289
BROCK, G. W. (1961) *Trans. A.I.M.E.*, **221**, 1055
BROOMFIELD, G. H., HARRIES, D. R. AND ROBERTS, A. C. (1965) *J. Iron Steel Inst.*, **203**, 502
BROWN, L. M. (1961) *Physica Status Solidi*, **1**, 585
BROWN, L. M. AND PRATT, P. L. (1963) *Phil. Mag.*, **8**, 717
BROWN, N. (1959) *Phil. Mag.*, **4**, 693
BROWN, N. AND LUKENS, K. F., Jnr. (1961) *Acta Metall.*, **9**, 106
BROWN, N. AND EKVALL, R. A. (1962) *Acta Metall.*, **10**, 1101
BRYDGES, W. T. (1969) *Scripta Metall.*, **3**, 271
BULLEN, F. P., HENDERSON, F., HUTCHISON, M. M. AND WAIN, H. L. (1964a) *Phil. Mag.*, **9**, 285
BULLEN, F. P., HENDERSON, F. AND WAIN, H. L. (1964b) *Phil. Mag.*, **9**, 803
BULLEN, F. P. AND ROGERS, C. B. (1964) *Phil. Mag.*, **9**, 401
BULLEN, F. P., HENDERSON, F., HUTCHISON, M. M. AND WAIN, H. L. (1967) *Scripta Metall.*, **1**, 29
BULLOUGH, R. AND NEWMAN, R. C. (1959) *Proc. R. Soc.*, **A249**, 427
BULLOUGH, R. AND NEWMAN, R. C. (1962a) *Proc. R. Soc.*, **A266**, 198
BULLOUGH, R. AND NEWMAN, R. C. (1962b) *Proc. R. Soc.*, **A266**, 209
BULLOUGH, R. AND NEWMAN, R. C. (1962c) *Acta Metall.*, **10**, 971
BURBACH, R., MORDIKE, B. L. AND HAASEN, P. (1966) *J. Iron Steel Inst.*, **204**, 390
BURKE, J. E. (1963) *Mater Sci. Research*, **1**, 69
BUSH, R. H. AND HUGGINS, R. A. (1964) *Acta Metall.*, **12**, 697
BUTLER, J. F. (1962a) *Acta Metall.*, **10**, 258

Butler, J. F. (1962b) *J. Mech. Phys. Solids*, **10**, 313

Butler, J. F. (1962c) *Trans. A.I.M.E.*, **224**, 89

Butler, J. F. (1966) *J. Iron Steel Inst.*, **204**, 127

Butler, R. D. and Wilson, D. V. (1963) *J. Iron Steel Inst.*, **201**, 16

Byrne, J. C. (1961) *Acta Metall.*, **9**, 1037

Cabrera, N. and Price, P. B. (1958) *Growth and Perfection of Crystals* (Wiley), pp. 204 ff

Cahn, J. W. (1957) *Acta Metall.*, **5**, 169

Caisso, J. (1959) *Revue Metall.*, **56**, 237

Caisso, J. and Guillot, J. (1962) *Mem. Scient. Revue Metall.*, **59**, 395

Campbell, J. D. (1953) *Acta Metall.*, **1**, 706

Campbell, J. D. and Marsh, K. J. (1962) *Phil. Mag.*, **7**, 933

Carlsen, K. and Honeycombe, R. W. K. (1954) *J. Inst. Metals*, **83**, 449

Carnahan, R. D., Arsenault, R. J. and Stone, G. A. (1967) *Trans. A.I.M.E.*, **239**, 1193

Carpenter, S. H. and Baker, G. S. (1965a) *Acta Metall.*, **13**, 917

Carpenter, S. H. and Baker, G. S. (1965b) *J. Appl. Phys.*, **36**, 1733

Carreker, R. P. and Guard, R. W. (1956) *Trans. A.I.M.E.*, **206**, 178

Carreker, R. P. and Hibbard, W. R. (1957) *Trans. A.I.M.E.*, **209**, 1157

Carreker, R. P. (1957) *Trans. A.I.M.E.*, **209**, 112

Carrington, W., Hale, K. F. and McLean, D. (1960) *Proc. R. Soc.*, **A259**, 203

Carrington, W. E. and McLean, D. (1965) *Acta Metall.*, **13**, 493

Chadwick, R. and Hooper, W. H. L. (1951) *J. Inst. Metals*, **80**, 17

Charnock, W. (1968) *Phil. Mag.*, **18**, 89

Charnock, W. (1969a) *Phil. Mag.*, **19**, 209

Charnock, W. (1969b) *Phil. Mag.*, **20**, 427

Chaudhuri, A. R., Patel, J. R. and Rubin, L. G. (1962) *J. Appl. Phys.*, **33**, 2736

Chiao, W. F. and Gordon, R. B. (1965) *Trans. A.I.M.E.*, **233**, 1164

Chossat, H. (1950a) *Revue Metall.*, **47**, 167

Chossat, H. (1950b) *Revue Metall.*, **47**, 343

Chow, J. G. Y., McRickard, S. B. and Gurinsky, D. H. (1963) *Proc. Amer. Soc. Testing Mater.*, S.T.P. No. 341

Christ, B. W. and Smith, G. V. (1967) *Acta Metall.*, **15**, 809

Christian, J. W. (1964) *Acta Metall.*, **12**, 99

Christian, J. W. and Masters, B. C. (1964) *Proc. R. Soc.*, **A281**, 223 and 240

Christian, J. W. (1967) *Acta Metall.*, **15**, 1257

Churchman, A. T. (1954) *Proc. R. Soc.*, **A226**, 216

Churchman, A. T. (1955) *Acta Metall.*, **3**, 22

Churchman, A. T., Mogford, I. and Cottrell, A. H. (1957) *Phil. Mag.*, **2**, 1271

Churchman, A. T. (1959) *J. Inst. Metall.*, **88**, 221

Clarebrough, L. M. (1957) *Acta Metall.*, **5**, 413

Clough, W. R. and Pavlovic, A. S. (1960) *Trans. Am. Soc. Metals*, **52**, 948

COCHARDT, A., SCHOECK, G. AND WIEDERSICH, H. (1955) *Acta Metall.*, **3**, 533
CODD, I. AND PETCH, N. J. (1960) *Phil. Mag.*, **5**, 30
COLEMAN, C. E. AND HARDIE, D. (1966a) *J. Less-Common Metals*, **11**, 168
COLEMAN, C. E. AND HARDIE, D. (1966b) *J. Inst. Metals*, **94**, 387
CONRAD, H. (1960) *Phil. Mag.*, **5**, 745
CONRAD, H. AND SCHOECK, G. (1960) *Acta Metall.*, **8**, 791
CONRAD, H. AND WIEDERSICH, H. (1960) *Acta Metall.*, **8**, 128
CONRAD, H. (1961) *J. Iron Steel Inst.*, **198**, 364
CONRAD, H. AND FREDERICH, S. (1962) *Acta Metall.*, **10**, 1013
CONRAD, H. (1963a) *Acta Metall.*, **11**, 75
CONRAD, H. (1963b) N.P.L. Symposium, No. 15, 'Relation between structure and mechanical properties of metals', p. 475 (HMSO)
CONRAD, H. (1963c) *J. Mech. Phys. Solids*, **11**, 437
CONRAD, H. AND HAYES, W. (1963a) *Trans. Am. Soc. Metals*, **56**, 125
CONRAD, H. AND HAYES, W. (1963b) *Trans. Am. Soc. Metals*, **56**, 249
CONRAD, H. AND STONE, G. (1964) *J. Mech. Phys. Solids*, **12**, 139
CONRAD, H., STONE, G. AND JANOWSKI, K. (1965) *Trans. A.I.M.E.*, **233**, 889
CONRAD, H. (1966) *Acta Metall.*, **14**, 1631
CONRAD, H. (1967) *Acta Metall.*, **15**, 147
CONTE, R. AND GROH, P. (1968) *Mem. Scient. Revue Metall.*, **65**, 93
COTTRELL, A. H. (1948) 'Report of a Conference on Strength of Solids' (Physical Society), p. 30
COTTRELL, A. H. AND GIBBONS, D. F. (1948) *Nature*, **162**, 488
COTTRELL, A. H. AND BILBY, B. A. (1949) *Proc. Phys. Soc.*, **A62**, 49
COTTRELL, A. H. AND CHURCHMAN, A. T. (1949a) *J. Iron Steel Inst.*, **162**, 271
COTTRELL, A. H. AND CHURCHMAN, A. T. (1949b) *Trans. A.I.M.E.*, **185**, 877
COTTRELL, A. H. AND CHURCHMAN, A. T. (1950) *Nature*, **167**, 943
COTTRELL, A. H. AND JASWON, M. A. (1949) *Proc. R. Soc.*, **A199**, 104
COTTRELL, A. H. AND LEAK, G. M. (1952) *J. Iron Steel Inst.*, **172**, 301
COTTRELL, A. H. (1953a) *Dislocations and Plastic Flow in Crystals* (O.U.P.)
COTTRELL, A. H. (1953b) *Phil. Mag.*, **44**, 829
COTTRELL, A. H., HUNTER, S. C. AND NABARRO, F. R. N. (1953) *Phil. Mag.*, **44**, 1064
COTTRELL, A. H. (1954) *Relation of Properties to Microstructure* (A.S.M., Cleveland), p. 131
COTTRELL, A. H. AND STOKES, R. J. (1955) *Proc. R. Soc.*, **A233**, 17
COTTRELL, A. H. (1957a) *J. Iron Steel Inst.*, **187**, 219
COTTRELL, A. H. (1957b) Conference on Properties of Materials at High Rates of Strain (London, Inst. of Mech, Eng.), p. 1
COTTRELL, A. H. (1958) *Trans. A.I.M.E.*, **212**, 192
COTTRELL, A. H. (1963) N.P.L. Symposium, No. 15, 'Relation between structure and mechanical properties of metals', p. 456 (HMSO)
COTTRELL, A. H. (1964) *Mechanical Properties of Matter* (Wiley)
COTTERILL, P. (1961) *Progress in Mater. Sci.*, **9**, 201
COULING, S. L. (1959) *Acta Metall.*, **7**, 133

COX, J. J., HORNE, G. I. AND MEHL, R. F. (1957) *Trans. Am. Soc. Metals*, **49**, 118
CRACKNELL, A. AND PETCH, N. J. (1955a) *Acta Metall.*, **3**, 186
CRACKNELL, A. AND PETCH, N. J. (1955b) *Acta Metall.*, **3**, 200
CRUSSARD, C. (1963a) N.P.L. Symposium, No. 15, 'Relation between structure and mechanical properties of metals', p. 547 (HMSO)
CRUSSARD, C. (1963b) *J. Aust. Inst. Metals*, **8**, 317
CRUSSARD, C. (1964) *Mem. Scient. Revue Metall.*, **61**, 231
CUPP, C. R. AND CHALMERS, B. (1954) *Acta Metall.*, **2**, 803
DAHL, W. AND LÜCKE, K. (1954) *Arch. für das Eisenhütt.*, **25**, 241
DALBY, W. E. (1913) *Proc. R. Soc.*, **A88**, 281
DAVENPORT, E. S. AND BAIN, E. C. (1935) *Trans. Am. Soc. Metals*, **23**, 1047
DAVIDSON, D. L., LINDHOLM, U. S. AND YEAKLEY, L. M. (1966) *Acta Metall.*, **14**, 703
DAVIES, R. G. AND STOLOFF, N. S. (1964) *Trans. A.I.M.E.*, **230**, 390
DAVIES, R. G. AND GILBERT, A. (1967) *Acta Metall.*, **15**, 665
DEWEY, M. A. P., SUMMER, G. AND BRAMMAR, I. S. (1965) *J. Iron Steel Inst.*, **203**, 938
DEW-HUGHES, D. AND ROBERTSON, W. D. (1960) *Acta Metall.*, **8**, 147 and 156
DINGLEY, D. J. AND HALE, K. F. (1966) *Proc. R. Soc.*, **A295**, 55
DINGLEY, D. J. AND MCLEAN, D. (1967) *Acta Metall.*, **15**, 885
DOLLINS, C. AND WERT, C. (1969) *Acta Metall.*, **17**, 711
DOREMUS, R. H. (1958) *Acta Metall.*, **6**, 674
DOREMUS, R. H. (1959) *Acta Metall.*, **7**, 399
DOREMUS, R. H. (1960) *Trans. A.I.M.E.*, **218**, 596
DOREMUS, R. H. AND KOCH, E. F. (1960) *Trans. A.I.M.E.*, **218**, 591
DORN, J. E. AND RAJNAK, S. (1964) *Trans. A.I.M.E.*, 230, 1052
DUFFAUT, F. AND LACOMBE, P. (1965) *C.R. Acad. Sci.*, **261**, 3398
DYSON, B. F., JONES, R. B. AND TEGART, W. J. MCG. (1958) *J. Inst. Metals.*, **87**, 340
EBORALL, R., LACK, M. AND PHILLIPS, V. A. (1952) *Bull. Inst. Metals*, **1**, 58
EDINGTON, J. W., LINDLEY, T. C. AND SMALLMAN, R. E. (1964) *Acta Metals*, **12**, 1025
EDINGTON, J. W. AND SMALLMAN, R. E. (1964) *Acta Metall.*, **12**, 1313
EDINGTON, J. W. AND SMALLMAN, R. E. (1965) *Acta Metall.*, **13**, 765
EDMONDS, D. V. AND BEEVERS, C. J. (1968) *Metal Science J.*, **2**, 228
EDWARDS, C. A. AND PFEIL, L. B. (1925) *J. Iron Steel Inst.*, **112**, 79
EDWARDS, C. A., JONES, H. N. AND WALTERS, B. (1939) *J. Iron Steel Inst.*, **140**, 341
EDWARDS, C. A., PHILLIPS, D. L. AND JONES, H. N. (1940) *J. Iron Steel Inst.*, **142**, 199
EDWARDS, C. A., PHILLIPS, D. L. AND LIU, Y. H. (1943) *J. Iron Steel Inst.* **147**, 45
EIKUM, A. AND THOMAS, G. (1964) *Acta Metall.*, **12**, 537
ELAM, C. F. (1927a) *Proc. R. Soc.*, **A115**, 133, 148 and 167

ELAM, C. F. (1927b) *Proc. R. Soc.*, **A116**, 694
ELAM, C. F. (1936) *Proc. R. Soc.*, **A153**, 273
ELAM, C. F. (1938) *Proc. R. Soc.*, **A165**, 568
ELLS, C. E. AND SAWATZKY, A. (1965) *Trans. A.I.M.E.*, **233**, 2041
EMBURY, J. D. AND NICHOLSON, R. B. (1963) *Acta Metall.*, **11**, 347
EMBURY, J. D. AND FISHER, R. M. (1966) *Acta Metall.*, **14**, 147
ENRIETTO, J. F. (1966) *J. Iron Steel Inst.*, **204**, 252
ENRIETTO, J. F., SINCLAIR, G. M. AND WERT, C. A. in *Columbium Metallurgy* (Interscience), p. 503
ENTWISTLE, K. N., FELL, J. H. AND KANG, I. K. (1962) *J. Inst. Metals*, **91**, 84
EPSTEIN, S., CUTLER, H. J. AND FRAME, J. W. (1950) *Trans. A.I.M.E.*, **188**, 830
ERGANG, R. (1952) *Metal Industry*, **81**, 261
ERGANG, R. AND WELZ, S. (1952) *Z. Metallkunde*, **43**, 45
ERGANG, R. AND WELZ, S. (1953) *Metall.*, **7**, 254
ESHELBY, J. D., FRANK, F. C. AND NABARRO, F. R. N. (1951) *Phil. Mag.*, **42**, 351
ESHELBY, J. D., NEWEY, C. W. A., PRATT, P. L. AND LIDIARD, A. B. (1958) *Phil. Mag.*, **3**, 75
ESHELBY, J. D. (1959) *Acta Metall.*, **7**, 631
EUSTICE, A. L. AND CARLSON, O. N. (1961a) *Trans. A.I.M.E.*, **221**, 238
EUSTICE, A. L. AND CARLSON, O. N. (1961b) *Trans. Am. Soc. Metals*, **53**, 501
EVANS, J. T. AND RAWLINGS, R. (1968) *Metal Science J.*, **2**, 221
EVANS, K. R. AND FLANAGAN, W. F. (1966) *Trans. A.I.M.E.*, **236**, 1381
EVANS, K. R. (1969) *Scripta Metall.*, **3**, 627
EVANS, P. R. V. (1962) *J. Less-Common Metals*, **4**, 78
EVANS, P. R. V., WEINBERG, A. F. AND VAN THYNE, R. J. (1963) *Acta Metall.*, **11**, 143
EVERS, M. (1959) *Z. Metallkunde*, **50**, 638
EVERS, M. (1961) *Z. Metallkunde*, **52**, 359
FARRELL, J. W. (1961) *Trans. Am. Soc. Metals*, **54**, 143
FARRELL, K. (1965) *J. Iron Steel Inst.*, **203**, 71
FARRELL, K., SCHAFFHAUSER, A. C. AND STIEGLER, J. O. (1967) *J. Less-Common Metals*, **13**, 548
FELBECK, D. K., GIBBONS, W. G. AND OVENS, W. G. (1965) *Trans. Am. Soc. Mech. Eng., J. Basic Eng.*, **87**, 319
FELL, E. W. (1937) *J. Iron Steel Inst., Carnegie Schol. Mem.*, **26**, 123
FELTHAM, P. AND MEAKIN, J. D. (1957) *Phil. Mag.*, **2**, 105
FELTHAM, P. AND COPLEY, G. J. (1960) *Acta Metall.*, **8**, 542
FENG, C. AND KRAMER, I. R. (1963) *J. Metals*, **15**, 674
FENG, C. AND KRAMER, I. R. (1965) *Trans. A.I.M.E.*, **233**, 1467
FERRISS, D. P., ROSE, R. M. AND WULFF, J. (1962) *Trans. A.I.M.E.*, **224**, 975
FERRO, A. AND MONTALENTI, G. (1963) *Phil. Mag.*, **8**, 105
FIORE, N. F. AND BAUER, C. L. (1964) *Acta Metall.*, **12**, 1329
FIORE, N. F. AND BAUER, C. L. (1967) *Progress in Mater. Sci.*, **13**, 85
FISHER, J. C. (1954) *Acta Metall.*, **2**, 9

FISHER, J. C. (1955) *Trans. Am. Soc. Metals*, **47**, 451
FISHER, J. C. AND ROGERS, H. C. (1956) *Acta Metall.*, **4**, 180
FISHER, R. M. (1961) Ph.D. Thesis, University of Cambridge
FLEISCHER, R. L. (1960) *Acta Metall.*, **8**, 32
FLEISCHER, R. L. (1961) *Acta Metall.*, **9**, 1034
FLEISCHER, R. L. (1962a) *Acta Metall.*, **10**, 835
FLEISCHER, R. L. (1962b) *J. Appl. Phys.*, **33**, 3504
FLEISCHER, R. L. (1967) *Acta Metall.*, **15**, 1513
FLEISCHER, R. L. (1968) *Scripta Metall.*, **2**, 113
FLINN, P. A. (1956) *Phys. Rev.*, **104**, 350
FLINN, P. A. (1958) *Acta Metall.*, **6**, 631
FLOREEN, S. AND SCOTT, T. E. (1964a) *Acta Metall.*, **12**, 758
FLOREEN, S. AND SCOTT, T. E. (1964b) *Acta Metall.*, **12**, 1459
FLOREEN, S. AND WESTBROOK, J. H. (1969) *Acta Metall.*, **17**, 1175
FORMBY, C. L. AND OWEN, W. S. (1965) *J. Less-Common Metals*, **9**, 25
FORMBY, C. L. AND OWEN, W. S. (1966) *Acta Metall.*, **14**, 1841
FOUNTAIN, R. W. AND MCKINSEY, C. R. (1963) In *Columbium and Tantalum* (Wiley)
FOURDEUX, A. AND BERGHEZAN, A. (1960) *J. Inst. Metals*, **89**, 31
FOURDEUX, A. AND WRONSKI, A. (1964) *J. Less-Common Metals*, **6**, 11
FRANK, W. (1967a) *Z. Naturforschung*, **22A**, 377
FRANK, W. (1967b) *Phys. stat. solidi*, **19**, 239
FRANZ, H., PFEIFFER, H. AND PFEIFFER, I. (1967) *Z. Metallkunde*, **58**, 87
FRIEDEL, J. (1964) *Dislocations* (Pergamon Press)
FUJITA, F. E. AND DAMASK, A. C. (1964) *Acta Metall.*, **12**, 331
GALLAGHER, C. J. (1952) *Phys. Rev.*, **88**, 721
GALLAGHER, P. C. J. (1967) *Phil. Mag.*, **15**, 51
GARLICK, R. G. AND PROBST, H. B. (1964) *Trans. A.I.M.E.*, **230**, 1120
GAROFALO, F., SMITH, G. V. AND MARSDEN, D. C. (1957) *Trans. Am. Soc. Metals*, **49**, 372
GAROFALO, F., VON GEMMINGEN, F. AND DOMIS, W. F. (1961) *Trans. Am. Soc. Metals*, **54**, 430
GARROD, R. I. AND WAIN, H. L. (1965) *Phil. Mag.*, **12**, 199
GAVERT, R. B. AND MACK, D. J. (1967) *Trans. A.I.M.E.*, **239**, 130
GEISELMAN, D. AND GUY, A. G. (1959) *Trans. A.I.M.E.*, **215**, 814
GELL, M. AND WORTHINGTON, P. J. (1966) *Acta Metall.*, **14**, 1265
GENSAMER, M., PEARSALL, E. B., PELLIN, W. S. AND LOW, J. R. (1942) *Trans. Am. Soc. Metals*, **30**, 983
GERBEREICH, W. W., MARTIN, C. F. AND ZACKAY, V. F. (1965) *Trans. Am. Soc. Metals*, **58**, 85
GILBERT, A., HULL, D., OWEN, W. S. AND REID, C. N. (1962) *J. Less-Common Metals*, **4**, 399
GILBERT, A., REID, C. N. AND HAHN, G. T. (1963) *J. Inst. Metals*, **92**, 351
GILBERT, A., WILCOX, B. A. AND HAHN, G. T. (1965) *Phil. Mag.*, **12**, 649
GLEN, J. (1957) *J. Iron Steel Inst.*, **186**, 21

GOLDMAN, K. M. (1960) In *The Metallurgy of Hafnium* (U.S. Govt. Printing Office), p. 229
GORMAN, J. A., WOOD, D. S. AND VREELAND, T., Jnr. (1969) *J. Appl. Phys.*, **40**, 833
GOUCHER, F. S. (1924) *Phil. Mag.*, **48**, 229
GOUZOU, J. (1964) *Acta Metall.*, **12**, 785
GRÅNÄS, L. AND AROUSSON, B. (1968) *Scripta Metals*, **2**, 541
GREEN, H. AND BROWN, N. (1953) *J. Metals*, **5**, 1240
GREETHAM, G. AND HONEYCOMBE, R. W. K. (1960) *J. Inst. Metals.*, **89**, 13
GREENMAN, W. F., VREELAND, T. Jnr. AND WOOD, D. S. (1967) *J. Appl. Phys.*, **38**, 3595
GREGORY, D. P. (1963) *Acta Metall.*, **11**, 455
GRIERSON, R. AND PARKINS, R. N. (1965) *Trans A.I.M.E.*, **233**, 1000
GRIMES, R. (1967) *Acta Metall.*, **15**, 953
GUARD, R. W. AND WESTBROOK, J. H. (1959) *Trans. A.I.M.E.*, **215**, 807
GUARD, R. W. (1961) *Acta Metall.*, **9**, 163
GUARD, R. W. AND FINE, M. E. (1965) *Trans. A.I.M.E.*, **233**, 1383
GUBERMANN, H. (1968) *Acta Metall.*, **16**, 713
GUIN, F. (1969) *Scripta Metall.*, **3**, 489
GUTMANAS, E. U., NADGORNY, E. M. AND STEPANOV, A. V. (1963) *Soviet Physics Solid St.*, **5**, 743
HAASEN, P. AND KELLY, A. (1957) *Acta Metall.*, **5**, 192
HAASEN, P. (1959) *Internal Stresses and Fatigue* (Elsevier), p. 205
HAASEN, P. AND KING, A. (1960) *Z. Metallkunde.*, **51**, 722
HADDRILL, D. M., YOUNGER, R. N. AND BAKER, R. G. (1961) *Acta Metall.*, **9**, 982
HAHN, G. T. (1962) *Acta Metall.*, **10**, 727
HAHN, G. T., COHEN, M. AND AVERBACH, B. L. (1962a) *J. Iron Steel Inst.*, **200**, 634
HAHN, G. T., REID, C. N. AND GILBERT, A. (1962b) *Acta Metall.*, **10**, 747
HALE, K. F. AND MCLEAN, D. (1963) *J. Iron Steel Inst.*, **201**, 337
HALL, E. O. (1950) *Proc. Phys. Soc.*, **B63**, 724
HALL, E. O. (1951a) *Proc. Phys. Soc.*, **B64**, 742
HALL, E. O. (1951b) *Proc. Phys. Soc.*, **B64**, 747
HALL, E. O. (1951c) *Proc. Phys. Soc.*, **B64**, 1085
HALL, E. O. (1952a) *J. Iron Steel Inst.*, **170**, 331
HALL, E. O. (1952b) Thesis, University of Cambridge
HALL, E. O. (1954) *Nature*, **173**, 948
HALL, E. O. (1962) *J. Aust. Inst. Metals*, **7**, 44
HALL, E. O. (1964) *J. Aust. Inst. Metals*, **9**, 264
HALL, E. O. (1968) *J. Inst. Metals*, **96**, 21
HAM, F. S. (1959) *J. Appl. Phys.*, **30**, 915 and 1518
HAM, R. K. AND JAFFREY, D. (1967) *Phil. Mag.*, **15**, 247
HARDING, H. J. AND HONEYCOMBE, R. W. K. (1966) *J. Iron Steel Inst.*, **204**, 259

HARDING, J. (1969) *Acta Metall.*, **17**, 949
HARPER, S. (1951) *Phys. Rev.*, **83**, 709
HARRIES, D. R. (1960) *J. Iron Steel Inst.*, **197**, 289
HARRIES, D. R., ARDY, A. F. AND BARTLETT, A. F. (1964) *J. Iron Steel Inst.*, **202**, 518
HARRIS, B. AND PEACOCK, D. E. (1965) *J. Less-Common Metals*, **8**, 78
HARRIS, I. R., DILLAMORE, I. L., SMALLMAN, R. E. AND BEESTON, B. E. P. (1966) *Phil. Mag.*, **14**, 325
HARRIS, S. G. (1958) Inst. of Metals Monograph No. 23, 'Vacancies and other point defects in metals and alloys', p. 220
HART, E. W. (1955) *Acta Metall.*, **3**, 146
HARTLEY, C. S. AND WILSON, R. J. (1963) *Acta Metall.*, **11**, 835
HARTLEY, S. (1966) *Acta Metall.*, **14**, 1237
HARTMANN, L. (1896) *Distribution des déformations dans les métaux soumis aux efforts* (Paris)
HAUSER, F. E., LONDON, P. R. AND DORN, J. E. (1956) *Trans. A.I.M.E.*, **206**, 589
HAUSER, J. J. (1961) *Trans. A.I.M.E.*, **221**, 305
HAYES, A. AND GRIFFIS, R. O. (1934) *Metals and Alloys*, **5**, 110
HEDGES, J. M. AND MITCHELL, J. W. (1953) *Phil. Mag.*, **44**, 233
HENDERSON, B. (1964) *Phil. Mag.*, **9**, 153
HENDRICKSON, A. A. AND FINE, M. E. (1961) *Trans. A.I.M.E.*, **221**, 103
HENDRICKSON, A. A. (1962) *Acta Metall.*, **10**, 900
HENDRICKSON, J. A., WOOD, D. S. AND CLARK, D. S. (1956a) *Acta Metall.*, **4**, 593
HENDRICKSON, J. A., WOOD, D. S. AND CLARK, D. S. (1956b) *Trans. Am. Soc. Metals*, **48**, 540
HENDRICKSON, J. A. AND WOOD, D. S. (1958) *Trans. Am. Soc. Metals*, **50**, 498
HESLOP, J. AND PETCH, N. J. (1956) *Phil. Mag.*, **1**, 866
HESLOP, J. AND PETCH, N. J. (1957) *Phil. Mag.*, **2**, 649
HEUBNER, U. AND SCHUMACHER, V. (1968) *Z. Metallkunde*, **59**, 337
HIRSCHHORN, J. S. (1963a) *Acta Metall.*, **11**, 1367
HIRSCHHORN, J. S. (1963b) *J. Less-Common Metals*, **5**, 493
HIRTH, J. P. AND COHEN, M. (1969a) *Scripta Metall.*, **3**, 107
HIRTH, J. P. AND COHEN, M. (1969b) *Scripta Metall.*, **3**, 311
HOLDEN, A. N. AND HOLLOMON, J. H. (1949) *Trans. A.I.M.E.*, **185**, 179
HOLDEN, A. N. AND KUNZ, R. W. (1952) *J. Appl. Phys.*, **23**, 799
HOLDEN, A. N. AND KUNZ, R. W. (1953) *Acta Metall.*, **1**, 495
HOLDEN, J. (1959) *Acta. Metall.*, **7**, 380
HOLDEN, J. (1960) *Acta Metall.*, **8**, 424
HOLLAND, J. R. (1967) *Acta Metall.*, **15**, 691
HOLLISTER, R. G. AND DARLING, A. S. (1967) *Platinum Metals Review*, **11**, 94
HOLZMANN, M. AND MAN, J. (1966) *J. Iron Steel Inst.*, **204**, 230
HONEYCOMBE, R. W. K., VAN ASWEGEN, J. S. T. AND WARRINGTON, D. H. (1963) N.P.L. Symposium, No. 15, 'Relation between structure and mechanical properties of metals', p. 380 (HMSO)

HOOK, R. E. AND ADAIR, A. M. (1963) *Trans. A.I.M.E.*, **227**, 151
HOOPER, W. H. L. (1952) *J. Inst. Metals*, **81**, 563
HORDON, M. J. (1963) *Acta Metall.*, **11**, 1006
HORNBOGEN, E. (1962) *Acta. Metall.*, **10**, 525
HORNBOGEN, E. (1963) *Trans. Am. Soc. Metals*, **56**, 16
HORNE, G. T., ROY, R. B. AND PAXTON, H. W. (1963) *J. Iron Steel Inst.*, **201**, 160
HOSFORD, W. F., FLEISCHER, R. L. AND BACHOFEN, W. A. (1960) *Acta Metall.*, **8**, 187
HOWE, L. M. AND THOMAS, W. R. (1960) *J. Nucl. Mater.*, **2**, 248
HULL, D. AND MOGFORD, I. M. (1958) *Phil. Mag.*, **3**, 1213
HULL, D. (1961) *Acta Metall.*, **9**, 191
HULL, D. AND MOGFORD, I. M. (1961) *Phil. Mag.*, **6**, 535
HULL, D., McIVOR, I. D. AND OWEN, W. S. (1963) N.P.L. Symposium, No. 15, 'Relation between structure and mechanical properties of metals', p. 596 (HMSO)
HULL, D. (1963) *Proc. R. Soc.*, **A274**, 5
HUME-ROTHERY, W. (1966) *The Structure of Alloys of Iron* (Pergamon)
HUNDY, B. B. (1954) *J. Inst. Metals*, **83**, 115
HUNDY, B. B. (1956) *Metallurgia*, **53**, 203
HUNDY, B. B. AND BOXALL, T. D. (1957) *Metallurgia*, **54**, 27
HUTCHISON, M. M. (1957) *J. Iron Steel Inst.*, **186**, 431
HUTCHISON, M. M. (1963) *Phil. Mag.*, **8**, 121
HUTCHISON, M. M. AND HONEYCOMBE, R. W. K. (1967) *Metal Science J.*, **1**, 129
HUTCHISON, M. M. AND PASCOE, R. T. (1969) *J. Aust. Inst. Metals*, **14**, 306
IMAI, Y. AND ISHIZAKI, T. (1962) *Sci. Rep. Res. Inst. Tohôku Univ.*, **14A**, 203
INGRAM, A. G., BARTLETT, E. S. AND OGDEN, H. R. (1963) *Trans. A.I.M.E.*, **227**, 131
IRANI, J. J. AND WEINER, R. T. (1965) *Nature*, **205**, 795
IRVINE, K. J., MURRAY, J. D. AND PICKERING, F. B. (1960) *J. Iron Steel Inst.*, **196**, 166
IRVINE, K. J. AND PICKERING, F. B. (1963) *J. Iron Steel Inst.*, **201**, 944
JACKSON, P. J. (1965) *Acta Metall.*, **13**, 1057
JAFFEE, R. I., MAYKUTH, D. J. AND DOUGLAS, R. W. (1961) In *Refractory Metals and Alloys* (Interscience, N.Y.)
JAFFREY, D. (1966) M.Sc. Thesis, McMaster Univ.
JAOUL, B. (1961) *J. Mech. Phys. Solids*, **9**, 66
JAOUL, B. AND GONZALES, D. (1961) *J. Mech. Phys. Solids*, **9**, 16
JEFFRIES, ZAY (1919) *Trans. A.I.M.E.*, **60**, 474
JENKINS, C. F. AND SMITH, G. V. (1969) *Trans. A.I.M.E.*, **245**, 2149
JENKINS, W. D. AND WILLARD, W. A. (1962) *Trans. Am. Soc. Metals*, **55**, 580
JINDAL, P. C. AND ARMSTRONG, R. W. (1967) *Trans. A.I.M.E.*, **239**, 1856
JINDAL, P. C. AND ARMSTRONG, R. W. (1969) *Trans. A.I.M.E.*, **245**, 623
JOHANSEN, H. A., GILBERT, H. L., NELSON, R. G. AND CARPENTER, R. L. (1954) U.S Bureau of Mines Rep. Inv. (5058)

Johnson, A. A. (1959) *Phil. Mag.*, **4**, 194
Johnson, A. A. and Peacock, D. E. (1959) *Phil. Mag.*, **4**, 528
Johnson, A. A. (1960a) *Acta Metall.*, **8**, 737
Johnson, A. A. (1960b) *Phil. Mag.*, **5**, 413
Johnson, A. A. (1962a) *Acta Metall.*, **10**, 975
Johnson, A. A. (1962b) *Phil. Mag.*, **7**, 177
Johnston, T. L., Davies, R. C. and Stoloff, N. S. (1965) *Phil. Mag.*, **12**, 305
Johnston, W. G. and Gilman, J. J. (1959) *J. Appl. Phys.*, **30**, 129
Johnston, W. G. (1962a) *J. Appl. Phys.*, **33**, 2050
Johnston, W. G. (1962b) *J. Appl. Phys.*, **33**, 2716
Johnston, W. G. and Stein, D. F. (1963) *Acta Metall.*, **11**, 317
Jolley, W. and Hull, D. (1964) *Acta Metall.*, **12**, 1337
Jolley, W. (1968) *Trans. A.I.M.E.*, **242**, 306
Jones, R. B. and Phillips, V. A. (1961) *Trans. Am. Soc. Metals*, **53**, 603
Jones, R. L. and Conrad, H. (1967) *Acta Metall.*, **15**, 649
Jones, R. L. and Conrad, H. (1968) *Scripta Metall.*, **2**, 239
Jones, R. L. and Conrad, H. (1969) *Trans. A.I.M.E.*, **245**, 779
Jordan, K. R. and Stoloff, N. S. (1969) *Trans. A.I.M.E.*, **245**, 2027
Kabler, M. N., Layer, H., Miller, M. G. and Slifkin, L. (1963) *Mater. Sci. Research*, **1**, 82
de Kazinczy, F. (1959a) *Acta Metall.*, **7**, 525
de Kazinczy, F. (1959b) *Acta Metall.*, **7**, 706
Keeler, J. H. (1955) *Trans. Am. Soc. Metals*, **47**, 157
Keh, A. S. and Wriedt, H. A. (1962) *Trans. A.I.M.E.*, **224**, 560
Keh, A. S. and Leslie, W. C. (1963) *Mater. Sci. Research*, **1**, 208
Keh, A. S. (1965) *Phil. Mag.*, **12**, 9
Keh, A. S., Nakada, Y. and Leslie, W. C. (1968) *Dislocation Dynamics* (McGraw Hill), p. 381
Kelly, A. and Chiou, C. (1958) *Acta Metall.*, **6**, 565
Kelly, A. and Nicholson, R. B. (1961) *Progress in Mater. Sci.*, **10**, 149
Kelly, A. (1966) *Strong Solids* (O.U.P.)
Kent, K. G. (1962) Ph.D. Thesis, University of Cambridge
Kent, K. G. and Kelly, A. (1963) *J. Inst. Metals*, **92**, 190
Kent, K. G. and Kelly, A. (1965) *J. Inst. Metals*, **93**, 536
Kenyon, R. L. and Burns, R. S. (1939) *Age Hardening of Metals* (A.S.M.), pp. 262 ff
Kieffer, R., Sedlatshek, K. and Braun, H. (1959) *J. Less-Common Metals*, **1**, 19
Kiessil, W. R. and Sinnott, M. J. (1953) *Trans. A.I.M.E.*, **197**, 331
King, G. W. and Spretnak, J. W. (1964) *Trans. A.I.M.E.*, **230**, 1481
Kirby, J. H. and Noble, F. W. (1967) *Phil. Mag.*, **16**, 1009
Klassen-Nekludova, M. (1929) *Z. Phys.*, **55**, 555
Klesnil, M., Holzmann, M., Lukáš, P. and Ryš, P. (1965) *J. Iron Steel Inst.*, **203**, 47
Klesnil, M. and Lukáš, P. (1967) *J. Iron Steel Inst.*, **205**, 746
Kloske, R. A. and Fine, M. E. (1969) *Trans. A.I.M.E.*, **245**, 217

KNIGHT, C. A. AND MURRAY, G. (1946) *Sheet Metal Industries*, **23**, 1741
KOCH, C. C. AND TROIANO, A. R. (1964) *Trans. Am. Soc. Metals*, **57**, 519
KOCHS, U. F. AND MADDIN, R. (1956) *Acta Metall.*, **4**, 91
KOCHS, U. F. (1959) *Acta Metall.*, **7**, 131
KOEHLER, J. S. AND SIETZ, F. (1947) *J. Appl. Mechanics*, **14**, A217
KOEHLER, J. S., LANGRETH, D. AND VON TURKOVICH, B. (1962) *Phys. Rev.*, **128**, 573
KOO, R. C. (1962) *J. Less-Common Metals*, **4**, 138
KOO, R. C. (1963a) *Trans. A.I.M.E.*, **227**, 280
KOO, R. C. (1963b) *Acta Metall.*, **11**, 1083
KOPPENAAL, T. J. AND FINE, M. E. (1961) *Trans. A.I.M.E.*, **221**, 1178
KOPPENAAL, T. J. (1964) *Acta Metall.*, **12**, 487
KOPPENAAL, T. J. (1965) *Phil. Mag.*, **11**, 1257
KOPPENAAL, T. J. (1966) *Acta Metall.*, **14**, 681
KOPPENAAL, T. J. (1968) *Acta Metall.*, **16**, 89
KOSSOWSKY, R. (1967) *Trans. A.I.M.E.*, **239**, 931
KOSSOWSKY, R. AND BROWN, N. (1966) *Acta Metall.*, **14**, 131
KÖSTER, W. (1927) *Z. Metallkunde*, **19**, 304
KÖSTER, W., BANGERT, L. AND HAHN, R. (1954) *Arch. für das Eisenhütt.*, **25**, 567
KÖSTER, W. AND KONZELMANN, W. (1963) *Z. Metallkunde*, **54**, 132
KÖSTER, W. AND SPEIDEL, M. O. (1965) *Z. Metallkunde*, **56**, 585
KÖTHE, A. (1968) *Acta Metall.*, **16**, 357
KOTVAL, P. S. (1968) *Trans. A.I.M.E.*, **242**, 1651
KRAFFT, J. M. AND SULLIVAN, A. M. (1959) *Trans. Am. Soc. Metals*, **51**, 643
KRAFFT, J. M. (1961) Report No. SSC-123, Ship Structure Cttee.
KRAFFT, J. M. (1962) *Acta Metall.*, **10**, 85
KRAFFT, J. M. AND SULLIVAN, A. M. (1962) *Trans. Am. Soc. Metals*, **55**, 101
KRAMER, I. R. AND MADDIN, R. (1952) *J. Metals*, **4**, 197
KRAMER, I. R. (1959) *Trans. A.I.M.E.*, **215**, 226
KRAMER, I. R. (1961) *Trans. A.I.M.E.*, **221**, 474
KRAMER, I. R. AND KUMAR, A. (1969) *Scripta Metall.*, **3**, 205
KRONBERG, M. L. (1962) *J. Am. Ceram. Soc.*, **45**, 274
KRAUSZ, A. S. (1968) *Acta Metall.*, **16**, 897
KRUPRIK, N. AND FORD, H. (1952) *J. Inst. Metals*, **81**, 601
KUMAR, A. (1968) *Acta Metall.*, **16**, 333
KUNZ, F. W. AND HOLDEN, A. N. (1954) *Acta Metall.*, **2**, 816
KURODA, M. (1938) *Sci. Pap. Inst. Phys. Chem. Research*, **34**, 1528
KURTZ, A. D. AND KULIN, S. A. (1954) *Acta Metall.*, **2**, 352
LACY, C. E. AND BECK, C. J. (1956) *Trans. Am. Soc. Metals*, **48**, 579
LANTANISION, R. M. AND STAEHLE, R. W. (1968) *Scripta Metall.*, **2**, 667
LAU, S. S. (1967) *Trans. A.I.M.E.*, **239**, 921
LAUTENSCHLAGER, E. AND BRITTAIN, J. O. (1962a) *Trans A.I.M.E.*, **224**, 48
LAUTENSCHLAGER, E. AND BRITTAIN, J. O. (1962b) *Trans. A.I.M.E.*, **224**, 606
LAWLEY, A., VIDOZ, E. A. AND CAHN, R. W. (1961a) *Acta Metall.*, **9**, 287
LAWLEY, A., LIEBMANN, W. AND MADDIN, R. (1961b) *Acta Metall.*, **9**, 841

Lawley, A., Van den Lype, J. and Maddin, R. (1962) *J. Inst. Metals*, **91**, 23
Lawley, A. and Gaigher, H. L. (1964) *Phil. Mag.*, **10**, 15
Lawley, A. (1966) *Intermetallic Compounds* (J. H. Westbrook (ed.)) (Wiley), Ch. 24
Laxar, F. H., Frame, J. W. and Blickwede, D. J. (1961) *Trans. Am. Soc. Metals*, **53**, 683
Leadbetter, M. J. and Argent, B. B. (1961) *J. Less-Common Metals*, **3**, 19
Lean, J. B., Plateau, J. and Crussard, C. (1959) *Mem. Scient. Revue Metall.*, **56**, 427
Lean, J. B. (1960) *Aust. J. Phys.*, **13**, 359
Leslie, W. C. and Rickett, R. L. (1953) *Trans. A.I.M.E.*, **197**, 1021
Leslie, W. C. (1961) *Acta Metall.*, **9**, 1004
Leslie, W. C. and Keh, A. S. (1962) *J. Iron Steel Inst.*, **200**, 722
Leslie, W. C. and Sober, R. J. (1967) *Trans. Am. Soc. Metals*, **60**, 99
Levine, E. D. (1966) *Trans. A.I.M.E.*, **236**, 1558
Lewis, M. H. (1966) *Phil. Mag.*, **13**, 777
Li, J. C. M. (1963) *Trans. A.I.M.E.*, **227**, 239
Li, J. C. M. and Michalak, J. T. (1964) *Acta Metall.*, **12**, 1457
Li, J. C. M. (1967) *Canadian J. Phys.*, **45**, 493
Li, J. C. M. (1969) *Phil. Mag.*, **19**, 189
Lindley, T. C. and Smallman, R. E. (1963a) *Acta Metall.*, **11**, 361
Lindley, T. C. and Smallman, R. E. (1963b) *Acta Metall.*, **11**, 626
Lindley, T. C. (1965) *Acta Metall.*, **13**, 681
Liss, R. B. (1957) *Acta Metall.*, **5**, 341
Liu, C. T. and Gurland, J. (1968) *Trans. A.I.M.E.*, **242**, 1535
Liu, T., Kramer, I. R. and Steinberg, M. A. (1956) *Acta Metall.*, **4**, 364
Logie, H. J. (1957) *Acta Metall.*, **5**, 106
Lomer, W. M. (1952) *J. Mech. Phys. Solids*, **1**, 64
Lothe, J. (1962) *Acta Metall.*, **10**, 663
Louat, N., (1956) *Proc. Phys. Soc.*, **B69**, 459
Low, J. R., Jnr. and Gensamer, M. (1944) *Trans. A.I.M.E.*, **158**, 207
Low, J. R., Jnr. and Guard, R. W. (1959) *Acta Metall.*, **7**, 171
Low, J. R., Jnr. (1963) *Iron and its Dilute Solid Solutions* (Interscience, N.Y.), p. 217
Lubahn, J. D. (1949) *Trans. A.I.M.E.*, **185**, 702
Lubahn, J. D. (1952) *Trans. Am. Soc. Metals*, **44**, 643
Luders, W. (1860) *Dinglers Polytech. J.*, **155**, 18
Ludwik, P. and Scheu, R. (1925) *Ber. Fach. Ver. deut. Eisen.*, **5**, 1
Lytton, J. L. and Teitz, T. E. (1964a) *Trans. A.I.M.E.*, **230**, 241
Lytton, J. L. and Teitz, T. E. (1964b) *Trans. A.I.M.E.*, **230**, 802
Macheraugh, E. and Vöhringer, O. (1963) *Acta Metall.*, **11**, 157
Maiden, C. J. (1959) *Acta Metall.*, **7**, 297
Makin, M. J. and Minter, F. J. (1956) *J. Inst. Metals*, **85**, 397
Makin, M. J. and Gillies (Mrs.) E. (1957) *J. Inst. Metals*, **86**, 108
Makin, M. J. (1958) *Phil. Mag.*, **3**, 287
Makin, M. J. (1959) *Acta Metall.*, **7**, 233

MAKIN, M. J. AND MINTER, F. J. (1959) *Acta Metall.*, **7**, 361
MAKIN, M. J. AND MINTER, F. J. (1960) *Acta Metall.*, **8**, 691
MAKIN, M. J. AND MANTHORPE, S. A. (1961) *Acta Metall.*, **9**, 886
MAKIN, M. J. AND BLEWITT, T. H. (1962) *Acta Metall.*, **10**, 241
MARCINKOWSKI, M. J. AND LIPSITT, H. A. (1962) *Acta Metall.*, **10**, 95
MARCINKOWSKI, M. J. AND FISHER, R. M. (1963) *J. Appl. Phys.*, **34**, 2135
MARCINKOWSKI, M. J. AND CHESSIN, H. (1964) *Phil. Mag.*, **10**, 837
MARCINKOWSKI, M. J. AND FISHER, R. M. (1965) *Trans. A.I.M.E.*, **233**, 293
MEAKIN, J. D. AND PETCH, N. J. (1963) ASP-TDR-63-324; Symposium on Role of Substructure, p. 243
MICHALAK, J. T. (1965) *Acta Metall.*, **13**, 213
MICHALAK, J. T. (1966) *Acta Metall.*, **14**, 1864
MIKLOVITZ, J. (1947) *J. Appl. Mechanics*, **14**, (1) A-31
MILES, G. D. (1964) *Acta Metall.*, **12**, 1241
MILKO, J. A., ADAMS, R. E. AND HARMS, W. O. (1958) *The Metal Thorium* (A.S.M., Cleveland), p. 186
MILNE, I. AND SMALLMAN, R. E. (1968) *Trans. A.I.M.E.*, **242**, 120
MINCHER, A. L. AND SHEELY, W. F. (1961) *Trans. A.I.M.E.*, **221**, 19
MINTZ, B. AND WILSON, D. V. (1965) *Acta Metall.*, **13**, 947
MITCHELL, T. E., FOXALL, R. A. AND HIRSCH, P. B. (1963) *Phil. Mag.*, **8**, 1895
MITCHELL, T. E. AND SPITZIG, W. A. (1965) *Acta Metall.*, **13**, 1169
MOGFORD, I. L. AND HULL, D. (1963) *J. Iron Steel Inst.*, **201**, 55
MOON, D. W. AND VREELAND, T., Jnr. (1968) *Scripta Metall.*, **2**, 35
MOON, D. W. AND VREELAND, T., Jnr. (1969) *Acta Metall.*, **17**, 989
MORDIKE, B. L. (1962) *Z. Metallkunde*, **53**, 586
MORDIKE, B. L. AND HAASEN, P. (1962) *Phil. Mag.*, **7**, 459
MORGAN, E. R. AND SHYNE, J. C. (1957a) *Trans. A.I.M.E.*, **209**, 65
MORGAN, E. R. AND SHYNE, J. C. (1957b) *J. Iron Steel Inst.*, **185**, 156
MORGAND, P., MOUTURAT, P. AND SAINFORT, G. (1968) *Acta Metall.*, **16**, 867
MORI, T. AND MESHII, M. (1969) *Acta Metall.*, **17**, 167
MORRISON, W. B. (1963) *J. Iron Steel Inst.*, **201**, 317
MORRISON, W. B. AND WOODHEAD, J. H. (1963) *J. Iron Steel Inst.*, **201**, 43
MORRISON, W. B. AND GLENN, R. C. (1968) *J. Iron Steel Inst.*, **206**, 611
MOTE, J., TANAKA, K. AND DORN, J. E. (1961) *Trans. A.I.M.E.*, **221**, 858
MURA, T. AND BRITTAIN, J. O. (1960) *Acta Metall.*, **8**, 767
MURA, T., LAUTENSCHLAGER, E. A. AND BRITTAIN, J. O. (1961) *Acta Metall.*, **9**, 453
MURA, T. AND BRITTAIN, J. O. (1962) *Acta Metall.*, **10**, 973
MCADAM, D. J. AND MEBS, R. W. (1943) *Trans. A.S.T.M.*, **43**, 661
MCEVILY, A. J. Jnr., BUSH, R. H., SCHADLER, F. W. AND SCHMATZ, D. J. (1963) *Trans. Am. Soc. Metals*, **56**, 753
MCLEAN, D. (1963) N.P.L. Symposium, No. 15, 'Relation between structure and mechanical properties of metals', p. 575 (HMSO)
MCLENNAN, J. E. AND HALL, E. O. (1963) *J. Aust. Inst. Metals*, **8**, 191
MCLENNAN, J. E. (1965) *Acta Metall.*, **13**, 1299

McReynolds, A. W. (1949) *Trans. A.I.M.E.*, **185**, 32
McReynolds, A. W., Augustyniak, W., McKeown, M. and Rosenblatt, D. B. (1955) *Phys. Rev.*, **98**, 418
McRickard, S. B. and Chow, J. G. Y. (1965) *Trans. A.I.M.E.*, **233**, 147
McRickard, S. B. (1968) *Acta Metall.*, **16**, 969
Nabarro, F. R. N. (1948) Report of a Conference on the Strength of Solids (London, Phys. Soc.), p. 38
Nacken, M. and Heller, W. (1960) *Arch. für das Eisenhütt.*, **31**, 103
Nakada, Y. and Keh, A. S. (1967) *Acta Metall.*, **15**, 879
Nakada, Y. and Keh, A. S. (1968) *Acta Metall.*, **16**, 903
Nakayama, Y., Weissmann, S. and Imura, T. (1962) *Direct Observations of Imperfections in Crystals* (Interscience), p. 573
Naybour, R. D. (1965) *Acta Metall.*, **13**, 1197
Naybour, R. D. (1966) *J. Iron Steel Inst.*, **204**, 1200
Nechay, Y. P. and Popov, K. V. (1965) *Physics of Metals and Metallog.*, **19**, (4) 117
Nichols, R. W. and Harries, D. R. (1962) A.S.T.M., S.T.P. No. 341, p. 162
Nishino, K. (1962) *Trans. J. Inst. Metals*, **3**, 63
Nishino, K. and Takahashi, K. (1962) *Trans. J. Inst. Metals*, **3**, 57
Nishino, K. (1963) *Sci. Rep. Tohôku Univ.*, **15**, 235
Noble, F. W. and Hull, D. (1964) *Acta Metall.*, **12**, 1089
Noble, F. W., Kirby, J. H. and Whitehead, R. S. (1968) *Scripta Metall.*, **2**, 425
Nye, J. F. (1949) *Proc. R. Soc.*, **A198**, 190, 1949
Okamoto, P. R. and Thomas, G. (1967) *Acta Metall.*, **15**, 1325
Ono, K., Meshii, M. and Kauffman, J. W. (1964) *Acta Metall.*, **12**, 361
Ono, K. and Meshii, M. (1965) *Acta Metall.*, **13**, 140
Ono, K. (1966) *Acta Metall.*, **14**, 1863
Orava, R. N. (1964) *Trans. A.I.M.E.*, **230**, 1614
Orava, R. N., Stone, G. and Conrad, H. (1966) *Trans. Am. Soc. Metals*, **59**, 171
Oren, E. C., Fiore, N. F. and Bauer, C. L. (1966) *Acta Metall.*, **14**, 245
Orowan, E. (1934) *Z. Phys.*, **89**, 634
Orowan, E. (1940) *Proc. Phys. Soc.*, **52**, 8
Osborne, P. W. (1962) *Acta Metall.*, **10**, 740
Owen, W. S., Cohen, M. and Averbach, B. L. (1958) *Trans. Am. Soc. Metals*, **50**, 517
Pansieri, C., Federighi, T. and Ceresara, S. (1963) *Trans. A.I.M.E.*, **227**, 1122
Parameswaran, V. R. and Weertman, J. (1969) *Scripta Metall.*, **3**, 477
Pascoe, R. T. and Newey, C. W. A. (1968) *Metal Science J.*, **2**, 138
Patel, J. R. and Chaudhuri, A. R. (1962) *J. Appl. Phys.*, **33**, 2223
Paxton, H. W. (1953) *J. Appl. Phys.*, **24**, 104
Paxton, H. W. and Churchman, A. T. (1953) *Acta Metall.*, **1**, 473
Paxton, H. W. and Bear, I. J. (1955) *Trans. A.I.M.E.*, **203**, 989

PAXTON, H. W., SHEEHAN, J. M. AND BABYAK, W. J. (1959) *Trans. A.I.M.E.*, **215**, 725
PEARSON, G. L., READ, W. T. AND FELDMANN, W. L. (1957) *Acta Metall.*, **5**, 181
PEIFFER, H. R. (1961) *Acta Metall.*, **9**, 385
PEIFFER, H. R. (1963) *J. Appl. Phys.*, **34**, 298
PENNING, P. AND DE WIND, G. (1959) *Physica*, **25**, 765
PETCH, N. J. (1953) *J. Iron Steel Inst.*, **174**, 25
PETCH, N. J. (1958) *Phil. Mag.*, **3**, 1089
PETCH, N. J. (1964) *Acta Metall.*, **12**, 59
PETERS, B. C., HENDRICKSON, A. A. AND RUNDMAN, K. B. (1965) *Acta Metall.*, **13**, 142
PETERS, B. C. AND HENDRICKSON, A. A. (1966) *Acta Metall.*, **14**, 1121
PETERSON, D. T. (1961) *Trans. Am. Soc. Metals*, **53**, 765 and 966
PETERSON, D. T. AND SKAGGS, R. L. (1968) *Trans. A.I.M.E.*, **242**, 922
PETERSON, J. L. (1963) *Trans. Am. Soc. Metals*, **56**, 304
PFEIL, P. C. L. AND HARRIES, D. R. (1965) A.S.T.M., S.T.P. No. 380, p. 202
PHILLIPS, R. AND CHAPMAN, J. A. (1965) *J. Iron Steel Inst.*, **203**, 511
PHILLIPS, V. A., SWAIN, A. J. AND EBORALL, R. (1952) *J. Inst. Metals*, **81**, 625
PHILLIPS, V. A. (1953) *J. Inst. Metals*, **81**, 649
PHILLIPS, V. A. AND JONES, R. B. (1961) *Trans. Soc. Metals*, **53**, 775
PHILLIPS, V. A. (1963) *Trans. Am. Soc. Metals*, **56**, 600
PHILLIPS, W. L. Jnr. (1963) *Trans. A.I.M.E.*, **227**, 84
PIERCY, G. R., CAHN R. W. AND COTTRELL, A. H. (1955) *Acta Metall.*, **3**, 331
PINK, E. (1968) *J. Less-Common Metals*, **16**, 119
PIOBERT, G., MORIN AND DIDION (1842) *Mém. de l'Artillerie*, **5**, 525
PITSCH, W. AND LUCKE, K. (1956) *Arch. für das Eisenhütt.*, **27**, 45
POLAKOWSKI, N. H. (1952) *J. Inst. Metals*, **81**, 617
POMEY, G., BRUMBACH, M. AND CRUSSARD, C. (1964) *Mem. Scient. Revue Metall.*, **61**, 243
POPE, D. P., VREELAND, T., Jnr. AND WOOD, D. S. (1967) *J. Appl. Phys.*, **38**, 4011
POPOV, L. Y. AND SUKHOVAROV, V. F. (1964) *Physics of Metals and Metallog.*, **17** (3), 428
PORTEVIN, A. AND LE CHATELIER, A. (1923) *C. R. Acad. Sci.*, **176**, 507
POWELL, G. W., MARSHALL, E. R. AND BACKOFEN, W. A. (1958) *Trans. Am. Soc. Metals*, **50**, 478
PREKEL, H. L. AND CONRAD, H. (1967) *Acta Metall.*, **15**, 955
PREKEL, H. L., LAWLEY, A. AND CONRAD, H. (1968) *Acta Metall.*, **16**, 337
PRICE, P. B. (1961) *Proc. R. Soc.*, **A260**, 251
PRICE, R. J. AND KELLY, A. (1963) *Acta Metall.*, **11**, 915
PRICE, R. J. AND KELLY, A. (1964a) *Acta Metall.*, **12**, 159
PRICE, R. J. AND KELLY, A. (1964b) *Acta Metall.*, **12**, 979
PUGH, J. W. (1955) *Trans. Am. Soc. Metals*, **47**, 984
PUGH, J. W. (1956a) *Trans. Am. Soc. Metals*, **48**, 593

Pugh, J. W. (1956b) *Trans. Am. Soc. Metals*, **48**, 677
Pugh, J. W. (1957) *Proc. A.S.T.M.*, **57**, 906
Pugh, J. W. (1958) *Trans. Am. Soc. Metals*, **50**, 1072
Quimby, R. M., Mote, J. D. and Dorn, J. E. (1962) *Trans. Am. Soc. Metals*, **55**, 149
Radcliffe, S. V. (1969) *Scripta Metall.*, **3**, 59
Raffo, P. L. and Mitchell, T. E. (1968) *Trans. A.I.M.E.*, **242**, 907
Ramaswami, B., Kocks, U. F. and Chalmers, B. (1965) *Trans. A.I.M.E.*, **233**, 1632
Rawlings, R. D. and Newey, C. W. A. (1967) *Acta Metall.*, **15**, 440
Raymond, L., Gerbereich, W. W. and Martin, C. F. (1965) *J. Iron Steel Inst.*, **203**, 933
Raynor, D., Whiteman, J. A. and Honeycombe, R. W. K. (1966) *J. Iron Steel Inst.*, **204**, 1114
Reid, C. N. and Owen, W. S. (1962) *J. Iron Steel Inst.*, **200**, 229
Reid, C. N. and Hahn, G. T. (1964) *Acta Metall.*, **12**, 757
Reid, C. N., Gilbert, A. and Hahn, G. T. (1967) *Trans. A.I.M.E.*, **239**, 467
Rhode, R. W. and Pitt, C. H. (1967) *J. Appl. Phys.*, **38**, 876
Richards, P. N. and Barratt, K. V. (1965) *Trans. Am. Soc. Metals*, **58**, 601
Riggs, B. A. and Demer, L. J. (1963) *Acta Metall.*, **11**, 1003
Roberts, C. S., Carruthers, R. C. and Averbach, B. L. (1952) *Trans. Am. Soc. Metals*, **44**, 1150
Roberts, J. M. and Brown, N. (1960) *Trans. A.I.M.E.*, **218**, 454
Robinson, P. M. and Bever, M. B. (1966) *Acta Metall.*, **14**, 693
Rogers, H. C. (1954) *Acta Metall.*, **2**, 167
Rogers, H. C. (1956) *Acta Metall.*, **4**, 114
Rogers, H. C. (1957) *Acta Metall.*, **5**, 112
Rogers, H. C. (1959) *Trans. A.I.M.E.*, **215**, 666
Rose, K. S. B. and Glover, S. G. (1966) *Acta Metall.*, **14**, 1505
Rose, R. M., Ferriss, D. P. and Wulff, J. (1962) *Trans. A.I.M.E.*, **224**, 981
Rosenfield, A. R. (1962) *J. Inst. Metals*, **91**, 104
Rosenfield, A. R. and Owens, W. S. (1963) *Trans. A.I.M.E.*, **227**, 603
Rosi, F. D. and Perkins, F. C. (1953) *Trans. Am. Soc. Metals*, **45**, 972
Rozner, A. G. and Wasilewski, R. J. (1966) *J. Inst. Metals*, **94**, 169
Rudee, M. L. and Huggins, R. A. (1964) *Acta Metals*, **12**, 501
Rudman, P. S. (1962) *Acta Metall.*, **10**, 253
Rudolph, G. and Mordike, B. L. (1967) *Z. Metallkunde*, **58**, 708
Russell, B. (1963) *Phil. Mag.*, **8**, 615
Russell, B. and Vela, P. (1963a) *Phil. Mag.*, **8**, 677
Russell, B. and Vela, P. (1963b) *J. Aust. Inst. Metals*, **8**, 328
Russell, B. (1965a) *Acta Metall.*, **13**, 11
Russell, B. (1965b) *Acta Metall.*, **13**, 999
Russell, B. (1965c) *Phil. Mag.*, **11**, 139
Russell, B. and Jaffrey, D. (1965) *Acta Metall.*, **13**, 1
Russell, B. (1966) *Acta Metall.*, **14**, 682

RUSSELL, T. L., WOOD, D. S. AND CLARK, D. S. (1961) *Acta Metall.*, **9**, 1054
SARGENT, G. A. AND SHAW, B. J. (1966) *Acta Metall.*, **14**, 909
SASTRI, A. S. AND MARCINKOWSKI, M. J. (1968) *Trans. A.I.M.E.*, **242**, 2393
SCHADLER, H. W. (1960) *Trans. A.I.M.E.*, **218**, 649
SCHADLER, H. W. (1964) *Acta Metall.*, **12**, 861
SCHMATZ, D. J. AND BUSH, R. H. (1968) *Acta Metall.*, **16**, 207
SCHNITZEL, R. H. (1965a) *Trans. Am. Soc.*, **58**, 74
SCHNITZEL, R. H. (1965b) *J. Less-Common Metals*, **8**, 81
SCHOECK, G. (1956) *Phys. Rev.*, **102**, 1458
SCHOECK, G. AND SEEGER, A. (1959) *Acta Metall.*, **7**, 469
SCHOECK, G. (1961) *Acta Metall.*, **9**, 382
SCHOECK, G. (1969) *Scripta Metall.*, **3**, 239
SCHRÖDER, K. (1958) *Proc. Phys. Soc.*, **72**, 33
SCHRÖDER, K. (1959) *Proc. Phys. Soc.*, **73**, 674
SCHUETZ, A. E. AND ROBERTSON, W. D. (1957) *Corrosion*, **13**, 437
SCHUSSLER, M. AND BRUNHOUSE, J. S., Jnr. (1960) *Trans. A.I.M.E.*, **218**, 893
SCHWARTZBART, H. AND LOW, J. R. (1949) *Trans. A.I.M.E.*, **185**, 637
SEEGER, A. (1956) *Phil. Mag.*, **1**, 651
SEGALL, R. L. (1961) *Acta Metall.*, **9**, 975
SEITZ, F. (1952) *Adv. in Physics*, **1**, 43
SHARPE, W. N. Jnr. (1966) *J. Mech. Phys. Solids*, **14**, 187
SHAW, B. J. AND SARGENT, G. A. (1964) *Acta Metall.*, **12**, 1225
SHEELY, W. F. (1962a) *J. Less-Common Metals*, **4**, 291
SHEELY, W. F. (1962b) *J. Less-Common Metals*, **4**, 487
SHEPHARD, L. A. AND DORN, J. E. (1956) *J. Metals*, **8**, 1229
SHERBY, O. D., ANDERSON, R. A. AND DORN, J. E. (1951) *Trans. A.I.M.E.*, **191**, 643
SIETHOFF, H. (1969) *Acta Metall.*, **17**, 793
SILCOCK, J. M. (1959) *Acta Metall.*, **7**, 359
SILCOCK, J. M. (1963) *J. Iron Steel Inst.*, **201**, 409
SILCOCK, J. M. AND TUNSTALL, W. J. (1964) *Phil. Mag.*, **10**, 361
SINCLAIR, G. M. AND CRAIG, W. J. (1952) *Trans. Am. Soc. Metals*, **44**, 929
SINGHAL, L. K. AND MARTIN, J. W. (1967) *J. Iron Steel Inst.*, **205**, 947
SLEESWYK, A. W. (1958) *Acta Metall.*, **6**, 598
SLEESWYK, A. W. (1960) *Acta Metall.*, **8**, 130
SMALLMAN, R. E., WILLIAMSON, G. K. AND ARDLEY, G. (1953) *Acta Metall.*, **1**, 126
SMALLMAN, R. E. AND TERRY, J. C. (1963) *Trans. A.I.M.E.*, **227**, 769
SMITH, C. L. (1947) *Nature*, **160**, 466
SMITH, E. (1960) *Acta Metall.*, **8**, 403
SMITH, E. (1962) in *Direct Observations of Imperfections in Crystals* (Interscience, N.Y.), p. 203
SMITH, W. H. AND SEYBOLT, A. U. (1956) *J. Electrochem. Soc.*, **103**, 347
SMITH, W. H. AND SEYBOLT, A. U. (1957) in *Ductile Chromium* (A.S.M.), p. 169
SNOEK, J. L. (1941) *Physica*, **8**, 734
SOLIE, K. E. AND CARLSON, O. M. (1964) *Trans. A.I.M.E.*, **230**, 480

SOMALOWSKI, M. (1962) *Hydrogen in Steel* (Pergamon Press)
SONON, D. E. AND SMITH, G. V. (1968) *Trans. A.I.M.E.*, **242**, 1527
SOO, P. AND HIGGINS, G. T. (1968a) *Acta Metall.*, **16**, 177
SOO, P. AND HIGGINS, G. T. (1968b) *Acta Metall.*, **16**, 187
SPERRY, P. R. (1962) *Trans. A.I.M.E.*, **224**, 191
SPERRY, P. R. (1963a) *Acta Metall.*, **11**, 152
SPERRY, P. R. (1963b) *Acta Metall.*, **11**, 1005
STANG, A. H., GREENSPAN, M. AND NEWMAN, S. B. (1946) *Bur. Stand. J. of Res.*, **37**, 211
STEFANOVIĆ, V. M. (1966) *J. Iron Steel Inst.*, **204**, 987
STEIGERWALD, E. A. AND HANNA, G. L. (1963) *Trans. Am. Soc. Metals*, **56**, 656
STEIN, D. F. AND LOW, J. R. Jnr. (1960) *J. Appl. Phys.*, **31**, 362
STEIN, D. F., LOW, J. R. Jnr. AND SEYBOLT, A. V. (1963) *Acta Metall.*, **11**, 1253
STEIN, D. F. (1966) *Acta Metall.*, **14**, 99
STEIN, D. F. AND LOW, J. R. Jnr. (1966) *Acta Metall.*, **14**, 1183
STEIN, D. F. (1967) *Acta Metall.*, **15**, 150
STEPHENSON, E. T. AND COHEN, M. (1961) *Trans. Am. Soc. Metals*, **54**, 72
STIEGLER, J. O., DUBOSE, C. K. H. AND MCHARGUE, C. J. (1964) *Acta Metall.*, **12**, 263
STIEGLER, J. O. AND DUBOSE, C. K. H. (1967) *Acta Metall.*, **15**, 953
STOKES, R. J. AND COTTRELL, A. H. (1954) *Acta Metall.*, **2**, 341
STOKES, R. J. (1962) *Trans. A.I.M.E.*, **224**, 1227
STOKES, R. J. AND LI, C. H. (1963) *Mater. Sci. Research*, **1**, 133
STOLOFF, N. S. AND DAVIES, R. G. (1964) *Trans. Am. Soc. Metals*, **57**, 247
STOLOFF, N. S., DAVIES, R. G. AND KU, R. (1965) *Trans. A.I.M.E.*, **233**, 1500
STOLOFF, N. S. AND DAVIES, R. G. (1966) *Progress in Mater. Sci.*, **13**, 1
STROH, A. N. (1957) *Phil. Mag. Supplement*, **6**, 418
SUGENO, T. AND TAKAGI, S. (1964) *J. Fac. Eng. Univ. Tokyo*, **A2**, 38
SUGIYAMA, A. (1966) *J. Phys. Soc. Japan*, **21**, 1873
SUITKIN, N. F. (1957) *Physics of Metals and Metallog.*, **5**, (3), 116
SUITS, J. C. AND CHALMERS, B. (1961) *Acta Metall.*, **9**, 854
SUKHOVAROV, V. F. (1962) *Physics of Metals and Metallog.*, **13**, (2), 109
SUKHOVAROV, V. F., ALEKSANDROV, N. A. AND KUDRYAVTSEVA, L. A. (1962) *Physics of Metals and Metallog.*, **14**, (6) 82
SUMERLING, R. AND NUTTING, J. (1965) *J. Iron Steel Inst.*, **203**, 398
SUZUKI, H. (1952) *Sci. Rep. Res. Inst. Tohôku Univ.*, **A4**, 455
SUZUKI, H. (1957) *Dislocations and Mechanical Properties of Crystals* (Wiley, N.Y.), p. 361
SUZUKI, H. AND BARRETT, C. S. (1958) *Acta Metall.*, **6**, 156
SUZUKI, H. (1962) *J. Phys. Soc. Japan*, **17**, 322
SUZUKI, H. (1963) N.P.L. Symposium, No. 15, 'Relation between structure and mechanical properties of metals', p. 517 (HMSO)
SUZUKI, T. AND KOJIMA, H. (1966) *Acta Metall.*, **14**, 913
SYLWESTROWICZ, W. AND HALL, E. O. (1951) *Proc. Phys. Soc.*, **B64**, 495

SYUTKINA, V. I. AND YAKOVLEVA, E. S. (1960) *Physics of Metals and Metallog.*, **10**, (3), 161
SZKOPIAK, Z. C. (1968) *Acta Metall.*, **16**, 381
TAKAGI, S. AND SUGENO, T. (1965) *Jap. J. Appl. Phys.*, **4**, 772
TAKAMURA, J. AND MIURA, S. (1962) *J. Phys. Soc. Japan*, **17**, 237
TAKEUCHI, S., FURUBAYSHI, E. AND TAOKA, T. (1967) *Acta Metall.*, **15**, 1179
TAMHANKAR, R., PLATEAU, J. AND CRUSSARD, C. (1958) *Revue Metall.*, **55**, (4) 383
TANKINS, E. S. AND MADDIN, R. (1961) in *Columbium Metallurgy* (Interscience), p. 343
TARDIFF, G. E. AND HENDRICKSON, A. A. (1962) *Acta Metall.*, **10**, 573
TARDIFF, H. P. AND BALL, C. S. (1956) *J. Iron Steel Inst.*, **182**, 9
TAYLOR, G. AND CHRISTIAN, J. W. (1965) *Acta Metall.*, **13**, 1216
TAYLOR, G. AND CHRISTIAN, J. W. (1967) *Phil. Mag.*, **15**, 873
TAYLOR, G. I. AND ELAM, C. F. (1926) *Proc. Roy. Soc.*, **A112**, 337
TEDMON, C. S. AND FERRISS, D. P. (1962) *Trans. A.I.M.E.*, **224**, 1079
TEKIN, E. AND KELLY, P. M. (1965) *J. Iron Steel Inst.*, **203**, 715
TERRY, J. C. AND SMALLMAN, R. E. (1963) *Phil. Mag.*, **8**, 1827
THEOCARIS, P. S. AND KORONEOS, E. (1963) *Phil. Mag.*, **8**, 1871
THOMAS, A. T. (1960) *Phil. Mag.*, **5**, 947
THOMAS, A. T. (1966) *Acta Metall.*, **14**, 1363
THOMAS, G. AND NUTTING, J. (1956) *J. Inst. Metals*, **85**, 1
THOMAS, G. AND NUTTING, J. (1959) *Acta Metall.*, **7**, 515
THOMAS, W. R. AND LEAK, G. M. (1955) *J. Iron Steel Inst.*, **180**, 155
THOMPSON, R. W. AND CARLSON, O. N. (1964) *J. Less-Common Metals*, **7**, 321
THOMSON, R. (1958) *Acta Metall.*, **6**, 23
THORNTON, P. H. (1963) *Phil. Mag.*, **8**, 2013
THORNTON, P. H. (1965) *Phil. Mag.*, **11**, 71
THORNTON, P. R. AND MITCHELL, T. E. (1962) *Phil. Mag.*, **7**, 361
TITCHENER, A. L. AND DAVIES, G. J. (1965) *Phil. Mag.*, **11**, 1225
TJERKSTRA, H. H. (1961) *Acta Metall.*, **9**, 259
TRECO, R. M. (1953) *Trans. Am. Soc. Metals*, **45**, 872
TSOU, A. L., NUTTING, J. AND MENTER, J. W. (1952) *J. Iron Steel Inst.*, **172**, 163
TUCKER, R. P. AND OHR, S. M. (1967) *Phil. Mag.*, **16**, 643
TURNER, N. G. AND ROBERTS, W. T. (1968) *J. Less-Common Metals*, **16**, 37
TÚRY, P. AND KRAUSZ, S. (1936) *Nature*, **138**, 331
TÚRY, P. AND KRAUSZ, S. (1937) *Nature*, **139**, 30
VAN ASWEGEN, J. S. T. AND HONEYCOMBE, R. W. K. (1962) *Acta Metall.*, **10**, 262
VAN ASWEGEN, J. S. T., HONEYCOMBE, R. W. K. AND WARRINGTON, D. H. (1964) *Acta Metall.*, **12**, 1
VAN BUEREN, H. G. (1960) *Imperfections in Crystals* (North Holland)
VAN FOSSEN, R. H. Jnr., SCOTT, T. E. AND CARLSON, O. N. (1965) *J. Less-Common Metals*, **9**, 437
VAN TORNE, L. I. AND THOMAS, G. (1963) *Acta Metall.*, **11**, 881

van Torne, L. I. and Thomas, G. (1966) *Acta Metall.*, **14**, 621
Vaughan, H. G. and de Morton, M. E. (1956a) *J. Iron Steel Inst.*, **182**, 389
Vaughan, H. G. and de Morton, M. E. (1956b) *Acta Metall.*, **4**, 224
Vaughan, H. G. and de Morton, M. E. (1957) *Brit. Welding J.*, **4**, 40
Verduzco, M. and Polakowski, N. H. (1966) *J. Iron Steel Inst.*, **204**, 1027
Violan, P., Simon, J., Bouchet, B. and de Fouquet, J. (1967) *Mem. Scient. Revue Metall.*, **64**, 37
Vidoz, A. E. and Brown, L. M. (1962) *Phil. Mag.*, **7**, 1167
Vidoz, A. E., Lazarević, D. P. and Cahn, R. W. (1963) *Acta Metall.*, **11**, 17
Vöhringer, O. and Macherauch, E. (1967a) *Z. Metallkunde*, **58**, 21
Vöhringer, O. and Macherauch, E. (1967b) *Z. Metallkunde*, **58**, 317
Votava, E. (1965) *J. Less-Common Metals*, **9**, 409
Votava, E. (1968) *Acta Metall.*, **16**, 285
Vreeland, T. Jnr., Wood, D. S. and Clark, D. S. (1953a) *Trans. Am. Soc. Metals*, **45**, 621
Vreeland, T. Jnr., Wood, D. S. and Clark, D. S. (1953b) *Acta Metall.*, **1**, 414
Wachtman, J. B. Jnr. and Maxwell, L. H. (1954) *J. Am. Ceram. Soc.*, **37**, 291
Wagenblast, H. and Damask, A. C. (1962) *J. Phys. Chem. Solids*, **23**, 221
Wagner, W. J. Jnr., Rosenfield, A. R. and Averbach, B. L. (1962) *Acta Metall.*, **10**, 256
Wain, H. L. and Cottrell, A. H. (1950) *Proc. Phys. Soc.*, **B63**, 339
Wain, H. L. (1952) *Proc. Phys. Soc.*, **B65**, 886
Wain, H. L., Henderson, F., Johnstone, S. T. M. and Louat, N. (1957) *J. Inst. Metals*, **86**, 281
Wainwright, C., Cook, A. J. and Hopkins, B. E. (1964) *J. Less-Common Metals*, **6**, 362
Waldron, G. W. J. (1965) *Acta Metall.*, **13**, 897
Warrington, D. H. (1963) *J. Iron Steel Inst.*, **201**, 610
Wasilewski, R. J. (1963) *Trans. Am. Soc. Metals*, **56**, 221
Wasilewski, R. J. (1967) *Scripta Metall.*, **1**, 45
Wasilewski, R. J., Butler, S. R. and Hanlon, J. E. (1967) *Trans. A.I.M.E.*, **239**, 1357
Weaver, C. W. (1957) *Nature*, **180**, 806
Weil, N. A., Bortz, S. A. and Firestone, R. F. (1963) *Mater. Sci. Research*, **1**, 291
Weinstein, D. (1965) U.S.A.E.C. Rep. COO-578-P27-4
Welter, G. and Gochowski, S. (1938) *Metallurgia*, **18** (8), 99
Welter, G. and Gochowski, S. (1939) *Metallurgia*, **19** (8), 143
Welter, G. (1945) *Metallurgia*, **31**, (1), 144
Wepner, W. (1957) *Acta Metall.*, **5**, 703
Wert, C. A. (1949) *J. Appl. Phys.*, **20**, 943
Wert, C. A. and Zener, C. (1950) *J. Appl. Phys.*, **21**, 5
Wessel, E. T. (1957a) *Trans. Am. Soc. Metals*, **49**, 149

WESSEL, E. T. (1957b) *Trans. A.I.M.E.*, **209**, 930
WESTBROOK, J. H. (ed.) (1960) *Mechanical Properties of Intermetallic Compounds* (Wiley, N.Y.)
WESTBROOK, J. H. (ed.) (1966) *Intermetallic Compounds* (Wiley, N.Y.)
WESTLAKE, D. G. (1964) *Acta Metall.*, **12**, 1373
WESTLAKE, D. G. (1965) *Trans. A.I.M.E.*, **233**, 368
WESTLAKE, D. G. (1967) *Trans. A.I.M.E.*, **239**, 1101
WESTLAKE, D. G. (1969) *Trans. A.I.M.E.*, **245**, 1969
WESTMACOTT, K. H., BARNES, R. S., HULL, D. AND SMALLMAN, R. E. (1961) *Phil. Mag.*, **6**, 929
WESTWOOD, A. R. C. AND BROOM, T. (1957a) *Acta Metall.*, **5**, 77
WESTWOOD, A. R. C. AND BROOM, T. (1957b) *Acta Metall.*, **5**, 249
WESTWOOD, A. R. C. (1963) *Mater. Sci. Research*, **1**, 114
WILCOX, B. A. AND HUGGINS, R. A. (1960) *J. Less-Common Metals*, **2**, 292
WILCOX, B. A. AND HUGGINS, R. A. (1961) *Columbium Metallurgy* (Interscience, N.Y.), p. 257
WILCOX, B. A., BRISBANE, W. A. AND KLINGER, R. F. (1962) *Trans. Am. Soc. Metals*, **55**, 179
WILCOX, B. A. AND SMITH, G. C. (1964) *Acta Metall.*, **12**, 371
WILCOX, B. A. AND SMITH, G. C. (1965) *Acta Metall.*, **13**, 331
WILCOX, B. A. AND GILBERT, A. (1967) *Acta Metall.*, **15**, 601
WILMS, G. R. (1964) *J. Less-Common Metals*, **6**, 169
WILSON, D. V. AND RUSSELL, B. (1959) *Acta Metall.*, **7**, 628
WILSON, D. V. AND RUSSELL, B. (1960a) *Acta Metall.*, **8**, 36
WILSON, D. V. AND RUSSELL, B. (1960b) *Acta Metall.*, **8**, 468
WILSON, D. V. (1961) *Acta Metall.*, **9**, 618
WILSON, D. V. (1967) *Metal Science J.*, **1**, 40
WILSON, D. V. (1968) *Acta Metall.*, **16**, 743
WILSON, D. V. AND OGRAM, G. R. (1968) *J. Iron Steel Inst.*, **206**, 911
WILSON, J. C. AND BERGGREN, R. G. (1955) *Proc. Amer. Soc. Testing Mater.*, **55**, 689
WILSON, J. C. AND BILLINGTON, D. S. (1956) *J. Metals*, **8**, 665
WINDLE, A. H. (1966) Ph.D. Thesis, University of Cambridge
WINDLE, A. H. AND SMITH, G. C. (1968) *Metal Science J.*, **2**, 187
WINLOCK, J. AND LEITER, R. W. E. (1937) *Trans. A.S.M.*, **25**, 163
WINLOCK, J. (1953) *J. Metals*, **5**, 797
WOOD, D. L. AND WESTBROOK, J. H. (1962) *Trans. A.I.M.E.*, **224**, 1024
WOOD, D. L. AND WESTBROOK, J. H. (1963) *Trans. A.I.M.E.*, **227**, 771
WOOD, D. S. AND CLARK, D. S. (1951) *Trans. Am. Soc. Metals*, **43**, 571
WOOD, D. S. AND CLARK, D. S. (1952) *Trans. Am. Soc. Metals*, **44**, 726
WOOD, T. W. AND DANIELS, R. D. (1965) *Trans. A.I.M.E.*, **233**, 898
WOOD, W. A., GARROD, R. I. AND WAIN, H. L. (1964) *Trans. A.I.M.E.*, **230**, 837
WORTHINGTON, P. J. AND SMITH, E. (1964) *Acta Metall.*, **12**, 1277
WORTHINGTON, P. J. AND SMITH, E. (1966) *Acta Metall.*, **14**, 35
WORTHINGTON, P. J. (1966) *Acta Metall.*, **14**, 1015
WORTHINGTON, P. J. (1967) *Acta Metall.*, **15**, 1795

WORTHINGTON, P. J. (1969) *Phil. Mag.*, **19**, 663
WRIGHT, J. C., TURNER, M. S. AND DAVIES, A. L. (1962) *J. Inst. Metals*, **90**, 369
WRONSKI, A. S. AND JOHNSON, A. A. (1962) *Phil. Mag.*, **7**, 213
WRONSKI, A. AND FOURDEUX, A. (1964) *J. Less-Common Metals*, **7**, 205
WYNBLATT, P., ROSEN, A. AND DORN, J. E. (1965) *Trans. A.I.M.E.*, **233**, 651
WYNBLATT, P. AND DORN, J. E. (1966) *Trans. A.I.M.E.*, **236**, 1451
YADA, H. AND KIRITANI, S. (1965) *J. Phys. Soc. Japan*, **20**, 2314
YADA, H. (1967) *Acta Metall.*, **15**, 1255
YAJIMA, M. AND ISHII, M. (1967) *Acta Metall.*, **15**, 651
YAJIMA, M. AND ISHII, M. (1969) *Scripta Metall.*, **3**, 65
YOKOBORI, T. (1954) *J. Appl. Phys.* **25**, 593
YOSHIDA, S., OHBA, Y. AND NAGATA, N. (1960) *J. Jap. Inst. Met.*, **1**, 50
YOUNGER, R. N. AND BAKER, R. G. (1960) *J. Iron Steel Inst.*, **196**, 188
ZENER, C. AND HOLLOMON, J. H. (1944) *J. Appl. Phys.*, **15**, 22
ZERWEKH, R. P. AND SCOTT, J. E. (1967) *Trans. A.I.M.E.*, **239**, 432

Index

Major discussions are indicated by *bold* type